Alloys and
Intermetallic Compounds
From Modeling to Engineering

Alloys and
Intermetallic Compounds
From Modeling to Engineering

Editor

Cristina Artini

Department of Chemistry and Industrial Chemistry
University of Genova
Genova
Italy

CRC Press
Taylor & Francis Group
Boca Raton London New York

CRC Press is an imprint of the
Taylor & Francis Group, an **informa** business

A SCIENCE PUBLISHERS BOOK

Cover credit

Image taken by electron microscope of a Ni/B alloy drop solidified on a ZrB_2 substrate. Reproduced by kind courtesy of Dr. F. Valenza, CNR-ICMATE, Genova, Italy (author of Chapter 2.1).

CRC Press
Taylor & Francis Group
6000 Broken Sound Parkway NW, Suite 300
Boca Raton, FL 33487-2742

First issued in paperback 2020

© 2017 by Taylor & Francis Group, LLC
CRC Press is an imprint of Taylor & Francis Group, an Informa business

No claim to original U.S. Government works

ISBN-13: 978-1-4987-4143-9 (hbk)
ISBN-13: 978-0-367-78220-7 (pbk)

Library of Congress Cataloging-in-Publication Data

`Names: Artini, Cristina, editor.
Title: Alloys and intermetallic compounds : from modeling to engineering/editor,
Cristina Artini, Department of Chemistry and Industrial Chemistry,
University of Genova, Genova, Italy.
Description: Boca Raton, FL : CRC Press, Taylor & Francis Group, 2017. | "A science
publishers book." | Includes bibliographical references and index.
Identifiers: LCCN 2016057607| ISBN 9781498741439 (hardback) |
ISBN 9781498741446 (e-book)
Subjects: LCSH: Alloys. | Alloys--Design. | Intermetallic compounds.
Classification: LCC TA483 .A47 2017 | DDC 669--dc23
LC record available at https://lccn.loc.gov/2016057607

Visit the Taylor & Francis Web site at
http://www.taylorandfrancis.com

and the CRC Press Web site at
http://www.crcpress.com

Preface

As I was first contacted by CRC Press with the invitation to propose a book focused on intermetallic compounds, I immediately came across the idea of describing the role of modeling in the design of alloys aimed at specific applications. Experimental difficulties in obtaining the desired compounds and the desired properties are a common experience among materials scientists; on the other side, theory is often deemed as a high and respected knowledge, yet meant to be well separated from experimental science. In this respect, modeling appears as the perfect bridge between theoretical and experimental knowledge: it is partly or totally generated by theory, but it is intended to provide scientists with information useful to take very practical decisions, such as the choice of the most proper composition to obtain a certain effect or to investigate a certain property.

The book is divided into two parts: in the first one several ab-initio and semi-empirical modeling techniques, able to predict thermodynamic and physical properties, are presented. In the second part some examples regarding the employment of such methods and of specific mathematical and engineering simulations for the design of intermetallic compounds with specific properties, are elucidated and widely discussed. Alloys aimed at very different applications, such as brazing alloys, biocompatible materials, metallic glasses, superconductors, thermoelectric compounds, are among the materials treated. Therefore, my hope is that this collection of review papers written by distinguished researchers can be of help for a wide scientific audience: not only for graduate students willing to approach the topic "modeling" for the first time, but also for scientists aiming to go in more depth into a particular issue.

Preface

Contents

PART 1
MODELING TECHNIQUES
AND PREDICTION OF PROPERTIES

Computational Thermodynamics
From Experiments to Applications

A. Watson,[1,*] *A. Kroupa,*[2] *A. Dinsdale,*[3] *J. Vrestal,*[4] *P. Broz*[4] and *A. Zemanova*[5]

INTRODUCTION

The properties of materials, and ultimately their application, are determined substantially by the composition and quantities of the different phases present in their microstructure. The nature of the phases and their relative amounts are determined by the chemical composition of the material and the processing routes employed. Considering the number of chemical elements available to the materials scientist, the possible permutations and the almost infinite variation in processing parameters, a simple method of trial and error is an impractical route for materials development. However, the *Phase Diagram*, and more recently, the advent of *Computational Thermodynamics*, have made the process of Materials Development much easier and more efficient in terms of time and effort.

Phase diagrams show, in a simple visual form, the state of chemical equilibrium between a number of components as a function of

[1] Institute for Materials Research, School of Chemical and Process Engineering, University of Leeds, Leeds, LS2 9JT.
[2] IPM ASCR, Žižkova, 22, Brno, 61662, Czech Republic.
[3] Hampton Thermodynamics, Jamnagar Clo. Staines, Middx. UK.
[4] Dept. of Chemistry, Masaryk University, Kamenice 753/5, Bohunice, Brno, Czech Republic.
[5] IPM ASCR, Žižkova, 22, Brno, 61662, Czech Republic.
* Corresponding author: a.watson@leeds.ac.uk

variables such as temperature, pressure or composition. One of the first published phase diagrams was by Heycock and Neville (Heycock and Neville 1904), who, following earlier work published by Roozeboom (Roozeboom 1899), combined the cooling curve analysis of liquid mixtures of Cu and Sn with metallographic study of the solidified alloys. By plotting the temperatures corresponding to thermal arrests on the cooling curves at the corresponding alloy compositions, they were able to produce a simplified version of the Cu-Sn phase diagram, correctly showing that for all compositions and temperatures above the 'liquidus curve' the alloys are completely liquid, those below the 'solidus curve' are completely solid. In between the two curves, the solid and liquid phases coexist.

This was essentially the first 'modeling' of phase equilibria resulting in the initiation of large experimental programmes in many countries around the world, as the importance and applicability of a pictorial representation of phase equilibrium became fully appreciated. The basic experimental techniques used by Heycock and Neville soon became standard for the determination of phase diagrams, particularly for equilibria involving the liquid phase in higher temperature regimes, and they are still in use today, albeit in a more refined form. For lower temperature equilibria involving solid state transformations, other techniques have evolved, generally relying on electron microscopy or X-ray analysis.

For more than 60 years, phase diagrams were determined by purely experimental methods. But with the advent of inexpensive but more powerful computers, it became relatively easy to calculate phase diagrams using thermodynamic properties; as phase diagrams show the state of thermodynamic equilibrium. As early as 1908, J.J. Van Laar (van Emmerik 1991) had shown how features of binary phase diagrams can be synthesised using a simple regular solution model for the thermodynamic properties, but it was the pioneering work of Larry Kaufman and co-workers (Kaufman and Bernstein 1970) who took these concepts and began to calculate real phase diagrams. The *Calphad* technique (CALculation of Phase Diagrams) has been under continuous development since the late 1960s and now plays a prominent role in materials development. Experimental observation of phase equilibrium is combined with experimental determination of thermodynamic properties to produce self-consistent thermodynamic model descriptions of all possible phases in a material system. By minimisation of the total Gibbs energy for the system as a function of temperature, pressure and composition, the state of chemical equilibrium can be calculated, and unlike traditional 'paper' phase diagrams, the calculation can, in principle, be conducted for many components, the limit being dictated by the array sizes in the software and the availability of suitable thermodynamic descriptions. But it is important to recognise that, despite recent developments in *ab initio* methods, the power of the *Calphad* technique lies in the fact that it is firmly rooted in experimental observation, and the accuracy of the calculations relies very much on the quality of the experimental data that are used in the modeling process.

Experimental Techniques

The experimental techniques that have and continue to be used in the determination of phase diagrams and thermodynamic properties are quite varied and need considerable expertise on the part of the user. The following is not intended to be an exhaustive critique but to just give an overview of the type of techniques that are available to the materials scientist. For a more detailed description, the reader should consult (Kubaschewski et al. 1993; Boettinger et al. 2006).

Phase Diagram Determination

Dynamic methods

It would seem to be a contradiction, but one of the most popular and successful experimental techniques to determine phase boundaries in an *equilibrium* phase diagram, is *thermal analysis*, which is a dynamic method. There have been a number of improvements and variations of the method over the years, but the premise is still the same. Heat flow to or from a sample is measured with respect to time, and as a phase boundary is crossed, there is a significant change in the rate of the heat flow owing to evolution of the enthalpy of transformation on the appearance (or disappearance) of a phase in the material. As this method is used mostly for determination of phase boundaries involving the liquid phase, this heat change is related to the enthalpy of fusion.

At its simplest, *thermal analysis* can be conducted using a thermocouple and a stop watch. For example, by measuring and plotting the temperature of a solidifying alloy with respect to time (Fig. 1), deflections in the cooling curve give the temperatures of the liquidus, solidus and eutectic reaction. The main experimental difficulty with this simple set up is that the heat effects often can be relatively small and thus have only negligible influence on the cooling curve. A major improvement was the advent of *Differential Thermal Analysis* (DTA) which incorporated the measurement of the temperature of an inert standard in parallel with that of the sample, and the signal recorded being the difference between the two signals (Fig. 2). Therefore, if the heat capacity of the sample under study and that of the reference are very similar, the net output will be zero until a phase

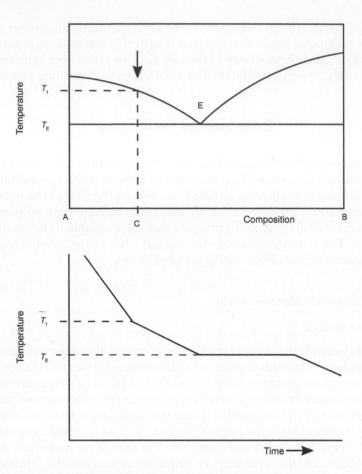

Fig. 1: Cooling curve analysis of a simple binary alloy system.

boundary is crossed, giving a well defined peak in the trace from the detector. Here, instead of plotting temperature *versus* time, the detector signal is plotted against sample temperature.

As Thermal Analysis is a dynamic measurement, it has to be said that equilibrium within the sample is unlikely to be reached. Kinetic constraints on phase transformation are an issue, but these can be reduced to a large degree by performing the measurements under slow cooling conditions (by fixing the cooling rate of the DTA furnace) or by performing the measurements on sample heating. However, heating/cooling rates need to be sufficiently fast in order to give a well defined peak; rates between 5 and 20°C per minute being the norm.

As a refinement of the DTA, the *Differential Scanning Calorimeter* measures the heat flow between the sample and the reference as a voltage

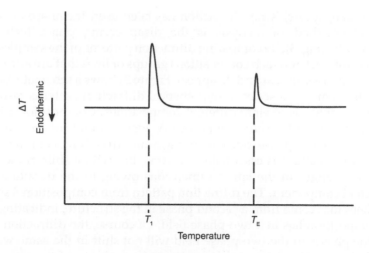

Fig. 2: The DTA Curve.

that is measured with respect to sample temperature. This not only leads to more accurate determination of transition temperature, but also makes the determination of the enthalpy change involved relatively straightforward.

Static methods

Dynamic methods of phase diagram determination require that the phase changes occur within a reasonable time. For high temperature work, particularly when the liquid phase is involved, these methods are acceptable and have been the staple techniques for many years, and continue to be used in the present day. But for lower temperature studies, particularly for solid state transformations, the times required to reach equilibrium are far too long for dynamic methods. Static methods involve preparation of alloys followed by their examination, which could involve microscopy (electron microscopy with microprobe analysis) or X-ray diffraction studies.

Typically, samples are prepared at a series of compositions within the system and then heat treated to equilibrium at a selected temperature. It is important that the heat treatment takes place in an inert atmosphere, which can involve sealing in silica ampoules under a reduced pressure of argon, or in a tube furnace under flowing inert gas. The time required for the samples to reach equilibrium will depend on the temperature of the heat treatment; the lower the temperature, the longer the time required. Samples are then cooled (rapidly quenched if there's a danger of not retaining the heat treated structure) for further examination.

For many years, X-ray diffraction has been used for phase analysis. A simple method of analysis is the disappearing phase technique (Cullity 1978) (Fig. 3). By comparing diffraction patterns of the samples, it is possible to detect at which compositions groups of lines that are attributed to specific phases appear and disappear. Figure 3 shows a schematic binary phase diagram and a series of hypothetical diffraction patterns resulting from X-ray analysis of eight different compositions. Composition 1 is the diffraction pattern for pure component A, whereas that for composition 2 is for the same phase but containing some dissolved component B. So, the X-ray pattern is essentially the same, but with a shift in the lines owing to changes in the lattice dimensions owing to the dissolution of the second component. The diffraction pattern from composition 3 shows extra lines that come from a second phase in the structure, indicating that this composition lies in a two-phase field. Of course, the diffraction lines from the phases in the two phase field will not shift in the same way as they do in a single phase region, and the compositions of the phases do not change, only their relative proportions. This is shown in the diagram of the hypothetical lattice parameters of the α and β phases. The lattice parameters are constant with respect to total composition within two phase fields. Composition 5 lies at the composition of the γ intermetallic compound, and hence the lines associated with the α phase disappear from the pattern. The patterns resulting from the remaining compositions show similar behaviour.

Although it is very good as a tool for phase identification, the drawback in using X-ray analysis is that the resolution of the technique can be as low as 5% with a corresponding uncertainty in composition of the location of a phase boundary. Therefore, it is now common practice to link X-ray studies with electron microscopy and microprobe analysis. Scanning electron microscopy not only reveals highly magnified detailed images of multiphase structures, but elemental maps produced by Energy Dispersive Analysis of X-rays emitted from the surface of the material reveal compositional variations between the different phases present. A recent development involves the incorporation of the analytical techniques into a single instrument. Electron Backscatter Diffraction (EBSD) involves the use of a special detector within a scanning electron microscope, which comprises, amongst other things, a phosphor screen and a CCD camera. The technique employs backscattered electrons that escape the sample, being diffracted by the crystal lattice of the material under study. The diffracted backscattered electrons excite the phosphor screen revealing bands, known as *Kikuchi bands*, which are then recorded by the CCD camera. With appropriate analysis, these bands lead to the identification of the crystal structure of the diffracting lattice.

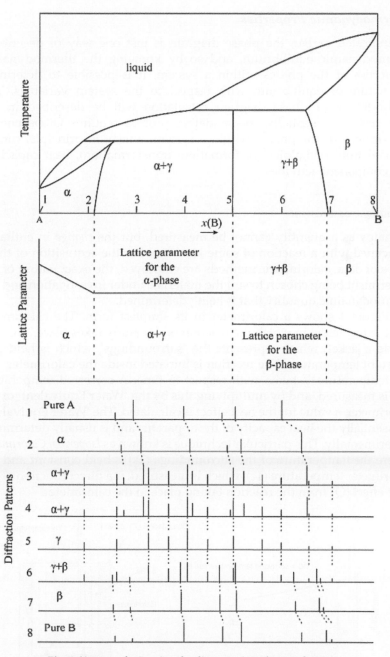

Fig. 3: X-ray analysis using the disappearing phase technique.

Thermodynamic Properties

As indicated earlier, the phase diagram is just one way of describing thermodynamic equilibrium, and so by knowing the thermodynamic properties of the phases within a system, it is possible to determine the nature of equilibrium with respect to the system variables. The methodology of phase diagram calculation will be described in the next section. Essentially, the modeling process requires Gibbs energy descriptions of the phases within a system, which, in principle, can be derived from enthalpies (of formation, transformation), heat capacities and component activities.

Enthalpy measurement

Enthalpy as a quantity cannot be measured, but the *change* in enthalpy associated with a reaction of some kind can. For the acquisition of these types of data, calorimetric methods are employed, the exact nature of the experiment being chosen to suit the materials under investigation and the thermodynamic quantity that is being determined.

Figure 4 shows a calorimeter in its simplest form. The calorimeter itself is the vessel in which the reaction under study takes place. This sits inside a jacket, which represents the 'surroundings', which is held at a constant temperature. The reaction is initiated inside the calorimeter and the temperature is taken with respect to time. A corrected temperature rise is measured, and by multiplying this by the 'Water Equivalent' of the calorimeter, a value for the heat effect is calculated. The 'Water Equivalent' is essentially the heat capacity of the apparatus and is usually determined experimentally. This particular technique is known as *Isoperibol Calorimetry* where the temperature of the surroundings (Ts) is held constant and the calorimeter temperature (Tc), which is measured, is a direct function of the heat effect (Q) from the reaction taking place in the calorimeter.

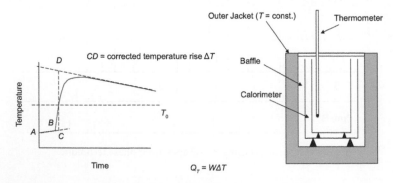

Fig. 4: Schematic of a simple calorimeter.

Other techniques are available, such as *Isothermal Calorimetry*, where Ts = Tc. For example, the Ice Calorimeter devised by Bunsen, captures the heat produced in the reaction melting a quantity of ice held in the calorimeter jacket, the weight of which can be measured, leading directly to the value of the heat effect, knowing the enthalpy of fusion of water.

The *Adiabatic Calorimeter* uses a completely different approach to the problem in that it aims to trap all of the heat produced in the reaction inside the calorimeter (Tc = Ts = f(Q)). The Adiabatic Calorimeter devised by Dench (Dench 1963) comprises a calorimetric cell suspended within a furnace. The furnace is maintained at a temperature equal to that inside the calorimetric cell by using a pair of differential thermocouples. A central heater supplies heat to the sample material inside the calorimetric cell. The heat required to form alloys from pressed compacts of the pure elemental powders can be determined by measuring the heat supplied directly by the sample heater, to raise the temperature of the material from a 'safe' temperature to a temperature at which alloying is complete (which is predetermined by a separate experiment). This is compared with the heat required to raise the temperature of the pure elements over the same temperature range, the difference between the two quantities being the enthalpy of formation of the alloy. As the furnace temperature outside the cell is the same as the temperature inside the cell, there is no net heat loss from the calorimeter and essentially all of the heat supplied to the specimen heater is used to raise the temperature of the sample and for the alloying process. Some minor heat losses can be expected *via* the heater leads themselves, but these can be compensated for in the final calculation.

However, the most common type of calorimeter used today is the *Heat Flow* calorimeter (Ts – Tc = constant). Typically, a pair of calorimetric cells, usually positioned side-by-side, are in contact with each other through many thermocouples which form a thermopile. The reaction under study takes place in the *reaction cell* and the heat produced flows through the thermopile to the *reference cell*. If the reaction taking place in the reaction cell is endothermic, then heat flows to the cell from the reference. This heat effect is realised as a voltage which is then plotted with respect to time for the duration of the experiment. The integral of this voltage time curve gives a value of the heat effect when multiplied by an appropriate calibration constant. Figure 5 shows a schematic of this calorimeter, set up in its usual configuration as a *drop calorimeter*.

Here, samples are dropped from a low temperature (typically room temperature) to a predetermined high temperature, and the heat effect measured, and with an appropriate calibration constant, a value for the enthalpy change can be determined. The drop technique can be used for different types of measurement. For example, by dropping successive samples of a pure element into a liquid metal held in an appropriate

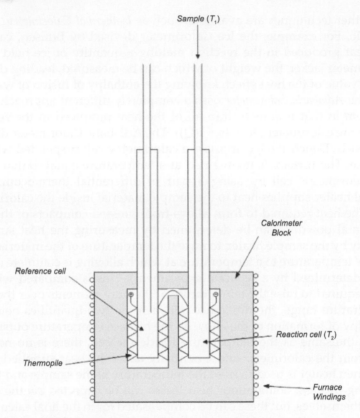

Fig. 5: Heat-flow calorimeter set up as a drop calorimeter.

crucible in the reaction cell, it is possible to determine enthalpies of mixing of liquid alloys. If alternatively, an appropriate liquid solvent is held in the crucible, and solid alloys and their component elements are dissolved in it, it is possible to determine enthalpies of formation of the solid alloys. The heat effect resulting from the dissolution of an alloy sample will result in a heat effect that comprises three parts. The enthalpy increment, namely the heat required to raise the temperature of the sample from the drop temperature to the calorimeter temperature, the heat effect on dissolving the sample, and finally, the enthalpy of formation of the alloy, as the dissolution results in the dispersion of the component elements to, in principle, infinite dilution; although this is in reality not achieved, but approached. The enthalpy of formation can be numerically much smaller than the other two heat effects, and so, if it is chemically possible, a direct reaction technique should be used to form the alloy within an empty reaction cell in the calorimeter from the individual components obviating the need for dissolution (for example (Watson and Hayes 1995)).

The high precision of this instrument, together with the versatility of the drop technique, has been put to great effect by a number of laboratories, most notably the group of Kleppa (for example (Meschel and Kleppa 2004)) and more recently, the group of Ipser in Vienna (for example (Plevachuk et al. 2014)).

In all types of calorimetric measurement, it is crucial to determine that the reaction under study goes to completion during the measurement. It is often better to reproduce the reaction outside the calorimeter, but replicating the measurement conditions. For example, a direct reaction measurement takes typically in the region of 20 minutes. Thus a sample of the same composition and size can be heated quickly to the calorimeter temperature inside a vertical tube furnace, held for 20 minutes and then quenched followed by X-ray and EPMA to check that the sample has been completely alloyed. If alloying is incomplete, a higher temperature is required for the reaction. This will result in an increase in the enthalpy increment portion of the heat effect, and hence adversely affect the uncertainty in the final result. Switching to a lower temperature dissolution technique may give better results in such a case.

Component activities—vapour pressure measurement

The determination of enthalpy is very useful in that it is relatively easy to do and it's an important component of the Gibbs energy. A second useful quantity that is relatively easy to determine is component activity. This can be done either as a vapour pressure measurement or *via* a galvanic cell.

The activity of a component in a mixture is defined as the ratio of its fugacity in the mixture to that of the component in the standard state. If the component is in its gaseous form, then this can be taken to be the ratio of its equilibrium vapour pressure above the mixture to that above the pure condensed component, assuming that the gas phase behaves ideally. Probably the most successful method of activity determination *via* vapour pressure measurement is using *Knudsen Effusion*.

The Knudsen Effusion cell (Fig. 6) is essentially a small container with a gas-tight lid with a knife-edged orifice, the diameter of which must be less than one-tenth the mean free path of the vapour inside the cell. The pressure of the effusing vapour in *vacuum* is given by:

$$\rho = \frac{m}{tA} \sqrt{\frac{2\pi RT}{M}} = 0.02256 \frac{m}{tA} \sqrt{\frac{T}{M}}$$

where m is the mass of the effusing species of molecular weight M, evaporating through an orifice of cross-sectional area A in time t. R is the Universal gas constant. Thus, by measuring the weight loss of the

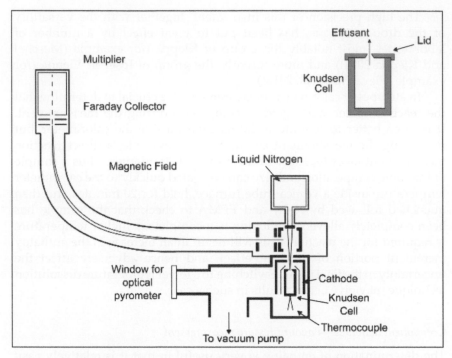

Fig. 6: Schematic of the Knudsen cell with the mass spectrometer.

cell during effusion it is possible to calculate the pressure of the effusing species. But by far the most popular and effective use of Knudsen Effusion is by combining it with mass spectrometry (Fig. 6). The effusing species are ionised by a beam of monoenergetic electrons. The ionised species are then subjected to an accelerating potential through a magnetic field that separates the different species according to charge/mass ratio, and the intensity of each is recorded. An alternative methodology involves non-magnetic quadrupole mass spectrometric separation via combination of direct-current (DC) and radio-frequency (RF) potentials or a pulsed accelerating potential with ions being separated by 'Time of Flight'. The advantage of combining Knudsen Effusion with mass spectrometry is the ability to identify the effusing species, which is particularly useful if there is more than one. The intensities of each are recorded, and the activity is simply the ratio of the ionic intensity of the species effusing from the mixture to that effusing from the pure component.

Gibbs energies—electrochemical methods

Gibbs energies of reaction can be determined directly if the reaction in question can be arranged in the form of an electrochemical cell. For example, the Gibbs energy of formation of an oxide of metal M, MO, can be determined electrochemically thus:

$$\text{Pt} < \text{M}, \text{MO} > \mid \text{electrolyte} \mid (\text{O}_2, 1 \text{ atm}) \text{ Pt}$$

A mixture of metal M and its oxide forms one electrode and is connected to the other electrode (in this case, oxygen gas) by a suitable electrolyte. Platinum connections complete the electrical circuit to an appropriate voltage measuring device. With the condition of zero current flow, the cell is in equilibrium, the reactions taking place at each electrode being reversible.

$$M + O^{2-} = MO + 2e^- \qquad \text{Left-hand side.}$$
$$\tfrac{1}{2}\,O_2 + 2e^- = O^{2-} \qquad \text{Right-hand side.}$$
$$\tfrac{1}{2}\,O_2 + M = MO \qquad \text{Overall reaction.}$$

The Gibbs energy for the reaction is given as:

$$\Delta G = -zFE$$

where z is the number of electrons taking part in the reaction, F is the Faraday constant (96485.3 C mol^{-1}) and E is the cell voltage.

The nature of the electrolyte depends very much on the current carrying species. Above, the current carrying species is the electron, and it's quite standard for liquid electrolytes to be used for this purpose. These can be aqueous salts, but often molten salts are used. Alternatively, positive ions can be the current carrying species. For example, in solid oxide electrolytes, such as zirconia-based materials, the current flows as a result of vacancies formed in the crystal lattice as a result of doping with an aliovalent ion.

Depending on the thermodynamic property required, the cell reaction can be quite complex. For example, Wierzbicka-Miernik et al. (Wierzbicka-Miernik et al. 2010) used the following cell to determine the activities of Sn in ternary Ag-Cu-Sn melts.

$$(-) \text{ Kanthal, Re} \mid \text{Ag-Cu-Sn(liq)}, \text{SnO}_2(\text{s}) \parallel \text{ZrO}_2 - \text{Y}_2\text{O}_3 \parallel \text{Ni(s)},$$
$$\text{NiO(s)} \mid \text{Pt} (+)$$

Cell reactions:

$$2NiO + 4e^- = 2Ni + 2O^{2-}$$
$$Sn + 2O^{2-} = SnO_2 + 4e^-$$

Overall:

$$2NiO + Sn = 2Ni + SnO_2$$

The Gibbs energy of the overall cell reaction is thus a sum of the chemical potentials of the products minus that of the reactants:

$$\Delta G = -4FE = 2\mu^\circ(Ni) + \mu^\circ(SnO_2) - 2\mu^\circ(NiO) - (\mu^\circ(Sn) + RT\ln(a_{Sn}))$$

and thus knowing the Gibbs energy of the formation of the Ni and Sn oxides, the activity of Sn in the ternary melt is given by:

$$\ln(a_{Sn}) = \frac{-2\Delta G^O_{f,NiO} + \Delta G^O_{f,SnO_2} + 4\,FE}{RT}$$

Modeling

Semi-empirical Approach to the Calculation of Phase Diagrams for Multicomponent Systems

Currently we can hardly imagine any work devoted to the development of new materials without intensive use of phase diagrams for the class of materials under study. But it has been the more recent explosion in phase diagram modeling and the sophisticated application to real materials problems that has made such an impact on Materials Science possible. As mentioned in the Introduction, the modern concept of Phase Diagram Modeling is based on the theoretical work of Gibbs and Van Laar (Gibbs 1876; van Emmerik 1991) in the late 19th and early 20th century, setting a clear thermodynamic foundation for the definition of phase equilibria and established the links between the thermodynamic properties of any system and the relevant phase diagram.

The thermodynamic properties of all phases potentially stable in the system under study have to be defined if the phase equilibrium for a particular set of conditions is to be calculated. The temperature and pressure are usually defined as independent conditions in the case of metallic materials and the relevant thermodynamic state function is the Gibbs energy. The phase equilibrium is then calculated as the state with the lowest Gibbs energy and by searching for the minima of this state function with respect to the independent conditions, we can calculate the whole phase diagram of the system. The phase diagram

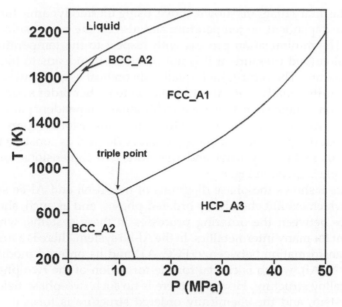

Fig. 7: The Calculated Phase Diagram for Iron.

for pure iron with respect to temperature and pressure is shown in Fig. 7.

There is an important difference between the phase diagram and any type of property diagram. Every point on the phase diagram has a specific meaning and gives information about the state of the system. In the case of the property diagram, only the curves on the diagram define any property of the material. The regions bound by the lines on the phase diagram are called phase fields and define the conditions under which the materials, in this case pure iron, exists in the indicated crystallographic structure. The lines themselves indicate the conditions under which the phases adjacent to the line coexist in thermodynamic equilibrium. The triple point (see Fig. 7) represents the only temperature and pressure under which the three phases are in equilibrium in accordance with Gibbs phase rule.

This *unary* (i.e., one component) phase diagram is the simplest example that can be used for the prediction of materials chemistry. The practical importance of these *theoretically modelled* phase diagrams increases with the complexity of the system as establishing an experimental phase diagram for complex systems is very difficult and only very limited regions can be known by experiment. Phase diagrams comprising of two components are known as *binary* phase diagrams, and they can be reasonably well established by experiment (see previous section) and consequently can be calculated in the same way as the unary diagram shown above, using thermodynamic functions, which are dependent on temperature and also on the composition of the system. The minimization process with respect to the temperature and composition (and pressure, if it is not constant) then leads to the binary phase diagram. The robust, high quality theoretical assessment serves as basis for further mathematical extrapolation to higher order systems. The composition of the system, which is an additional independent parameter of the system, would theoretically lead to a three-dimensional representation of the phase diagram, but the pressure can be defined as constant for vast majority of metallic systems and therefore two dimensions are used. An example is given in Fig. 8.

Figure 8 shows the phase diagrams of the Al-Ni and Al-Fe systems, both of which exhibit chemically ordered phases, and as such, shows the difference between the ordering processes in these systems, which are important for many intermetallics. In the Al-Ni system, there is a first order phase transformation between γ (FCC_A1) and its ordered modification γ' (L1$_2$ – Ni$_3$Al), which is crucial for the formation of the two-phase (γ + γ') superalloy structure. However, there is no such two-phase field in the Al-Fe system, and the chemically ordered structure is formed through a second order phase transformation between BCC_A2 and BCC_B2 phases. This means that the morphology of both alloys will be completely different, influencing significantly their materials properties. In addition, the NiAl phase (BCC_B2) is formed at around 50 mol% Al in the Al-Ni system, exhibiting a very high melting point and the presence of many defects in the structure leading to wide composition range for the phase. Robust and precise thermodynamic descriptions of these important binary systems, together with knowledge of other binary Ni-X (Fe-X) systems (X being key alloying elements), allows the prediction of the behaviour of more complex materials.

Computational Thermodynamics, and in particular the *Calphad* method, requires a complete description of the thermodynamic state functions of all phases in the system and their dependence on temperature, pressure and composition. Ever increasing computer capacity and power has enabled this method to be applied to the modeling of phase diagrams of highly complex systems, and its versatility and extreme usefulness was soon "discovered" not only by those working in basic science but also in applied science and industrial laboratories.

The *Calphad* method is based on a sequential (bottom-up approach) modeling of thermodynamic functions, and in particular the Gibbs energy, starting from the simplest systems (e.g., unaries), constructing robust and

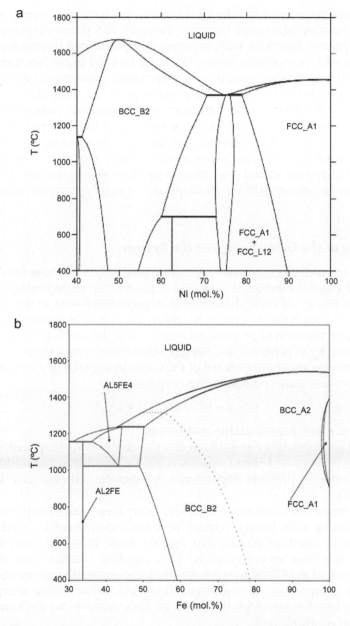

Fig. 8: Phase Diagram for the Al-Ni (a) and Al-Fe (b) Systems at Atmospheric Pressure.

reliable descriptions of the data for binary and ternary systems, where the best possible agreement between the modelled phase diagrams and thermodynamic functions with experimental phase and thermodynamic data is sought. Such robust descriptions can be used in conjunction with well tested and reliable models for extrapolating into systems with more and more components. The combination of powerful computer software and reliable thermodynamic databases (often available commercially— see the section "Thermodynamic databases") give the materials scientist a powerful tool in alloy design and process engineering, enabling a limitless number of "virtual experiments" to be conducted, saving time and the expense associated with huge experimental programmes, making it possible to obtain detailed descriptions of highly complex materials systems.

Modeling of the Gibbs Energy of the System

In order to produce high-quality, accurate predictions of phase equilibria using the *Calphad* method, it is crucial to have an accurate description of the Gibbs energy of each phase that may potentially exist in the system under study. Such Gibbs energy functions are stored in thermodynamic databases in the form of polynomial expressions, the coefficients of which are obtained by experiment and/or from theoretical evaluations.

The temperature dependence of the Gibbs energy (G) of a component (element or any pure species) is often expressed as:

$$G = a + bT + cT\ln(T) + \Sigma\, d_i T^n$$

where a, b, c & d_i are adjustable coefficients.

With appropriate differentiation, it is possible to derive expressions for Enthalpy, Entropy and Heat Capacity (Cp). It's possible to add additional terms to cater for pressure dependence or magnetic contributions (Lukas et al. 2007).

This treatment is adequate to model unary (one component) systems, but in dealing with binary, ternary and multicomponent systems, it is necessary to consider phases that contain more than one component, such as solutions or compounds. This requires consideration of the combination of the Gibbs energies for the components plus a contribution from the composition dependence. Thus, the molar Gibbs energy of a substitutional solution phase Φ can be considered to be the sum of a number of contributions:

$$G_m^\Phi = G_{ref}^\Phi + G_{id}^\Phi + G_e^\Phi + G_{mag}^\Phi + G_p^\Phi + \cdots \tag{1}$$

where G^{Φ}_{ref} refers to the weighted sum of the Gibbs energy of the components of the phase relative to the chosen reference state (generally, the Standard Element Reference—SER), and is defined as

$$G^{\Phi}_{ref} = \sum_{i=1}^{n} x_i \cdot {}^0G^{\Phi}_i \tag{2}$$

where x_i is the mole fraction of component i in the system and ${}^0G_i^{\Phi}$ is the Gibbs energy of the pure component i in the crystallographic structure Φ.

The contribution to the Gibbs energy from ideal random mixing of the components in the phase is given by the second term.

$$G^{\Phi}_{id} = R \cdot T \cdot \sum_{i=1}^{n} x_i \cdot ln(x_i) \tag{3}$$

The deviation in the Gibbs energy from this ideal behaviour is described by the third term, the excess term.

$$G^{\Phi}_E = \sum_{\substack{i,j=1 \\ i \neq j}}^{n} x_i \cdot x_i \cdot \sum_{z=0}^{m} {}^zL_{ij} \cdot (x_i - x_j)^z + \sum_{\substack{i,j,k=1 \\ i \neq j \neq k}}^{n} L_{ijk} \tag{4}$$

This is the Muggianu extension of the Redlich-Kister formalism (Muggianu et al. 1975; Redlich and Kister 1948), where ${}^zL_{i,j}$ is the interaction parameter describing the mutual interaction between the components i and j. This parameter can be temperature dependent—usually a linear dependence is adequate. The parameter L_{ijk} describes the ternary interaction between any three components in the system (Hillert 1980).

Additional terms may be necessary for a more accurate description of the Gibbs energy. For example, G_{mag} is the magnetic contribution described in detail by Hillert and Jarl (Hillert and Jarl 1978) and G_p is an additional term to account for variation of Gibbs energy with pressure.

This model is used for phases with a disordered structure, from liquids to solid solutions. For modeling data for intrinsically ordered phases, such as *intermetallic phases*, a different type of model is used, which will be briefly described below. More details can be found, for example, in (Lukas et al. 2007).

The simplest type of ordered phase is one that exists over a negligible composition range; a line compound. As there is no composition range, the expression that is used to describe the thermodynamic properties of pure components can be used. More often however, this is simplified further if only the enthalpy and entropy of formation of the compound (i.e., the Gibbs energy of formation, G_f) are known:

$$G^{\Phi}_m = G^{\Phi}_{ref} + G^{\Phi}_f \tag{5}$$

However, this assumes that the Cp of the compound is equal to the weighted sum of those of the pure components (Neumann-Kopp), which may not be appropriate and can lead to unexpected results (Lukas et al. 2007). The Gibbs energy of formation can be also evaluated during the process of assessment (see the next section), based on any experimental data.

The modeling of data for ordered phases which have a range of homogeneity presents much more of a challenge. In principle, the simple substitutional solution model described above can be used (and has been used) to model the thermodynamic properties of such phases and can lead to good phase diagrams. However, it is much more useful to employ a model that also reflects the ordered nature of the phase and is capable of modeling the preferential site occupancy within the crystal lattice. So, the modeling of ordered intermetallic phases usually involves the application of the *Compound Energy Formalism* (CEF) (Sundman and Ågren 1981; Andersson et al. 1986). This can be considered as an amalgamation of the model for pure phases and that for substitutional solutions.

In this case the crystal lattice of the intermetallic phase is considered in terms of a collection of *sublattices* that will show preferential occupancy for one of the constituents of the phase. Therefore, depending on the number of sublattices in the phase, there are a number of identifiable 'virtual' compounds, the stoichiometry of which is given by considering that each sublattice is completely occupied by a single constituent (It is worth mentioning here that a vacant lattice site (Va) is also considered as a constituent in the phase). This then gives for the reference Gibbs energy:

$$^0G^\Phi_{ref} = \sum_{\substack{i,j,\ldots,k=1 \\ i \neq j \neq \ldots \neq k}}^{n} y_i^1 \cdot y_j^2 \cdot \ldots \cdot y_k^s \cdot {}^0G^\Phi_{(i:j:\ldots:k)} \tag{6}$$

The $^0G^\Phi_{(i,j,\ldots,k)}$ terms are the Gibbs energies of the 'virtual' compounds in the structure of the phase Φ, s is the number of sublattices and the y_i^l terms are the site fractions of constituents i occupying the lth sublattice. Of course, in order to satisfy the range of homogeneity that the intermetallic phase may have, it is necessary to introduce mixing of all components on these sublattices as well, and this is satisfied by including a term similar to the substitutional solution model. By analogy with the model for the substitutional solution, the ideal Gibbs energy is given by:

$$G^\Phi_{id} = R \cdot T \cdot \sum_{l=1}^{s} f_l \sum_{i=1}^{n} y_i^l \cdot ln(y_i^l) \tag{7}$$

where f_p is the stoichiometric coefficient for a given sublattice and the second sum describes the ideal mixing within the sublattice p.

The excess term is then given by (for the simpler case of two sublattices):

$$G_E^{\Phi} = \sum_{\substack{i,j,k=1 \\ i \neq j}}^{n} y_i^1 \cdot y_j^1 \cdot y_k^2 \cdot \sum_{z=0}^{m} {}^{z}L_{i,j:k} \cdot (y_j^1 - y_j^2)^z \tag{8}$$

where ${}^{z}L_{i,j:k}$ is the interaction parameter describing the mutual interaction between components i and j on the first sublattice when the second is fully occupied by component k.

As mentioned above, we generally need an expression for the variation in the Gibbs energy of every phase potentially stable in that system, and to evaluate this, we must be able to evaluate the Gibbs energies of all the pure components in all phases (${}^{0}G_i^{\Phi}$), and the Gibbs energies of all 'virtual' compounds (${}^{0}G^{\Phi}_{(I,j,...,k)}$). We also have to describe the crucial mutual interaction between components in the phases (${}^{z}L_{i,j'}$, $L_{ijk'}$, ${}^{z}L_{i,j:k}$). Here only the important interactions are necessary and most of them can be neglected.

This description can be obtained for binary and ternary systems with very good precision and it is possible to model the system if reliable experimental data are available describing both the thermodynamic properties (e.g., heat capacity, enthalpies of formation or mixing, activities) and phase equilibria (e.g., temperatures of invariant reactions, compositions and amounts of phases in equilibria). At least one set of experimental thermodynamic properties is very important; it can be shown easily that almost identical phase diagrams of a system can be calculated using different sets of Gibbs energy expressions, but which will give quite different values for the thermodynamic properties. Experimental thermodynamic data allows us to select a realistic set of model parameters.

An intrinsic feature of the *Calphad* method is the need to evaluate Gibbs energies both for components in meta- or unstable structures and for 'virtual' compounds which cannot exist. Here the data can be estimated or obtained by extrapolation from higher order systems (Lukas et al. 2007). The second possibility is to derive the data during the assessment process (the optimization of parameters to the thermodynamic functions) for a higher order system where experimental data for the solution of the element in the phase in question are available. Recently, it has been possible to perform calculations using quantum mechanics based on *density functional theory (ab initio)* as a result of the rapid development in computer power and mathematical methods. It allows the evaluation of the differences in energy for an element between two phases (Sob et al. 2009), for example. These *ab initio* values are becoming more and more accurate so that it is possible to use them (with necessary care) as 'computer experimental values' together with other experimental values in the critical assessment of thermodynamic data. The disadvantages of such data (the calculations are valid for a temperature of 0 K), can be compensated for by including the results of phonon calculations to provide heat capacities.

The approach described above can be applied to a wide variety of materials and phases, from simple solid solutions to phases with complex crystallography, or liquid phases with a strong tendency to form molecular species. It is also possible to model complex ordering processes in systems. A detailed description for interested readers can be found, e.g., in (Hillert and Staffansson 1970; Saunders and Miodownik 1998; Kroupa et al. 2007a; Lukas et al. 2007).

Critical Assessment of Phase Equilibria and Thermodynamic Properties

The process which leads to a reliable set of thermodynamic parameters G and L, and to the robust and correct mathematical description of the phase diagram of any system is called *Thermodynamic Assessment*. Generally, it can be applied to binary and ternary systems where a reasonable amount of experimental data is available.

The assessment is based on a scientific analysis of all relevant experimental data followed by a mathematical process which leads to the best possible agreement between experiment properties and those calculated from the models. The least squares method is used as the process to provide such agreement. The scientific analysis is critically dependent on the knowledge and experience of the assessor as the most reasonable experimental data has to be selected from an often extensive number of measurements of the same thermodynamic or phase equilibrium quantity. Data in the scientific literature can often contradict each one another and the selection of the 'correct' set is the key task in the assessment process.

The selected experimental data are used as a basis for the optimization of model parameters G and L. Such parameters should represent the best theoretical thermodynamic description of the system under investigation.

The choice of the most appropriate models (see above—solid solution, CEF, etc.) is at the discretion of the assessor, but it should respect, for example, the crystallography, degree of stoichiometricity and ordering properties. Taking the physical properties into the account during the assessment usually makes the extrapolation from binary and ternary systems to higher order systems easier.

After the data selection, the final optimization is carried out using the assessment modules that are available in all of the main commercial software packages for the modeling of phase diagrams using the *Calphad* method. ThermoCalc with the Parrot module (Andersson et al. 2002), MTDATA with its Assessment module (Davies et al. 2002),

Pandat with the Pan-optimizer (Chen et al. 2002) and FactSage (Bale et al. 2002) are the main software packages used for this purpose.

It is necessary to test the obtained dataset after the optimization as mathematical artefacts may appear as the result of the process, the agreement with the selected experimental dataset must be verified and it is also important to check that the extrapolation to high and low temperatures and in terms of composition is reasonable (Schmid-Fetzer et al. 2007).

The final verified dataset can serve as a building block for the construction of a thermodynamic database, which in turn allows reliable predictions of phase diagram sections in very complex systems (e.g., advanced steels and superalloys, sometimes with more than 10 components). The way in which the database is constructed is described in next part of the chapter.

Thermodynamic Databases—Key Conditions for Successful Modeling of Phase Diagrams and Thermodynamic Properties of Complex Systems

The software available for the thermodynamic modeling of phase diagrams and thermodynamic properties of multicomponent system using the *Calphad* method is just an empty shell without proper thermodynamic data. Therefore, the existence of robust thermodynamic datasets for simpler systems that use models capable of reliable extrapolation of existing data to higher order systems is necessary. These datasets are available in the form of large thermodynamic databases, either general ones or those specialized for particular types of materials (Ni-based superalloys, steels, solders, oxide systems, etc.).

As mentioned above, the completed theoretical assessment has to be subjected to a whole set of tests to prove the quality of the dataset. Nevertheless, this is not the final test of the quality of the dataset as every assessment is bound to become part of larger database, either created by the author of the assessment or (very often) created by different scientific groups or teams of producers of commercial software, using the datasets already published in the scientific literature. In all cases, the datasets must be tested to ensure that they obey an additional set of conditions and rules.

Creation of thermodynamic databases

The thermodynamic databases that are used in conjunction with the available software packages are not just a "mechanical" collection of binary and ternary thermodynamic datasets published in the literature or assessed by the authors themselves. In particular, if the data originate from various sources, additional tests and checks are necessary in order to ensure that the complex thermodynamic database is correct, consistent and applicable for the prediction of thermodynamic properties and phase diagrams of unknown higher order systems.

Some of the problems which can appear without proper testing of particular assessments and the complete database have been summarized in the work of Schmid-Fetzer et al. (Schmid-Fetzer et al. 2007). The need for very detailed and meticulous database checks can be nicely illustrated by the example shown in the Figs. 9a and b. It shows the phase diagram of the Ag-Bi system, calculated using the specialized solder database before and after the necessary set of tests had been carried out. Just a quick comparison of Fig. 9a with the experimental phase diagram reveals the presence of the HCP phase in the calculated diagram, which is not actually shown as stable in the experimentally determined phase diagram. The presence of this phase in the calculated diagram is due to the large solder database containing the Gibbs energies for both elements in the HCP and other ("hypothetical") crystallographic structures, which are needed as 'virtual structures (compounds)' for the *Calphad* method. When the theoretical assessment of the Ag-Bi system was carried out, there was no need to take the HCP phase into consideration and therefore no interaction parameters were defined for this phase. However, once the assessed dataset for this system was incorporated into the larger database, the software was able to calculate the Gibbs energy of HCP phase for this system as the values for the pure elements were present. And as interaction parameters for this phase in this system were missing, the software then treated this phase as an ideal solution (Fig. 9a). The result is that the HCP phase erroneously appears as stable in the system. The only solution to this problem is by the inclusion of the missing interaction parameters for the HCP phase in the Ag-Bi system into the database. As a general rule in cases such as this, it is sufficient to assign to the problem phase the same interaction parameters as used for one of the stable phases. The correctly calculated phase diagram is shown in Fig. 9b.

The example given above shows a typical situation, when the binary theoretical assessment is correct and reproduces well the phase diagrams and thermodynamic properties of system; however, problems appear after introducing the assessed thermodynamic dataset into a larger database. At least three other tests should be carried out to ensure correctness and consistency of any large database.

A basic criterion for the construction of the database is that a single set of unary data is used for all constituents (elements and species). That is to say that only one unique thermodynamic description is used to represent

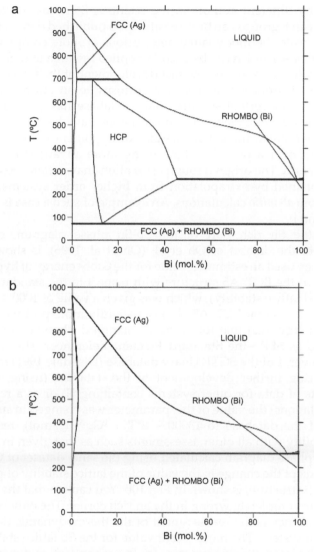

Fig. 9: (a) Incorrect phase diagram of the Ag-Bi system with the redundant HCP phase in the central part. (b) Correct phase diagram of the Ag-Bi system.

the Gibbs energy of each element, not only for their stable structures, but also for all other hypothetical (non-existent) structures, which are defined for that particular element. The majority of assessors now use the SGTE Unary database version 1 as a source of data for the elements (Dinsdale 1991), the creation of which was crucial in order to realise the full potential of universal use of the thermodynamic assessments published by different authors in the scientific literature. Because of it, various research groups can build easily upon published data in their own work and create thermodynamic descriptions of more complex systems or even databases which can be used in applied research and in industrial laboratories. This stimulated the fast development of the *Calphad* method and its growth from a basic research curiosity to an important tool for everyday use. Nevertheless, this unary database has been the subject of continuous development and new versions have been introduced, in which, in particular, the Gibbs energies of hypothetical structures have been refined over a period of time using more advanced methods for their evaluation. The original rough estimations have often been replaced by data obtained by extrapolation from higher order systems, or more recently, from *ab initio* calculations. An example of such a case is shown in Figs. 10a and b.

Here, the Sn rich part of the Sb-Sn phase diagram, calculated according to the dataset of Oh et al. (Oh et al. 1996), is shown in the Fig. 10a. They used an estimated value for the Gibbs energy of hypothetical state of Sb in the BCT_A5 structure with respect to the standard state of Sb (so called lattice stability), which was given a value of $1000 + {}^0G_{Sb}{}^{A7}$ (J/mol). This means that BCT_A5 Sb has a slightly more positive value for its Gibbs energy than that for Sb in the standard Rhombo_A7 structure (referred to as SER—the Standard Element Reference). This value was taken from ver. 1 of the SGTE Unary database (Dinsdale 1991) mentioned above. During further development of the database (using, e.g., new assessments of data for other systems containing Sb or as a result of *ab initio* calculations) the value of this parameter was changed in subsequent versions of the database to $13000 - 8*T + {}^0G_{Sb}{}^{A7}$ (J/mol), resulting in incompatibility with all older assessments such as that given in (Oh et al. 1996). The phase diagram calculated using the same dataset of Oh (Oh et al. 1996), except the change in the value of the lattice stability of pure Sb in the BCT_A5 structure, is shown in Fig. 10b. You can see that the invariant reactions are completely wrong in the Sn rich corner. The only solution to this inconsistency is the reassessment of the thermodynamic description of the Sb-Sn system. The more recent value for the Sb lattice stability was used and the interaction parameters for the concentration dependence of the Gibbs energy of Sb in the BCT_A5 phase were changed to maintain agreement with the experimental data available in the literature (Kroupa and Vizdal 2007b).

The second important issue relates to the models used to represent the temperature, pressure and composition dependence of the Gibbs energy of a given phase including any additional contributions to it (e.g., related to magnetic properties, surface energy contribution, etc.). The most common method used today for the expressing the concentration dependence of

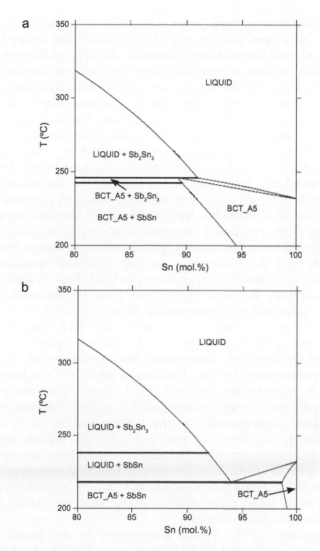

Fig. 10: Sn rich part of the binary Sb-Sn phase diagram: (a) correct phase diagram using the unary data from (Dinsdale 1991). (b) Phase diagram after substitution of the unary Gibbs energies of Sb into the structure of tin using the value from the SGTE unary database 4.4 (Unary 4.4 2001).

the thermodynamic properties is the Redlich-Kister expression (Redlich and Kister 1948) for binary systems, which was extended by Muggianu (Muggianu et al. 1975) to ternary systems. Other formalisms can be used for the extrapolation into ternary systems, for example, the Kohler equation or Toop equation (Saunders and Miodownik 1998). Similarly, several models can be used to express the temperature dependence of the excess Gibbs energy (Kaptay 2014; Schmid-Fetzer et al. 2015). In a similar way different approaches have been used to model Chemical/Short Range Ordering (SRO) (Ansara and Dupin 1998). Therefore, the type of the polynomials used for the expression of excess Gibbs energy has to be checked and the same models applied to the same types of phase throughout the database.

The last important consistency check is related more to the formal structure of the database—firstly, one must assure that the names assigned to all unique phases in the database are exclusive for that phase but at the same time identical for it in all systems in the database, where that phase exists. Secondly, the models assigned in the Compound Energy Formalism (Sundman and Ågren 1981; Andersson et al. 1986) for the phases with the same or similar crystallographic structure, existing in different systems (binary, ternary, etc.), has to be identical if we want to model, for example, the experimentally confirmed complete solubility of alloying elements between terminal phases in a ternary system. If this condition is not fulfilled, a two phase field always appears between otherwise identical phases. As an example, the Ni-Sb-Sn system can be used (Figs. 11 and 12). Here complete solubility between Ni_3Sn_2 and NiSb phases exist and was confirmed experimentally (Kroupa et al. 2014).

A further complication is caused by the fact, that the phases have different solubility ranges in the binary systems and even different crystallographic structures. The structure of the Ni_3Sn_2 phase is characterized by the Pearson symbol $hP6$ (Ni_2In-type), the structure of NiSb phase by $hP4$ (NiAs-type). Nevertheless, both structures are very closely related, which allows complete solubility to be observed experimentally in the ternary system. As is shown in Figs. 12a and b, two layers of atoms (Ni and As/In) are identical in both crystallographic structures, whereas the third layer is empty in the crystallographic structure corresponding to the NiAs prototype but fully occupied in the structure corresponding to the Ni_2In prototype. In the particular case of Ni_3Sn_2 phase existing in the Ni-Sb-Sn system, this third layer contains structural vacancies.

Unfortunately, in the recent assessments of the binary systems Ni-Sn (Liu et al. 2004) and Ni-Sb (Zhang et al. 2008) both of the above mentioned phases had been modelled using different and incompatible models. Therefore, in order to be able to model the mutual solubility of the NiSb and Ni_3Sn_2 phases, proper modeling of the Ni-Sb-Sn ternary phase diagram requires the reassessment of at least one of the subsystems, and careful selection of the model is necessary. Such a model should allow the modeling of not only these two phases, but also all other phases with such crystallographic structures in any other system of interest (Zemanova et al. 2012; Kroupa et al. 2014).

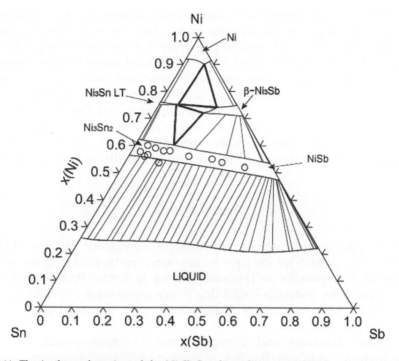

Fig. 11: The isothermal section of the Ni-Sb-Sn phase diagram at 900°C, where complete solubility exists between the Ni_3Sn_2 and NiSb phases. The solubility was confirmed by (Kroupa et al. 2014) (o—experimental points).

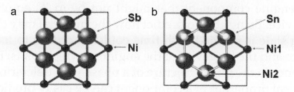

Fig. 12: The crystallographic structure of (a) NiSb phase (prototype NiAs) and (b) Ni_3Sn_2 phase (prototype Ni_2In).

This consistency condition is the most difficult to fulfil and careful checks are necessary for every new dataset inserted into the larger database. There are no general rules how the type of the phase (terminal solid solution, intermediate phases, SRO) and its characteristics (crystallographic structure, composition, solubility range) should be described by the model. Those conducting the assessment are free to choose the model for each phase and to allocate any name to it. Therefore this has to be analysed by creators of the database.

The above mentioned rules have to be implemented for any database, if we want to obtain an accurate phase diagram and thermodynamic property modeling, and especially in order to exploit fully the predictive power of the *Calphad* method. Without detailed analysis of all datasets, the reliability of a complex thermodynamic database is questionable. Often, the database is deliberately limited to some concentration and/or temperature region and its authors should properly define the conditions under which it can be used. This is true especially for the specialized databases covering only certain types of materials. A more detailed description of the database consistency rules is available in Kroupa (2013).

Applications

The properties of materials are strongly dependent on their phase composition, which for the equilibrium state, can be read from the phase diagram. Computational Thermodynamics is a tool that can provide information for materials containing many components, something that the traditional 'paper' phase diagram cannot do. The principles of both are, however, the same.

Phase diagrams and Computational Thermodynamics consider equilibrium conditions. But in Computational Thermodynamics, having the ability to remove individual phases from consideration in a calculation can give an insight into non-equilibrium processing. But in any case, knowing the equilibrium condition is useful. For example, consider the case of an engine component that might not be in its equilibrium state when in service. Being able to predict how the material may change as the equilibrium state is reached with time could be crucial to its mechanical performance and the longevity of the engine in which it sits. For example, the appearance in the microstructure of a phase with deleterious effects on the mechanical properties of a component can be easily predicted by these computational techniques.

There are numerous examples of applications of Computational Thermodynamics in the literature. Below are just a few cases.

Steels

In alloyed steels, the occurrence in the microstructure of the sigma-phase as hard, brittle precipitates has a deleterious effect on the mechanical and corrosion properties of the material. Careful control of its presence in multicomponent systems in the equilibrium state can be achieved with the aid of calculation of phase equilibria using the *Calphad* method.

The sigma-phase is stable in 43 different binary systems and has a Topologically Close-Packed (TCP) structure (Frank-Kasper phases). The sigma-phase is characterised crystallographically with Space Group $P4_2/mnm$ (Pearson symbol $tP30$) consisting of 30 atoms on 5 non-equivalent positions in the unit cell, given as ($2a$, $4g$, $8i_1$, $8i_2$, $8j$), where the subscripts to the numbers refer to the Wyckoff positions in the unit cell (a, g, i_1, i_2, j) and the numbers refer to the number of atoms per position, or sublattice. Therefore, in a two-component sigma phase, there are 32 possible non-equivalent combinations of the components on the 5 positions of the structure.

From the earlier discussion of the Compound Energy Formalism, which is the model of choice for ordered intermetallic phases where there is preferential site occupation by the components, it is apparent that in order to describe the structure fully, parameters are required for each of the 32 combinations, or 'virtual compounds'. In addition, parameters describing the mixing of the components on the individual sublattices may be required as well. The fitting of such a large number of parameters in the least-squares optimisation procedure requires very many experimental data in order to do this unambiguously. In order to make the modeling of this complex phase a simpler task, the complicated five sublattice model has been simplified in a number of ways. Early attempts at modeling this phase considered the simplest of approaches, that is, as a stoichiometric compound. This was employed in the Al-Nb system (Nb_2Al) (Kaufman and Nesor 1978) and in the Re-Ta system (Re_3Ta_2) (Cui and Jin 1999). Also, the phase has been treated as a substitutional solid solution (A,B), essentially considering the phase as comprising a single sublattice (Watson 1987; Vrestal 2001). But neither of these models have been particularly successful. The sigma-phase often appears with a wide range of homogeneity, making a simple stoichiometric model completely inappropriate. On the other hand, a substitutional solution model, while allowing a wide range of homogeneity, does not give any information about sublattice occupancy.

The most successful simplification has been to reduce the number of sublattices in the model from 5 to 3. This has been achieved by consideration of the site occupancy and the coordination number of the lattice sites. This has enabled a simplified model to be proposed by Andersson and Sundman (Andersson et al. 1986) of $(A,B)_{18}(A)_4(B)_8$, where A and B are component elements. Here, the sublattices representing the a, i_1 and i_2 positions are combined. The model shows preferential occupation of A and B in sublattices 2 and 3, with mixing taking place on the first sublattice allowing an appreciable homogeneity range to be described. This is not the best combination of the sublattices, a more appropriate approach combining the a and i_1 sublattices, and the i_2 and j sublattices, giving the model $(A,B)_{10}(A)_4(A,B)_{16}$ (Ansara et al. 1997). Further extensions to the modeling have been made, for example using the 'modified CEF' using the descriptions such as $(A,B)_4(A,B)_{16}(A,B)_{10}$ (Watson and Hayes 2001) and $(A,B)_{20}(A,B)_{10}$. But recent progress in DFT calculations has allowed the full five sublattice modeling of $(A,B)_2(A,B)_4(A,B)_8(A,B)_8(A,B)_8$ to be explored, with thermodynamic parameters being supplied by *ab initio* calculations. The advantages and disadvantages of the different simplifications have been discussed in detail in Joubert (2008). But at present, the model of choice for the sigma phase is the $A_4(A,B)_{18}B_8$ CEF model, and has been used in many of the databases that are available, e.g., STEEL16 (Kroupa et al. 2001) or the Solution database provided by SGTE, SSOL (SSOL 2015).

The application of calculations of phase equilibria involving the sigma-phase in corrosion resistant steels has been shown, for example in work by Svoboda et al. (Svoboda et al. 2004) and Kraus et al. (Kraus et al. 2010). As an example, Nicrofer 3127 (27.0 wt.% Cr, 31.0 wt.% Ni, 6.4 wt.% Mo, 1.6 wt.% Mn, 1.3 wt.% Cu, 0.30 wt.% Si, 0.009 wt.% C, 0.20% wt.% N, Fe balance) was chosen.

Calculations using Thermocalc with either the STEEL16 or SSOL databases predict an austenite matrix in equilibrium with $M_2(C,N)$ carbonitride and the sigma-phase. The calculated phase fraction of the sigma-phase in the equilibrium state of Nicrofer 3127 at 700°C is 0.186 when using STEEL16 and 0.169 when using SSOL (Kraus et al. 2010).

Experimentally, during heating of Nicrofer 3127 at 700°C, the volume fraction of the sigma-phase determined on foils by transmission electron microscopy (TEM) increased from 0 to (37 ± 5) volume % at 6170 hours of annealing (Kraus et al. 2010), see Fig. 13.

The calculation, in this case, gives valuable assistance in identifying the completeness of the process of sigma-phase precipitation.

Ni-based Superalloys

Intermetallic compounds with transition metals involving primarily nickel and aluminium as the main components are characterized by their resistance to high temperatures. Alloys containing such intermetallic phases, known as Ni-based superalloys, play a key role in high temperature engineering applications, such as for materials for power plant turbine engines or gas turbine engines, and are therefore important for the power, car and aerospace industries. Because of their high-temperature material properties, their inclusion in gas turbines, for instance, allows higher engine operating temperatures with an accompanying improvement in efficiency, resulting in a reduction in fuel consumption and a subsequent

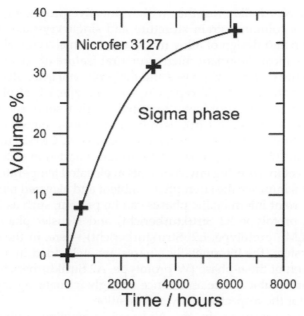

Fig. 13: Time dependence of the precipitation of sigma-phase in Nicrofer 3127 at 700°C.

saving of natural resources. Over recent years, attention has been focused on the study of model systems that are able to simulate the behaviour of real materials containing a large number of alloying elements. The alloying additions have a dramatic effect on the material properties, with each chosen for a specific purpose; thus combinations of elements can give improved high temperature properties at the same time as oxidation resistance, for example. The presence of intermetallic compounds in the microstructure of the alloy has been shown to help in improving the mechanical properties. Strength can be increased by presence of the intermetallic particles that impede dislocation movement. On the other hand, the presence of certain intermetallic phases can promote material damage owing to their negative effect on properties through embrittlement.

Over the past 50 years, materials scientists have been able to raise the operating temperature of turbine blades by around 400°C so that blades can now operate at temperatures over 1100°C. As the performance of a turbine engine improves with increasing operating temperature, the demand for further development of superalloys is ongoing. The improvement of the high temperature creep strength of advanced Ni-based superalloys has been achieved by a combination of complex alloying of the basic Ni-Al alloy together with innovative processing techniques (the use of a columnar grain structure and single crystals). The careful investigation and design of alloy microstructure has a crucial role in this process. The most important microstructural feature of Ni superalloys is the coexistence of the $\gamma + \gamma'$ phases and the positional geometry of ordered intermetallic phase γ' (Ni_3Al type, Cu_3Au prototype, $L1_2$ Strukturbericht) with respect to the disordered terminal Ni-rich solid solution γ (Cu prototype, A1 Strukturbericht), both having a face-centered cubic (FCC) lattice structure. Advanced superalloys (CMSX 4, CMSX 10, RR 3010) contain up to 70% of the γ' phase. Chromium is added in order to reduce embrittlement in oxidising environments at elevated temperatures. Cobalt and titanium improve the strength at ambient and elevated temperatures. Other important intermetallic phases can be present, such as the β (NiAl type, CsCl prototype B2 Strukturbericht) and Heusler phases ($AlNi_2Ti$ type, $AlCu_2Mn$ prototype, $L2_1$ Strukturbericht)—here, in the case of the Al-Ni-Ti system, the intermetallic phases are embedded in a disordered solid solution of the α-phase (W prototype, A2 Strukturbericht) having a body-centered cubic lattice and hence strongly influencing the materials properties for the respective alloy composition.

The phase relations in the Ni-based superalloy multicomponent systems can be successfully predicted using the modeling approaches described above. Owing to the crystallographic sublattice symmetry of the intermetallic phases, a simplified two sublattice form of the compound energy formalism (CEF) can be used to describe their thermodynamic properties; $(A,B)_3(A,B)_1$ for the γ' phase and $(A,B)(B,Va)$ or $(A,B,Va)(A,B,Va)$ for the β-phase (Dupin 1995; Ansara et al. 1997; Dupin and Sundman 2001), where A and B are metallic elements and Va represents a vacant sublattice site. The Heusler phase can modelled separately as $(A,B,C)(A,B,C)(B,Va)_2$ (Dupin 1995) with C being a third element. However, in order to describe these phases correctly it is important to include structural vacancies in the thermodynamic modeling. For other alloy systems, e.g., Al-Fe, the vacancies can be ignored (Sundman et al. 2009).

The intermetallic phases discussed above have ordered crystal structures that are closely related to their respective disordered matrix phases. For this reason, a strategy for the thermodynamic modeling of these intermetallic phases is to use a single Gibbs energy function which describes both the ordered and the disordered state of the FCC or the BCC phase, respectively. In order to enable a simultaneous thermodynamic modeling of the processes of ordering of the β and the Heusler phase from the α matrix in a ternary or higher order system using the one Gibbs energy function only, a four sublattice description of the same form $(A,B,C,Va)(A,B,C,Va)(A,B,C,Va)(A,B,C,Va)$ for both the β (Dupin 2001) and the Heusler phase (Sundman et al. 2009) is necessary. Similarly, the previous sublattice description for the γ´ phase can be modified resulting in (A,B,C)(A,B,C)(A,B,C)(A,B,C) (Dupin 2001), which leads to a complete unification of the thermodynamic models for the description of order-disorder transformations of the BCC and FCC structures in Ni-based superalloys.

A particular example of the technically important Ni-based alloys is the Al-Ni-Ti system. This system combines the above mentioned properties of the basic binary Al-Ni binary system with those of the Al-Ti alloys, which are known to contribute an improvement in room temperature ductility while simultaneously maintaining good creep and oxidation resistance. Figure 14 shows a calculated isothermal section of the Al-Ni-Ti system at 1050°C (Buršík and Brož 2009), which yields valuable information for engineers and materials scientists, indicating which phase structures can be expected at specific compositions at elevated temperatures. This enables the material scientist to select alloy compositions that will give the required properties for the application.

Calculation of Thermophysical Properties

Calculations of phase equilibria for a multicomponent alloy may be carried out for a fixed composition over a range of temperatures to identify where a material solidifies and to calculate such properties as, the liquidus temperature, solidus temperature, fraction solid, phase compositions, heat capacity and enthalpy change. This can be extended to other properties such as molar volume (or density) if the necessary volumetric data are also available. Such calculations assume that the material is at equilibrium throughout the solidification (the so-called lever-rule approximation). Such information can be fed into software packages for the simulation of entire casting processes of large engineering components such as engine blocks thereby improving product quality, energy control and materials usage.

Such an approach is particularly useful for aluminium alloys. In reality equilibrium is rarely maintained during solidification because of the finite diffusion rates of atoms or species within phases. Equilibrium solidification can be seen as one extreme case where atoms or species are assumed to diffuse very rapidly as compared to the time scales associated with solidification. An alternative extreme case can be calculated where it is assumed that diffusion in the liquid phase is extremely rapid while diffusion in the crystalline phases is very slow. Assuming that equilibrium is still maintained at the interface between the solid and liquid this gives rise to the so-called Scheil approximation. In reality neither model is

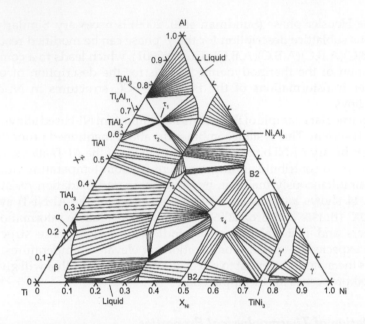

Fig. 14: Calculated phase diagram of Al-Ni-Ti system for 1050°C (Buršík and Brož 2009) [τ₁: Ti₅Ni₂Al₁₃ (π), τ₂: TiNiAl₂ (μ), τ₃: TiNiAl (γ), τ₄: TiNi₂Al (Heusler phase)].

correct although solidification of different classes of materials may be closer to one extreme than to the other.

Figure 15a shows the calculated fraction of phase formed during solidification of a very commonly used aluminium alloy, LM25. The specification of LM25 can vary a little but will always contain Si with minor additions of Fe, Mg and Mn as well as Si. A large number of intermetallic phases can potentially form in the system in addition to the liquid phase and the solid solution phases in which the pure elements may crystallise. Calculations indicate that at low temperatures the phases Mg₂Si, and ternary phases emanating from the Al-Fe-Si and Al-Mn-Si systems may crystallise out. Both of these latter phases are complicated requiring use of a model with four sublattices with mixing taking place on at least one of these sublattices.

Other important properties may be calculated at the same time such as the variation of the heat capacity (Fig. 15b), the enthalpy released and, if the database contains volumetric information, the density (Fig. 16). These are properties which are vital to a successful simulation of a casting process.

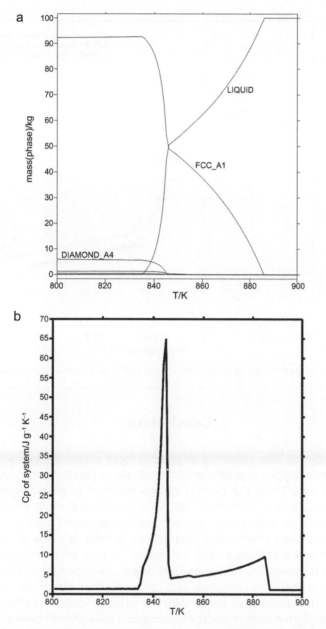

Fig. 15: (a) Calculated phase fraction of aluminium alloy (LM25) (b) calculated heat capacity.

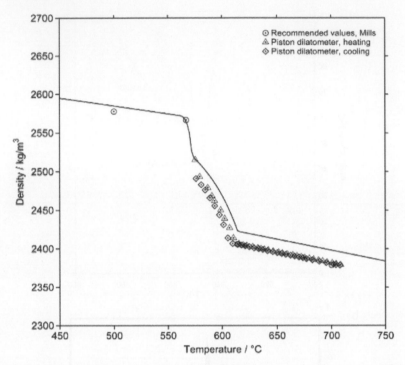

Fig. 16: Calculation of density of aluminium alloy (LM25) compared with experimental data.

Conclusions

The mysteries of how elements interact to form useful materials has been with us since copper was first mixed with tin to produce bronze, around 5000 years ago. From the earliest representations of phase equilibria in diagrammatic form, more than 100 years ago, to the most sophisticated mathematical modeling of complex multicomponent phase equilibria of today, the basic science has changed very little. It's one of the oldest fields in materials science and yet today is one of the most vibrant. As interest in some of the simpler yet highly precise experimental techniques fades, they are replaced by computer-power-hungry simulation software; although calculations will never take the place of good experiments and the continued provision of good experimental data should be encouraged. However, one thing is sure: even though some may think that the study of phase equilibria and thermodynamics is 'old-hat', it remains at the very basis of materials science.

Acknowledgements

The authors wish to acknowledge the support of the Grant Agency of the Czech Republic (Project No. GA14-15576S).

Keywords: Phase Diagram, Experimental Thermodynamics, Calphad, Thermodynamic Database, Computational Thermodynamics, Thermodynamic Assessment

References

Andersson, J.-O., A. Fernandez Guillermet, M. Hillert, B. Jansson and B. Sundman. 1986. A compound energy model of ordering in a phase with sites of different coordination numbers. Acta Metall. 34: 437–445.

Andersson, J.-O., T. Helander, L. Höglund, P. Shi and B. Sundman. 2002. Thermo-Calc & DICTRA, computational tools for materials science. Calphad 26: 273–312.

Ansara, I., T.G. Chart, A. Fernandez Guillermet, F.H. Hayes, U.R. Kattner, D.G. Pettifor, N. Saunders and K. Zeng. 1997. Workshop on thermodynamic modeling of solutions and alloys, group 2: alloys systems I, thermodynamic modeling of selected topologically close-packed intermetallic compounds. Calphad 21: 171–218.

Ansara, I., N. Dupin, H.L. Lukas and B. Sundman. 1997. Thermodynamic assessment of the Al-Ni system. J. Alloys Compd. 247: 20–30.

Ansara, I. and N. Dupin. 1998. Thermodynamic modelling. pp. 1–15. *In*: I. Ansara, A. Dinsdale and M.H. Rand [eds.]. Thermodynamic Database for Light Metal Alloys, Vol. 2. European Communities, Luxembourg.

Bale, C.W., P. Chartrand, S.A. Degterov, G. Eriksson, K. Hack, R. Ben Mahfoud, J. Melançon, A.D. Pelton and S. Petersen. 2002. FactSage thermochemical software and databases. Calphad 26: 189–228.

Boettinger, W.J., U.R. Kattner, K.-W. Moon and J.H. Perepezko. 2006. DSC Measurements of Alloy Melting and Freezing, NIST Special Publication 960-15. US Government Printing Office, Washington.

Buršík, J. and P. Brož. 2009. Constitution of Ni–Al–Ti system studied by scanning electron microscopy. Intermetallics 17: 591–595.

Chen, S.-L., S. Daniel, F. Zhang, Y.A. Chang, X.-Y. Yan, F.-Y. Xie, R. Schmid-Fetzer and W.A. Oates. 2002. The PANDAT software package and its applications. Calphad 26: 175–188.

Cui, Y. and Z. Jin. 1999. Assessment of the Re-Ta binary system. J. Alloy Compd. 285: 150–155.

Cullity, B.D. 1978. Elements of X-ray Diffraction. 2nd ed. Addison-Wesley, Reading, Mass., USA.

Davies, R.H., A.T. Dinsdale, J.A. Gisby, J.A.J. Robinson and S.M. Martin. 2002. MTDATA—Thermodynamics and phase equilibrium software from the National Physical Laboratory. Calphad 26: 229–271.

Dench, W.A. 1963. Adiabatic high-temperature calorimeter for measurement of heats of alloying. Trans. Faraday Soc. 59: 1279–1292.

Dinsdale, A.T. 1991. SGTE data for pure elements. Calphad 15: 317–425.

Dupin, N. 1995. Contribution a l'evaluation Thermodynamique des Alliages Polyconstitutes a Base de Nickel. Ph.D. Thesis, LTPCM, Institut National Polytechnique de Grenoble, France.

Dupin, N. and B. Sundman. 2001. A thermodynamic database for Ni-base superalloys. Scand. J. Metall. 30: 184–192.

Gibbs, J.W. 1876. On the equilibrium of heterogeneous substances. Trans. Connect. Acad. III: 108–248.

Heycock, C.T. and F.H. Neville. 1904. On the constitution of the copper-tin series of alloys. Phil. Trans. Roy. Soc. London Ser. A. 202: 1–70.

Hillert, M. and L.-I. Staffansson. 1970. Regular solution model for stoichiometric phases and ionic melts. Acta Chem. Scand. 24: 3618–3626.

Hillert, M. and M. Jarl. 1978. A model for alloying in ferromagnetic materials. Calphad 2: 227–238.

Hillert, M. 1980. Empirical methods of predicting and representing thermodynamic properties of ternary solution phases. Calphad 4: 1–12.

Joubert, J.-M. 2008. Crystal chemistry and Calphad modelling of the sigma phase. Prog. Mater. Sci. 53: 528–583.

Kaptay, G. 2014. On the abilities and limitations of the linear, exponential and combined models to describe the temperature dependence of the excess Gibbs energy of solutions. Calphad 44: 81–94.

Kaufman, L. and H. Bernstein. 1970. Computer Calculation of Phase Diagrams. Academic Press, New York.

Kaufman, L. and H. Nesor. 1978. Coupled phase diagrams and thermochemical data for transition metal binary systems—V. Calphad 4: 325–348.

Kraus, M., A. Kroupa, P. Miodownik, M. Svoboda and J. Vřešťál. 2010. Microstructure of super-austenitic steels after long-term annealing. Int. J. Mat. Res. 101: 729–735.

Kroupa, A., J. Havránková, M. Coufalová, M. Svoboda and J. Vřešťál. 2001. Phase diagram in the iron-rich corner of the Fe-Cr-Mo-V-C system below 1000 K. J. Phase Equilib. 22: 312–323.

Kroupa, A., A.T. Dinsdale, A. Watson, J. Vrestal, J. Vizdal and A. Zemanova. 2007a. The development of the COST 531 lead-free solders thermodynamic database. JOM 59: 20–25.

Kroupa, A. and J. Vizdal. 2007b. The thermodynamic database for the development of modern lead-free solders. Defect and Diffusion Forum. 263: 99–104.

Kroupa, A. 2013. Modelling of phase diagrams and thermodynamic properties using *Calphad*—development of thermodynamic databases. Comput. Mater. Sci. 66: 3–13.

Kroupa, A., R. Mishra, D. Rajamohan, H. Flandorfer, A. Watson and H. Ipser. 2014. Phase equilibria in the ternary Ni–Sb–Sn system: Experiments and calculations. Calphad 45: 151–166.

Kubaschewski, O., C.B. Alcock and P.J. Spencer. 1993. Metallurgical Thermochemistry, 6th Edition. Pergamon Press, Oxford.

Liu, H.S., J. Mang and Z.P. Jin. 2004. Thermodynamic optimization of the Ni–Sn binary system. Calphad 28: 363–370.

Lukas, H.L., S.G. Fries and B. Sundman. 2007. Computational Thermodynamics—The Calphad Method. Cambridge University Press, Cambridge.

Meschel, S.V. and O.J. Kleppa. 2004. Thermochemistry of some binary alloys of silver with the lanthanide metals by high temperature direct synthesis calorimetry. J. Alloy Compd. 376: 73–78.

Muggianu, Y.-M., M. Gambino and J.-P. Bros. 1975. Enthalpies of formation of liquid alloys bismuth-gallium-tin at 723K—Choice of an analytical representation of integral and partial thermodynamic functions of mixing for this ternary system. J. Chim. Phys. Physicochim. Biol. 72: 83–88.

Oh, C.-S., J.-H. Shim, B.-J. Lee and D.-N. Lee. 1996. A thermodynamic study on the Ag-Sb-Sn System. J. Alloys Compd. 238: 155–166.

Plevachuk, Yu., A. Yakymovych, S. Fuertauer, H. Ipser and H. Flandorfer. 2014. The enthalpies of mixing of liquid Ni-Sn-Zn alloys. J. Phase Equilib. Diff. 35: 359–368.

Redlich, O. and A. Kister. 1948. Algebraic representation of thermodynamic properties and the classification of solutions. Indust. Eng. Chem. 40: 345–348.

Roozeboom, H.W.B. 1899. Solidification of mixtures of two substances. Z. Phys. Chem. 30: 385–412.

Saunders, N. and A.P. Miodownik. 1998. CALPHAD (A Comprehensive Guide). Pergamon Press, Oxford.

Schmid-Fetzer, R., D. Andersson, P.Y. Chevalier, L. Eleno, O. Fabrichnaya, U.R. Kattner, B. Sundman, C. Wang, A. Watson, L. Zabdyr and M. Zinkevich. 2007. Assessment techniques, database design and software facilities for thermodynamics and diffusion. Calphad 31: 38–52.

Schmid-Fetzer, R., P. Wang and S.-M. Liang. 2015. Inherently consistent temperature function for interaction parameters demonstrated for the Mg-Si assessment. Presented at *Calphad* XLIV, Loano, Italy, June 2015, Book of abstracts p. 29.

Sob, M., A. Kroupa, J. Pavlu and J. Vrestal. 2009. Application of *Ab initio* electronic structure calculations in construction of phase diagrams of metallic systems with complex phases. Solid State Phenomena 150: 1–28.

SSOL. 5.2 2015. Version 5.2 of the SGTE Solution database. SGTE, St. Martin d'Heres, France.

Sundman, B. and J. Ågren. 1981. A regular solution model for phases with several components and sub-lattices, suitable for computer-applications. J. Phys. Chem. Solids 42: 297–301.

Sundman, B., I. Ohnuma, N. Dupin, U.R. Kattner and S.G. Fries. 2009. An assessment of the entire Al–Fe System including D0$_3$ ordering. Acta Mater 57: 2896–2908.

Svoboda, M., A. Kroupa, J. Sopoušek, J. Vřešťál and P. Miodownik. 2004. Phase changes in superaustenitic steels after long-term annealing. Z. Metallkde 95: 1025–1030.

Unary 4.4 2001. Version 4.4 of the SGTE Unary database. SGTE, St Martin d'Heres, France.

Van Emmerik, E.P. 1991. J.J. Van Laar, a Mathematical Chemist. Ph.D. Thesis, The University of Delft, Delft, Netherlands.

Vrestal, J. 2001. Recent progress in modelling of sigma-phase. Arch. Metall. 46: 239–247.

Watson, A. 1987. A Thermodynamic Study of the Chromium-Nickel, Chromium-Nickel-Silicon and Nickel-Vanadium Systems. Ph.D. Thesis, UMIST, Manchester, UK.

Watson, A. and F.H. Hayes. 1995. Enthalpies of formation of solid Ni-Cr and Ni-V alloys by direct reaction calorimetry. J. Alloy Compd. 220: 94–100.

Watson, A. and F.H. Hayes. 2001. Some experiences modelling the sigma-phase in the Ni-V system. J. Alloy Compd. 320: 199–206.

Wierzbicka-Miernik, A., G. Garzel and L.A. Zabdyr. 2010. Emf measurements in the liquid Au-Cu-Sn lead-free solder alloys. J. Phase Equilib. Diff. 31: 34–36.

Zemanova, A., A. Kroupa and A. Dinsdale. 2012. Theoretical assessment of the Ni–Sn system. Monatsh Chem. 143: 1255–1261.

Zhang, Y., C. Li, Z. Du and C. Guo. 2008. A thermodynamic assessment of the Ni-Sb system. Calphad 32: 378–388.

1.2

Thermophysical Properties of Metallic Alloys from *Ab Initio* Methods and Applications to Thermodynamic Modeling

Mauro Palumbo

INTRODUCTION

The aim of *ab initio* methods is to calculate properties of (condensed) matter with minimal empirical information. To achieve this goal, the interaction of atoms is described by the fundamental laws of physics based on quantum mechanics. The traditional formulation uses Schrödinger's equation involving many-body wavefunctions. The large number of interacting particles that must be dealt with in solids limits the applicability of this equation, and it is necessary to introduce several approximations in order to apply the laws of quantum mechanics to real materials. A key problem is that these approximations must still allow us to obtain results which are reasonably accurate.

One of the formulations which has proven very useful in materials science is density functional theory (DFT), where the many-body wavefunctions in the Schrödinger equation are recast into single-particle

International School for Advanced Studies (SISSA), via Bonomea, 265, 34136 Trieste, Italy; ICAMS, Ruhr-Universität Bochum, Universitätsstr. 150, 44801, Bochum, Germany. Email: mpalumbo@sissa.it wavefunctions based on the electronic density. The complications of many-body interactions are not lost, but moved into the exchange-correlation functional, whose exact form is unknown. Although DFT is in principle an exact theory, only approximate formulations for the exchange-correlation functional have been derived to date. This is the greatest source of discrepancy between DFT and experimental results and among different DFT results. Different functionals lead to properties which may be significantly different and only an *a posteriori* evaluation of the accuracy of the results is possible, by comparison with experimental data or more accurate methods. Nonetheless, DFT has, in recent years, allowed material scientists to obtain reasonably accurate results in a short-enough timeframe, provided that the system is not too large. Among the quantities that can be calculated with DFT-based methods are energies of formation, electronic densities, lattice parameters, elastic constants, free energies, heat capacities and thermal expansions.

Although most DFT calculations are carried out at 0 K, a detailed knowledge of lattice dynamics is essential for understanding and quantitative prediction of many thermophysical properties of solids. Some of them, such as thermal expansion and thermal conductivity, are intrinsically determined by atomic vibrations. Therefore, we will also present DFT-based methods to treat vibrational properties of crystalline solids.

The application of *ab initio* methods as described in the next sections greatly benefits from translational and other symmetry constraints originating from the ordered nature of compounds. This is not true, however, for disordered alloys, amorphous and liquid phases. Disordered systems can still be treated with *ab initio* methods but they require special approaches. For instance, thermophysical properties of liquids can be obtained from *ab initio* molecular dynamics simulations. A detailed treatment of liquid and amorphous phases is beyond the scope of this chapter and the interested reader can refer to a recent review (Becker 2014) and references therein. Instead, in Section 5 we will briefly describe some of the techniques used to treat solid disordered alloys.

Finally, in the last section we will present an active field of research where *ab initio* results are used in a multiscale approach together with computational thermodynamic modeling (cf. Chapter on Computational Thermodynamics). We will show how this approach allows us to extend the range of systems for which *ab initio* methods can be applied to very complex multicomponent alloys with up to 10 or 15 components. For systems of such complexity, it is still a daunting task to determine thermophysical properties using only *ab initio* methods spanning the entire compositional space.

Theoretical Background

We describe in this section the main ideas behind *ab initio* calculations (and in particular DFT) used today in materials science. Since the field is very wide, however, for a detailed treatment the interested reader is referred to the book by Martin (Martin 2004).

Limiting our considerations to time-independent cases, i.e., stationary ground states, and neglecting relativistic effects and magnetic fields, a system of interacting electrons and nuclei can be described by the time-independent Hamilton operator $\hat{H}$:

$$\hat{H} = -\sum_{I} \frac{\hbar^2}{2M_I} \nabla_I^2 + \frac{1}{2} \sum_{I \neq J} \frac{Z_I Z_J e^2}{|\vec{R}_I - \vec{R}_J|} \tag{1}$$
$$-\frac{\hbar^2}{2m_e} \sum_{i} \nabla_i^2 - \sum_{i,I} \frac{Z_I e^2}{|\vec{r}_i - \vec{R}_I|} + \frac{1}{2} \sum_{i \neq j} \frac{e^2}{|\vec{r}_i - \vec{r}_j|}$$

where the first term is due to the kinetic energy of the nuclei, the second to the Coulomb interaction among the nuclei, the third to the kinetic energy of the electrons, the fourth to the Coulomb interaction between nuclei and electrons and the last term to the Coulomb interaction among the electrons. The time-independent Schrödinger equation for this system is an eigenvalue equation:

$$\hat{H} \Psi_i = E_i \Psi_i \tag{2}$$

where $\Psi_i(\dots \vec{r}_i \dots \vec{R}_I \dots)$ is the many-body wavefunction corresponding to the eigenstate with E_i energy.

In principle, every property of our system can be obtained by solving Eq. 2 with the Hamilton operator (1) for the system under investigation. An exact solution of this equation is, however, only possible for very simple cases. Several approximations need to be introduced to obtain equations which are numerically suitable for describing real solid systems.

A first approximation that is excellent for many purposes and represents a starting point for other cases is the Born-Oppenheimer or adiabatic approximation. According to this formulation, the electronic and nuclear motions are decoupled since they occur on different time-scales. Therefore, one can solve the "electronic problem" assuming the nuclei are not in motion, and then separately solve the nuclear motion. The wavefunction for a system of N electrons now depends only on the $3N$ coordinates of the electrons $\Psi_i(\dots \vec{r}_i \dots)$. Although simplified, the resulting "electronic problem" is still a many-body one whose solutions are in practice restricted to rather small molecules and systems and are not applicable to solids with $\approx 10^{23}$ electrons. Several approaches have been developed to overcome this shortcoming and apply the Schrödinger equation to complex systems. In the next section we present one of these approaches, namely density functional theory.

Density Functional Theory

Most of the *ab initio* methods in use today for calculating the properties of solids are based on DFT as derived by Hohenberg and Kohn (Hohenberg and Kohn 1964) and Kohn and Sham (Kohn and Sham 1965).

The fundamental tenet of DFT is that any property of a system of many interacting particles can be viewed as a *functional* of the ground-state electronic density $\rho(\vec{r})$. In other words, the role of the wavefunction $\Psi_i(\ldots \vec{r}_i \ldots)$ is substituted by the electronic density $\rho(\vec{r})$, the latter being a substantially simpler function of position $\vec{r}$. The existence of such functionals is proved in a surprisingly simple way in the original work by Hohenberg and Kohn (1964), but no clue is given for constructing functionals for systems of interacting electrons. The total Hohenberg-Kohn energy for the system can be expressed as:

$$E_{HK}[\rho] = \int v_{ext}\rho dx + F_{HK}[\rho] \tag{3}$$

where v_{ext} is the external potential generated by the nuclei of the system and $F_{HK}[\rho]$ is a universal functional depending only on the electronic density.

DFT would probably remain only a curiosity if it was not for the *ansatz* by Kohn and Sham (Kohn and Sham 1965), which has provided a practical way to obtain approximate ground-state functionals for real systems of interacting electrons. The ansatz replaces the interacting problem with an auxiliary independent-particle problem with all many-body effects included in the exchange-correlation functional of the density. The universal Hohenberg-Kohn functional is then reformulated as a Kohn-Sham functional:

$$F_{KS}[\rho] = T_s[\rho] + E_{Hartree}[\rho] + E_{xc}[\rho] \tag{4}$$

That is, as a sum of the kinetic energy of the single-particle electrons $T_s[\rho]$, the classical Coulomb interaction $E_{Hartree}[\rho]$ between electronic charge densities and a remaining term $E_{xc}[\rho]$ which include all the complications of quantum many-body interactions. In other words, the exchange-correlation term is by definition given by $E_{xc}[\rho] = F_{KS}[\rho] - (T_s[\rho] + E_{Hartree}[\rho])$. Furthermore, the original many-body electronic Schrödinger equation is substituted by a system of N single-particle eigenvalue equations:

$$\hat{H}_{KS}\phi_i = \epsilon_i\phi_i \tag{5}$$

with the Kohn-Sham Hamiltonian given by:

$$\hat{H}_{KS} = -\frac{\hbar^2}{2m_e} \nabla^2 + v_{ext} + v_{Hartree} + v_{xc} \tag{6}$$

where $v_{Hartree}$ and v_{xc} are the potentials corresponding to the functionals already introduced (being the potential equal to the functional derivative $v = \partial E/\partial \rho$). The auxiliary Kohn and Sham single-particle wavefunctions ϕ_i (or Kohn-Sham orbitals) must build up the true ground-state electronic density by summing over the number of states:

$$\rho(\vec{r}) = \sum_i \phi_i^*(\vec{r})\phi_i(\vec{r}) \tag{7}$$

Equations 5 and 6 are usually solved numerically with different possible schemes; several codes are available to perform such calculations. The most basic and fundamental properties that can be obtained from a DFT calculation are the ground-state electronic density $\rho(\vec{r})$ and the total energy $E[\rho]$.

We remark that DFT as formulated above is an exact theory. The uncertainties in actual calculations for real systems originate from the choice of the exchange-correlation functional, whose exact form is unknown. Several approximate forms of this functional have been proposed, with different degrees of success. Some additional sources of uncertainty are also related to the numerical schemes adopted in actual calculations, as will be described in the following section.

The first approximate exchange-correlation functionals adopted were based on the properties of the homogeneous electronic gas and are called local-density approximation (LDA) functionals:

$$E_{xc}^{LDA}[\rho] = \int \rho(\vec{r})\,\varepsilon_{xc}(\rho)d\vec{r} \tag{8}$$

The energy $\varepsilon_{xc}(\rho)$ is the local exchange-correlation energy per particle of the homogeneous electron gas of density ρ. Several LDA functionals exist that give rather similar results.

Considering that the LDA is based on a very simple assumption, these functionals lead to surprisingly accurate results for many systems, but they can also fail dramatically. The typical example of the latter case is the ground-state structure of pure Fe, which is wrongly predicted by LDA as being the nonmagnetic state of the f.c.c. phase. To overcome the shortcomings of LDA functionals, several functionals based not only on the local density but also on the (semi-local) gradient of the electronic density $\vec{\nabla}\rho$ have been proposed. These generalized gradient approximation (GGA) functionals lead to a significant improvement in the calculation of several properties, although this is not always true. One of the most popular GGA functionals is that of Perdew-Burke-Ernzerhof (PBE) (Perdew et al. 1996). An example of typical differences between results calculated with LDA and with GGA functionals is shown in Fig. 1 for the calculated total

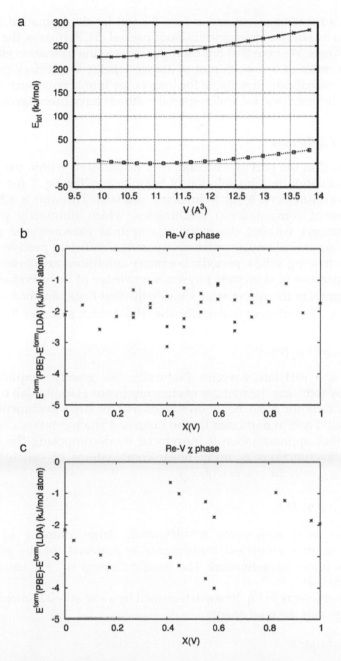

Fig. 1: (a) Calculated total energy E as a function of volume for f.c.c. Ni using Quantum Espresso and LDA and PBE functionals. (b) Formation energy differences GGA-LDA for every possible hypothetical configuration (see Section 6) for the sigma phase in the binary Re-W system (calculations performed with the VASP code). (c) As (b) for the chi phase.

energy as a function of volume in f.c.c. Ni and for the formation energy (cf. Section 3.1) of two topologically close-packed (TCP) phases, the sigma and chi phases. We note that typical differences in the formation energies are of the order of a few kJ/mol of atoms. Finally, we remark that the above functionals are unsuitable for long-range interactions such as Van der Waals interactions, for which special methods have been developed.

Numerical Schemes

As discussed in the previous section, DFT calculations within the Kohn-Sham formulation require solution of Schrödinger-like Eq. 5 for single-particle wavefunctions ϕ_i (or Kohn-Sham orbitals). This aim is achieved with different computational machineries, which ultimately provide useful numbers without the input of empirical parameters if proper technical/numerical implementations guarantee sufficient precision.

When treating solids, periodic boundary conditions are chosen, as is standard practice in solid state physics. Knowledge of the wavefunctions $\psi_{\vec{k}}^v(\vec{r})$ in terms of the wavevector $\vec{k}$ inside the first Brillouin zone and the band index v is sufficient to describe the whole solid, since for the Bloch theorem

$$\psi_{\vec{k}}^v(\vec{r}+\vec{R}) = e^{i\vec{k}\cdot\vec{R}}\psi_{\vec{k}}^v(\vec{r}) \tag{9}$$

where $\vec{R}$ is a real lattice vector. Naturally, this greatly simplifies the problem by reducing the number of electrons in our Hamiltonian to those in the first Brillouin zone. Equation 9 holds true for any wavefunction in a periodic solid and in particular for the Kohn and Sham orbitals.

A further approximation is introduced by decomposing the single-particle wavefunctions $\phi_{\vec{k}}$ using linear combinations of suitable basis functions $\varphi_{\vec{k},i}$ with the coefficients $c_{\vec{k},i}^v$

$$\psi_{\vec{k}}^v(\vec{r}) = \sum_i c_{\vec{k},i}^v \varphi_{\vec{k},i}(\vec{r}) \tag{10}$$

where the sum runs over a sufficiently large number of basis functions so that numerical convergence is achieved for the physical properties under investigation. The basis functions $\varphi_{\vec{k},i}$ are usually an orthonormal set.

The coefficients in Eq. 10 are determined by a variational procedure of minimization of the total energy

$$\langle \psi_{\vec{k}}^v(\vec{r}) \,|\, \hat{H}_{\vec{k}} \,|\, \psi_{\vec{k}}^v(\vec{r}) \rangle \tag{11}$$

In the previous equation, $\hat{H}_{\vec{k}}$ is the (translational invariant) Kohn-Sham Hamiltonian describing the interactions among electrons and nuclei in the unit cell.

Combining Eqs. 10 and 11, one obtains the following matrix eigenvalue equation

$$\sum_j H_{ij} c^v_{\vec{k},j} = \epsilon^v_{\vec{k}} \sum_j S_{ij} c^v_{\vec{k},j} \tag{12}$$

where $H_{ij} = \langle \varphi_{\vec{k},i} | \hat{H}_{\vec{k}} | \varphi_{\vec{k},j} \rangle$ are the matrix elements of the Hamiltonian, $S_{ij} = \langle \varphi_{\vec{k},i} | \varphi_{\vec{k},j} \rangle$ are the elements of the overlap matrix formed by basis functions $\varphi_{\vec{k},i}$, $\epsilon^v_{\vec{k}}$ are the eigenvalues and $c^v_{\vec{k},j}$ the components of the eigenvectors to be calculated. For many applications, most of the computational effort is related to solving the above matrix equation. Once the coefficients $c^v_{\vec{k},j}$ are determined, the electronic density is calculated by summing over the squares of the Kohn and Sham orbitals. However, the electronic density is needed to calculate the Kohn-Sham Hamiltonian $\hat{H}_{\vec{k}}$ and thus the matrix elements H_{ij} in Eq. 12. This is a self-consistency problem whose typical solution is to start from an initial "reasonable" guess of the electronic density ρ_0 (for example a simple overlap of atomic electronic densities), compute the matrix elements H_{ij} and solve (numerically) Eq. 12, then compute back the electronic density ρ. The procedure is iterated until some convergence criterion is satisfied.

It remains to describe the exact form of the basis functions $\varphi_{\vec{k},i}$ in Eq. 10. A popular choice, in particular for metallic systems, is to use plane waves. The functions $\varphi_{\vec{k},i}$ thus have the form:

$$\varphi_{\vec{k},i}(\vec{r}) = \frac{1}{\sqrt{\Omega}} e^{i(\vec{k}+\vec{G}_j)\vec{r}} \tag{13}$$

with $\vec{G}_j$ a vector in the reciprocal lattice and Ω a normalization factor.

This choice has the advantage of a simple mathematical form and the quality of the expansion can be systematically improved by increasing the number of plane waves. Nonetheless, it presents difficulties in describing the rapid changes in the potential and wavefunctions near the nuclei. To overcome this problem, pseudopotentials are introduced by which a simplified potential, screened by the inner-shell electronic states, is introduced to describe the region near the nuclei, while the full potential and plane waves are used in the regions between the atoms (where bonding is important). Examples of software packages implementing this approach are the Vienna *ab initio* simulation package (VASP)[1] and Quantum Espresso.[2] There are several recipes for constructing pseudopotentials, some of them differing only for very technical details which are beyond the scope of this chapter. Nonetheless, it is important to mention that two kinds of pseudopotentials are nowadays emerging as

---

[1] www.vasp.at

[2] www.quantumespresso.org

the most suitable (and accurate) for many applications, namely ultrasoft and Projector Augmented Waves (PAW) pseudopotentials. Compared with the more traditional family of "norm-conserving" pseudopotentials, they guarantee a comparable accuracy but using pseudofunctions that are as "smooth" as possible. This goal is achieved by re-formulating the problem using a smooth function and an auxiliary function around ion cores, where the density is rapidly varying. In addition, in the PAW approach the full all-electron wavefunction is kept and all integrals are evaluated as a combination of smooth functions extending in the whole space plus localized contributions evaluated near the nuclei. Both kinds of pseudopotentials are available in the above mentioned software packages.

We remark that the choice of plane waves is not the only possibility as a basis set; a variety of different approaches (and codes) have been developed. As an example, localized atomic-like orbital functions can also be used.

The higher the number of plane waves, the more precise will be the expansion in Eq. 10. It is customary to define the cut-off energy $E_{cut-off}$ as a single parameter which determines the number of plane waves in the expansion:

$$\frac{\hbar}{2m_e}|\vec{k}+\vec{G}_j|^2 < E_{cut-off} \tag{14}$$

Finally, we remark that the evaluation of the electronic density as in Eq. 7 is carried out with some numerical method on a grid of points (k-points) in the first Brillouin zone. Some additional technical parameters, besides the number of plane waves, are introduced (k-point density and distribution scheme, smearing factor, etc.) which need to be tested for convergence in order to obtain accurate DFT results for different properties.

Ground-State Properties

By applying the theory described in the previous sections, several materials properties can be derived either directly from the electronic density or with some additional modeling. In this section we will present an overview of some materials properties that can be obtained from ground-state DFT calculations. Although these properties are limited to 0 K, they are already remarkably useful in many applications where the temperature-dependence can be neglected as a first approximation. We also have to remark, however, that not every material's property can be satisfactory calculated using DFT. A well-known example is the so-called "band gap problem", that is, the fact the band gaps are strongly underestimated by DFT calculations (up to 50%).

Formation Energies and Lattice Parameters

The most fundamental property that can easily be calculated by virtually any DFT software is the ground-state electronic total energy E_{TOT} or simply E. In a thermodynamic context and later in this chapter, we also refer to this quantity as the static energy E^{st} to underline the difference with other energy terms related to electronic excitations (E^{el}), phonons (E^{vib}), etc. This quantity depends on the actual distribution of the atoms in space and hence on the crystal structure, volume and atomic coordinates. A self-consistent DFT calculation results in a particular value of E_{TOT} for a given geometry. Different volumes, lattice parameters and atomic positions will yield a different E_{TOT} and, for a given crystal structure, the minimum energy geometry has to be found. One way of finding this minimum value for E_{TOT} and the corresponding geometry for simple isotropic crystal structures (simple cubic, f.c.c., b.c.c.) is to compute the total energy for different volumes and fit the resulting points with some kind of equation of state (EOS) $E_{TOT}(V)$. A typical EOS is that of Murnaghan, which comes from the assumption of linear behaviour of the bulk modulus of a solid compressed to a finite strain with respect to pressure. At the temperature of absolute zero, the link between pressure P and volume V reads

$$P(V) = \frac{B_0}{B_0'}\left[\left(\frac{V_0}{V}\right)^{B_0'} - 1\right] \tag{15}$$

where B_0 is the isothermal bulk modulus, B_0' its derivative with respect to volume and V_0 is the equilibrium volume (corresponding to the minimum of the energy). From $P(V) = -\,dE(V)/dV$ the related expression for the energy is obtained as[3]

$$E(V) = \frac{B_0 V}{B_0'}\left[\left(\frac{V_0}{V}\right)^{B_0'}\frac{1}{B_0'-1} + 1\right] + K \tag{16}$$

Using an EOS has the advantage of determining not only the minimum energy E_0 but also other important properties such as the bulk modulus and equilibrium volume and the relationship between pressure and volume (around the equilibrium volume). Sometimes, polynomials of different degrees are used instead of an EOS. We note that the minimum energy corresponds to $P = 0$.

For complex crystal structures, however, volume is not the only variable to be considered. In general, the energy E depends on lattice constants and angles ($a, b, c, \alpha, \beta, \gamma$) as well as atomic coordinates for the atoms in general positions in the unit cell and the above fitting procedure

---

[3] The expression currently used is slightly different with an equivalent form for the integration constant K which has advantages in fitting energy vs. volume data. See V.G. Tyuterev, N. Vast, Computational Materials Science 38 (2006) 50.

becomes cumbersome or unfeasible. Finding the minimum-energy geometry (or relaxed geometry or relaxed structure) in these cases is done by applying some automatic algorithms which are implemented in many DFT codes (for instance in the already mentioned VASP and Quantum Espresso). The automatic procedure is called "relaxation" of the crystal structure and must start from initial values of lattice constants and angles which are not too far from the equilibrium ones. A possible shortcoming of this automatic procedure is the occurrence of local minima different from the absolute minimum.

We note here that absolute values of the total energy E are usually not meaningful, since different DFT codes use different reference states, pseudopotentials and technical implementation details. However, comparison of DFT energy differences, choosing suitable reference states, is meaningful and results reported in the literature are always in this form. For pure elements in different (hypothetical) crystal structures, the differences in energy are also called *lattice stabilities* (Saunders and Miodownik 1998).

A common choice for comparing the stability of different crystal structures in binary and higher order systems is to report their energy of formation E_δ^{form} (or ΔE_δ^{form}, δ referring to the crystal structure), using the energy of the pure components in their stable structures at 0 K as reference states. For a binary ordered compound $A_x B_y$ with crystal structure δ, where A and B are its components, x and y define its stoichiometry, the energy of formation is given by

$$E_\delta^{form} = E_{0,\delta}(A_x B_y) - x E_{0,\lambda}(A) - y E_{0,\mu}(B) \tag{17}$$

with λ and μ the stable structures of A and B, respectively, at 0 K.

On the right-hand side of the above equation all E_0 are ground-state total energies at the equilibrium ($P = 0$) geometry and hence the left-hand side is also referred to as an enthalpy of formation (H_δ^{form} or ΔH_δ^{form}). For the same stoichiometry, the energies of formation of different structures E^{form}, E_θ^{form}, etc. can be calculated. A thorough comparison of several hypothetical compounds in different crystal structures allows us to assess the phase stability in binary or higher-order systems.

As an example, Fig. 2 shows the formation energies calculated with DFT (VASP, PBE exchange-correlation functional) for several ordered structures potentially competing in stability in the binary Re-V system.

For the TCP structures (A15, sigma, chi, mu, C14, C15, C36) all hypothetical ordered compounds obtained by permutations of Re and V atoms in the Wyckoff positions were considered (cf. Section 6). For f.c.c., b.c.c. and h.c.p. phases, some typical related ordered structures were considered. For example, for the b.c.c. phase, the formation energies of the ordered B2, B32 and D03 were calculated and, as shown in Fig. 2, the most stable structures in this system were found to be B2 (at 50 at% V) and D03 (at 75 at% V). Each set of straight-line segments represents the convex hull for a given group of phases (for example for all configurations of the sigma phase). This kind of calculation is very useful for assessing the relative stability of different structures in a given system. Furthermore, DFT energies for metastable structures are also useful in modeling approaches such as phase-field or other solidification simulations.

The calculation of energies of formation in a high-throughput environment is becoming very popular in materials science as a tool for assessing phase stability in binary and multicomponent systems. A huge effort is underway in the framework of the Materials Genome Initiative in the US and several similar programmes in other countries and open databases are becoming available on the web.[4] Because of the computational effort involved, however, most of these calculations are still limited to 0 K (cf. Section 4).

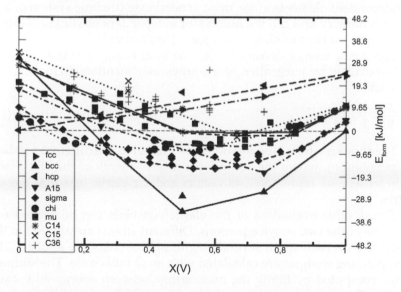

Fig. 2: Calculated DFT formation energies of several structures in the Re-V system. Calculations were carried out with the VASP code. For the TCP phases (A15, sigma, chi, mu, C14, C15, C36), the formation energies of all possible hypothetical configurations obtained by permutation of the atoms in the different Wyckoff positions were evaluated (see also Section 6). For the b.c.c., f.c.c. and h.c.p. phases the following ordered structures were considered, respectively: A2, B2, B32, D0$_3$; A1, L1$_2$, L1$_0$; A3, D0$_{19}$, B$_h$. For details, see (Palumbo et al. 2014).

---

[4] www.materialsproject.org, www.aflowlib.org, www.nomad-repository.eu, www.aiida.net

Elastic Properties

Among the properties of materials of high technological relevance, mechanical properties such as elastic constants and elastic moduli play a major role. Knowledge of these data enables the prediction of the material behaviour under different circumstances.

Ab initio methods based on DFT can be reliably applied to compute elastic constants of single (perfect) crystals. The calculation essentially involves the evaluation of the energy (or forces) of a number of distorted geometries produced by a hypothetical applied stress. The relationship between stress and strain is assumed to be linear as in Hooke's law

$$S_{mn} = C_{mnpq}\varepsilon_{pq} \tag{18}$$

where (in Einstein notation) S_{mn} is the second rank stress tensor, ε_{pq} the second rank strain tensor and C_{mnpq} the fourth rank elastic tensor, whose elements are the elastic constants. The stress and strain tensors are always symmetric and have six independent elements, while the elastic tensor has 21 independent elements in the most general case (triclinic systems). In a higher symmetry unit cell the number of independent elastic constants is further reduced (to a minimum of 3 for cubic systems).

The elastic energy density U, i.e., the total energy per unit volume, can be obtained by integration of the previously mentioned Hooke's law and reads

$$U = \frac{E}{V} = \frac{1}{2}C_{mnpq}\varepsilon_{mn}\varepsilon_{pq} \tag{19}$$

The above relationships are sometime presented in a different notation (Voigt notation), where the six independent elements of the stress and strain tensors are represented as vectors and the elastic tensor as a 6 x 6 matrix.

The *ab initio* evaluation of the elastic constants can be carried out using one of the two above equations. Different strains are applied to the crystal and for each strain several magnitudes of stress are applied. The corresponding energies are calculated with an *ab initio* code. The stiffness is then computed by fitting the relationship between energy and strain (Eq. 19). This scheme can be applied with all DFT codes. Alternatively, using DFT codes which can directly produce the stress tensor for a given strained geometry, one can apply Eq. 18, which in principle requires only one evaluation for each geometry. To enhance the accuracy of the results, however, at least three magnitudes of strain are usually applied in this case too.

For polycrystalline materials, some kind of macroscopic averaging scheme must be applied to produce the elastic moduli using the single-crystal results. The averaging procedure runs over all possible orientations of the single crystals in the polycrystalline system.

Temperature-Dependent Properties

Ab initio calculations are often viewed as being restricted to 0 K ground-state properties without the possibility of including temperature effects. At finite temperatures, a solid is brought into exited states and it becomes necessary to consider the entropy and free energy. Although it is true that the calculation of temperature-related excitations requires a significantly higher computational effort compared to ground-state calculations, routine methods have become available nowadays to perform such calculations for many systems.

Temperature effects physically originate from three main kinds of excitations:

1. *Electronic excitations.* When the temperature is raised above 0 K, electrons start to populate energy states above the Fermi level giving a contribution to the thermophysical properties of the system. These effects can be calculated using the Fermi-Dirac distribution and with a knowledge of the electronic energy states.

2. *Vibrational excitations or phonons.* Atoms vibrate around their equilibrium positions even at 0 K according to quantum mechanics. At higher temperatures the intensity of these vibrations increases according to Bose-Einstein statistics. Phonons provide the main contribution to temperature-dependent properties.

3. *Magnetic excitations or magnons.* For magnetic systems, an increase in temperature results in a progressive departure from the original ordered magnetic state (parallel spins, antiparallel spins or more complex situations) at 0 K. At a high enough temperature, the spins become randomly oriented and the magnetic ordering is completely lost. The contribution of magnetic excitations is important in some systems, while in other cases it can be safely neglected.

Considering the Helmholtz energy as a reference property in thermodynamics (from which other properties can be derived), the different temperature-dependent contributions can be expressed as:

$$F(T,V) = E^{st}(0K,V) + F^{vib}(T,V) + F^{el}(T,V) + F^{magn}(T,V) \qquad (20)$$

To simplify the problem, as is usual in physics, different excitations are normally treated separately, assuming they are independent. This assumption is based on the different timescales of physical processes involving electrons, phonons and magnons and is reasonable for many applications. Nonetheless, coupling effects such as electron-phonon interactions are fundamental for properties such as electrical resistivity and superconductivity, for which special methods have to be adopted. While magnetic ordering still represents a challenge for modern solid-state physics, electronic and vibrational excitations can be tackled using DFT-based methods with different degrees of approximation and corresponding computational effort. In the following, we will describe in detail how to treat electronic and vibrational excitations, but we will not examine magnetic transitions. The interested reader is referred to (Koermann 2014) for a recent review.

Electronic Excitations

Computing the electronic energy states is the aim of any *ab initio* code, as described in Section 2 for the DFT single-particle approach. This information is often represented as a density of states (DOS) $N(E)$, that is, the number of states in a given energy interval is reported for the range of energies of interest. At 0 K, the electrons occupy the lowest energy states up to the Fermi energy (or level) E_F. On increasing the temperature, more and more electrons acquire the necessary energy to occupy higher energy states. The underlying statistics is given by the Fermi-Dirac distribution $f_{FD}(E,T)$.

The energy of the system is then given by

$$E_{el}(T) = 2 \int_{-\infty}^{+\infty} EN(E)f_{FD}(E,T)dE \tag{21}$$

The corresponding expressions for the electronic entropy and heat capacity are

$$S_{el}(T) = -2k_B \int_{-\infty}^{+\infty} \left\{ \begin{array}{c} f_{FD}(E,T) \ln(f_{FD}(E,T)) \\ +(1-f_{FD}(E,T)) \ln(1-f_{FD}(E,T)) \end{array} \right\} N(E)dE \tag{22}$$

$$C_{el}(T) = 2 \int_{-\infty}^{+\infty} (E - E_F)N(E)\frac{\partial f_{FD}(E,T)}{\partial T}dE \tag{23}$$

In the low temperature limit, the heat capacity expression can be simplified to

$$C_{el}(T) \approx \frac{2\pi^2}{3} N(E_F) k_B^2 T \tag{24}$$

which is reported in many textbooks on solid-state physics.

The electronic heat capacity for some elements calculated from Eq. 23 and the DOS obtained from the Quantum Espresso code is shown in Fig. 3.

We remark that for many elements the electronic heat capacity represents a significant contribution to the total heat capacity, the latter being of the order of $3R$. Furthermore, it can be noticed that, as in Eq. 24, for every element at low temperature ($T < 0.2T_M$) the C_{el} depends linearly on temperature and for Al and Cu this holds true up the melting point. For the other elements, however, significant deviations from linearity occur, which ultimately demonstrate differences in the corresponding electronic DOS around the Fermi energy.

We finally note that the above single-particle model fails completely for systems where electron-phonon coupling is important (that is, superconductors). Besides, when increasing the temperatures, atomic vibrations can also significantly modify the electronic DOS, which is calculated by DFT codes assuming atoms remain still. Because of atomic motion, the DOS is smeared out at high temperatures (Asker et al. 2008), but little is known about this effect.

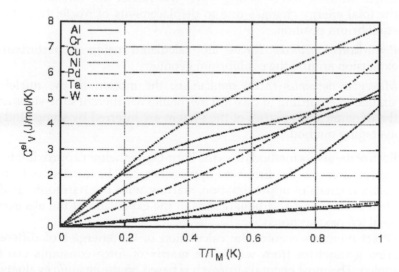

Fig. 3: Comparison of DFT-based electronic contribution to the heat capacity of several pure elements (Al f.c.c., Cr b.c.c., Cu f.c.c., Ni f.c.c., Pd f.c.c., Ta b.c.c., W b.c.c.). The electronic DOS was calculated using Quantum Espresso, including spin polarization when appropriate. The heat capacity was evaluated from the DOS using Eq. 23 as described in the text. The x axis is the normalized temperature (T_M is the melting point of each element).

Vibrational Excitations

Fundamental thermodynamic properties of solid phases such as Helmholtz and Gibbs energies, entropy and heat capacity as well as nonlinear properties such as thermal expansion and heat conduction are mostly determined by the vibrations of the constituent atoms in the lattice. Fortunately, the theory of lattice vibrations, both classical and quantum, is well developed and probably one of the most successful in solid-state physics. The early development of the theory took place in the 1930s and was reviewed by Born and Huang (Born and Huang 1954). The main focus of these early studies was on understanding of properties of the dynamical matrix, and less attention was devoted to atomic interactions. The latter were described with empirical potentials such as that of Lennard-Jones where the parameters were obtained by fitting the experimental data. The advent of DFT and the progress in *ab initio* methods for numerical solution of the quantum mechanical equations, together with the increasing power of available computers, have made it possible today to study lattice dynamics without any experimental input.

There are three main *ab initio* methods for the study of lattice-dynamic problems:

1. *Direct methods*: forces acting upon the nuclei are calculated from the total energy changes due to displacements of nuclei from their equilibrium position.

2. *Perturbative methods*: forces are calculated using a perturbative expansion around the equilibrium geometry.

3. *Molecular dynamics (MD) simulations*: the motion of the nuclei is simulated according to classical mechanics at different time-steps and the dynamical properties of the system are inferred by some kind of averaging method.[5]

Each of the above methods uses the Born-Oppenheimer approximation. The calculations in the first two approaches can be carried out with increasing degrees of approximation, starting with the harmonic model and then considering additional terms in the expansion of the electronic potential (cf. next sections).

Direct methods require the calculation of the energies of different distorted geometries from which the matrix of force constants can be assembled. The most general approach is based on the *ab initio* evaluation

---

[5] Although in principle the motion of nuclei could be determined by quantum laws, molecular dynamic simulations to date are carried out using classical laws.

of forces on all atoms, produced by a set of finite displacements of a few atoms in an otherwise perfect crystal. The perfect crystal surrounding the displaced atoms has to be large enough to ensure that interactions of the perturbation with its translation symmetry equivalent are small. This is normally achieved by constructing sufficiently large supercells. This approach is, for example, described in (Kresse et al. 1995; Parlinski et al. 1997). Some of the available codes for generating the necessary finite displacements and assembling the force-constant matrix are: PHON,[6] PHONON[7] and PHONOPY.[8] Phonon frequencies calculated by this method are very accurate provided that the supercells are sufficiently large. The main advantage of direct methods is that any *ab initio* code capable of calculating reasonably accurate forces can be used. Their main shortcoming is the necessity of considering large supercells for simple crystal structures.

Perturbative methods are quite accurate, efficient and elegant; they require considerable effort for their implementation, but have been implemented in computer codes such as Quantum Espresso. They are based on the assumption that the lattice distortion associated with nuclear vibrations is a perturbation acting on the electrons and causing a linear response of the electron density. This response determines the energy variation up to the third order and can be determined from the Kohn-Sham orbitals using density functional perturbation theory (DFPT) (Baroni et al. 1987; Baroni et al. 2001).

The third class of methods based on molecular dynamics simulations offers the possibility of exploring the space-time trajectories of the atoms and deriving vibrational properties from suitable averaging procedures. For instance, vibrational frequencies can be obtained as the Fourier transform of the atomic velocity autocorrelation function (Sampoli 1998). The main advantage of MD simulations is that they can capture the full anharmonic behaviour of the system without the need to consider further expansion terms in the electronic potential (cf. next section). Their shortcomings are the great computational effort necessary, the difficulties in reaching equilibrium within a reasonable time-scale (in particular at low temperatures) and the accuracy of the averaging procedures to obtain vibrational properties. In addition, most implementations are based on classical equations for the atomic motion and hence they cannot describe the quantum behaviour of the system at low temperatures (for example, the zero point energy, ZPE).

---

[6] www.homepages.ucl.ac.uk/~ucfbdxa/phon/
[7] www.wolf.ifj.edu.pl/phonon/
[8] www.phonopy.sourceforge.net/

Harmonic approximation

We consider now a three-dimensional lattice of N unit cells and n nuclei per unit cell. The position of a unit cell, indexed by $l = 1, \ldots, N$, is determined by the vector $\vec{r_l} = l_1\vec{a_1} + l_2\vec{a_2} + l_3\vec{a_3}$, with primitive translational vectors $\vec{a_1}, \vec{a_2}, \vec{a_3}$. The equilibrium position of the kth nucleus ($k = 1,\ldots, n$) in the lth unit cell is given by $\vec{r_{l,k}} = \vec{r_l} + \vec{r_k}$, where $\vec{r_k}$ is the relative position of the nucleus inside the unit cell. If the nucleus is now vibrating around its equilibrium position, at time t it will be displaced by a relative amount $\vec{u_{l,k}}(t)$ and its coordinates are finally given by

$$\vec{R_{l,k}} = \vec{r_{l,k}} + \vec{u_{l,k}}(t) \tag{25}$$

Let us now derive the equation of motion of this nucleus. In the following, time derivatives are denoted with dots, while Greek subscripts represent one of the Cartesian components of a vector. Within the Born-Oppenheimer approximation, the nucleus moves in a potential energy landscape given by the total energy of the electronic system calculated assuming fixed nuclei (for instance using DFT). We refer to this potential energy as $E(\vec{R_{l,k}})$. The electrons are assumed to be in the ground state for each nuclear configuration. For small displacements $\vec{u_{l,k}}(t)$, we can expand the energy in a Taylor series:

$$E = E_0 + \sum_{lk\alpha} \frac{\partial E}{\partial u_{lk\alpha}} u_{lk\alpha} + \frac{1}{2} \sum_{lk,l'k'} \sum_{\alpha\beta} \frac{\partial^2 E}{\partial u_{lk\alpha} \partial u_{l'k'\beta}} u_{lk\alpha} u_{l'k'\beta} + \cdots \tag{26}$$

where $E_0 = E(\vec{r_{l,k}})$ (no displacements) and the derivatives are calculated at $u = 0$. The first term E_0 is unimportant for the dynamical problem while the first derivatives are zero since we are expanding around the minimum. If the nuclear displacements are small enough, we can truncate the series at the third (quadratic) term which hence determine the dynamics of the system. This is by definition the harmonic approximation. In this case, the potential energy of the system is thus reduced to

$$E = + \frac{1}{2} \sum_{lk,l'k'} \sum_{\alpha\beta} \Phi_{lk\alpha,l'k'\beta} u_{lk\alpha} u_{l'k'\beta} \tag{27}$$

where we have conveniently introduced the 3 x 3 force-constant matrices $\Phi_{lk,l'k'}$ whose elements are given by

$$\Phi_{lk\alpha,l'k'\beta} = \frac{\partial^2 E}{\partial u_{lk\alpha} \partial u_{l'k'\beta}} \tag{28}$$

Each force constant matrix relates the displacement of nucleus $l'k'$ to the force exerted on nucleus lk:

$$F_{lk\alpha} = \sum_{l'k'\beta} \Phi_{lk\alpha,l'k'\beta} u_{l'k'\beta} \tag{29}$$

The force-constant matrix satisfies a number of symmetry requirements. For instance, from the translational invariance of the crystal, it can be derived that

$$\sum_{lk} \Phi_{lk\alpha,l'k'\beta} = 0 \tag{30}$$

Furthermore, the crystal symmetry operations leave not only the potential energy and its derivatives but also the force constant elements unchanged. These properties of the force constants are technically important since they reduce the necessary calculation effort.

With M_{lk} being the mass of nucleus lk, we can now derive the equations of motion in the absence of an external field:

$$F_{lk\alpha} = M_{lk}\ddot{u}_{lk\alpha} = -\frac{\partial E}{\partial u_{lk\alpha}} = -\sum_{l'k'\beta} \Phi_{lk\alpha,l'k'\beta} u_{l'k'\beta} \tag{31}$$

which are usually rewritten in the equivalent form

$$\sqrt{M_{lk}}\ddot{u}_{lk\alpha} = -\sum_{l'k'\beta} D_{lk\alpha,l'k'\beta} \sqrt{M_{l'k'}} u_{l'k'\beta} \tag{32}$$

where the real and symmetric matrices D are called dynamical matrices. The above set of equations constitutes a system of $3nN$ coupled differential equations.

The corresponding Hamiltonian (using the force constants) is

$$\hat{H} = \frac{1}{2}\sum_{lk} M_{lk}[\dot{u}_{lk}]^2 + \frac{1}{2}\sum_{lk,l'k'} \sum_{\alpha\beta} \Phi_{lk\alpha,l'k'\beta} u_{lk\alpha} u_{l'k'\beta} \tag{33}$$

where the first term is the kinetic energy given by the momenta of the nuclei and the second one is the potential energy due to the force acting on each atom displaced from its equilibrium (minimum energy) position.

To solve the system of Eq. 32, we can try a solution in the form of plane waves (phonons):

$$u_{lk\alpha} = \frac{1}{\sqrt{M_{lk}}} e_{k\alpha}(\vec{q}) e^{i(\vec{q}\cdot\vec{r}_l - \omega_q t)} \tag{34}$$

In the above equation, $\vec{q}$ is a vector in the first Brillouin zone, the time-dependence is given by the simple exponential $e^{-i\omega_q t}$ and the displacements of the atoms in each unit cell (identified by the vector $\vec{r}_l$) can be obtained from the displacements in any particular unit cell, for instance at $t = 0$ and $\vec{r}_l = 0$: $\frac{1}{\sqrt{M_{lk}}} e_{k\alpha}(\vec{q})$.

Inserting the plane-wave solution into the system of Eqs. 32, we obtain

$$\omega_q^2 e_{k\alpha}(\vec{q}) = \sum_{l'k'\beta} D_{lk\alpha,l'k'\beta} e^{i\vec{q}\cdot(\vec{r}_l - \vec{r}_{l'})} \tag{35}$$

which are N systems of $3n$ linear equations. For a crystal with n atoms per unit cell and for each value of $\vec{q}$, there are $3n$ solutions of the eigenproblem above, with eigenvalues $\omega^2(\vec{q}, v)$ and eigenvectors $\vec{e}_k(\vec{q}, v)$. An eigenmode with eigenvalue $\omega^2(\vec{q}, v)$ and eigenvector $\vec{e}_k(\vec{q}, v)$ is called the vth normal mode (or phonon) of the system. A complete solution of the dynamical problem is obtained when for each normal mode the dispersion relationship for eigenvalues $\omega = \omega(\vec{q})$ and the corresponding dispersion relationship for eigenvectors $\vec{e} = \vec{e}(\vec{q})$ are known. The graphical representation of the $3n$ functions $\omega(\vec{q}, v)$ as a function of $\vec{q}$ for $v = 1, ...3n$ is called a vibrational (or phonon) dispersion spectrum and each $\omega(\vec{q}, v)$ is called a phonon branch. Usually the phonon spectrum is represented in a plot with all ω_i along some particular directions in the first Brillouin zone, especially along characteristic symmetry points of the crystal. There are always three acoustic branches, so that $\omega(\vec{q}) \rightarrow 0$ when $\vec{q} \rightarrow 0$, and $3n - 3$ optical branches, so that $\omega(\vec{q}) \rightarrow constant$ when $\vec{q} \rightarrow 0$. Vibrations corresponding to any mode can be longitudinal ($\vec{e}_{i,k} \parallel \vec{q}$) or transverse ($\vec{e}_{i,k} \perp \vec{q}$) or a combination of longitudinal and transverse. Another convenient way to express the vibrational properties is using phonon density of states $g(\omega)$, that is, the frequency distribution of all normal modes.

As an example, Fig. 4 presents the phonon dispersion (a) and DOS (b) for f.c.c. Ni as calculated with Quantum Espresso and the density functional perturbation theory.

As one atom is present in the unit cell, we have a total of three acoustic modes (sometimes overlapping in some parts of the plot) in the dispersion graph. The phonon dispersion is plotted starting from the Γ point (0,0,0) and along the X (1,0,0), W(1,0.5,0), X(1,1,0), K(0.75, 0.75,0) and L(0.5,0.5,0.5) points of the reciprocal cell. As a further example, a significantly more complex phonon dispersion (Fig. 5) is obtained for the sigma phase, containing 30 atoms in the unit cell.

A total of 90 vibrational modes (3 acoustic and 87 optic) are present. The calculations were performed with the VASP and PHON codes. As discussed for example in (Palumbo et al. 2014), both the direct method and the density functional perturbation theory lead to essentially the same results if properly converged.

Knowledge of phonon frequencies allows us to calculate the thermodynamic properties of the harmonic system using statistical thermodynamics. We consider the phonons as a set of particles which can

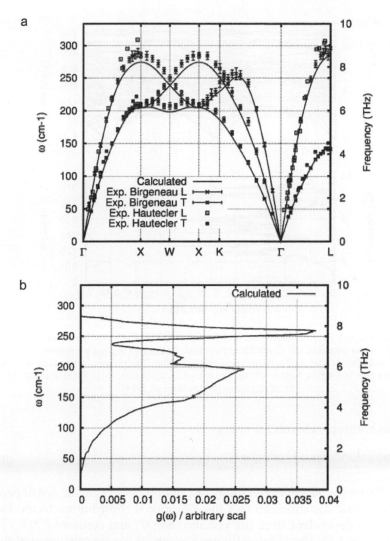

Fig. 4: Calculated phonon dispersion and DOS for Ni f.c.c. compared with available experimental data at 293 K. Calculations were performed using Quantum Espresso and the Density Functional Perturbation Theory as described in the text using the harmonic approximation. The path in the Brillouin zone is along the points $\Gamma(0,0,0)$, $X(1,0,0)$, $W(1, 0.5, 0)$, $X(1,1,0)$, $K(0.75,0.75,0)$ and $L(0.5,0.5,0.5)$.

occupy any possible eigenstate $(\vec{q}, \nu)$ with energy $\hbar\omega(\vec{q}, \nu)$. The partition function of the system is then given by

$$Z(T) = \prod_{\vec{q}\nu} \frac{e^{-\hbar\omega(\vec{q}, \nu)/2k_BT}}{1-e^{-\hbar\omega(\vec{q}, \nu)/k_BT}} \tag{36}$$

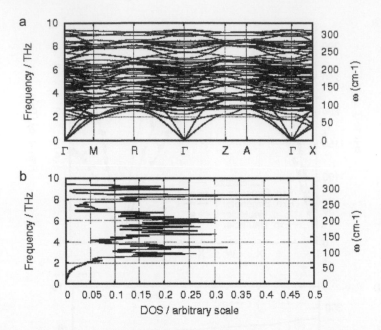

Fig. 5: Calculated phonon dispersion and DOS for CrReReCrCr configuration (cf. Section 6) of the sigma phase in the Cr-Re binary system. Calculations were performed using the PHON and VASP codes in the harmonic approximation. The phonon dispersion is shown along the points Γ(0,0,0), M(0.5, 0.5, 0), R(0.5,0.5,0.5), Z(0,0,0.5), A(0.5,0,0.5) and X(0.5,0,0). More details in (Palumbo et al. 2014).

and the harmonic Helmholtz energy is

$$F^{harm}(T) = \frac{1}{2}\sum_{\vec{q},v} \hbar\omega(\vec{q},v) + k_B T \sum_{\vec{q},v} \ln\left(1 - e^{-\hbar\omega(\vec{q},v)/k_B T}\right) \tag{37}$$

In the harmonic approximation, the phonon frequencies do not depend on interatomic distances, so that the vibrational contribution to the free energy is independent from the volume (Eq. 37) and constant $F^{vib}(V,T) \approx F^{harm}(V = const, T)$. The phonon frequencies are of course calculated at this constant volume, which is given by $V = N\Omega$, where Ω is the volume of the unit cell. Harmonic phonons are non-interacting particles, with infinite lifetime. Consequently, thermal expansion is zero, thermal conductivity is infinite, elastic constants are independent of temperature, etc. Additionally, the heat capacity at constant pressure is equal to that at constant volume. These are some of the shortcomings of the harmonic approximation.

The quasi-harmonic approximation

A proper treatment of anharmonicity should account for phonon-phonon interactions for all vibrational modes. Although the calculation of such interactions is possible from first principles, the task is challenging from the computational point of view.

A simple extension of the harmonic scheme, which corrects most of its deficiencies without requiring an explicit evaluation of anharmonic interaction coefficients, is the quasi-harmonic approximation (QHA) (Baroni et al. 2010). In this approach, the Helmholtz energy of a crystal is still assumed to be determined as in the harmonic case

$$F^{qha}(X,T) = \frac{1}{2} \sum_{\vec{q},v} \hbar\omega(\vec{q}, v, X) + k_B T \sum_{\vec{q},v} \ln\left(1 - e^{-\hbar\omega(\vec{q}, v, X)/k_B T}\right) \tag{38}$$

but here the phonon frequencies (and hence the free energy) are also dependent on X, which represents any global static constraint on which they may depend. Most commonly, $X = V$ with V the volume of the crystal, but in general X can represent internal distortions of the unit cell, anisotropic components of the strain tensor or other thermodynamic constraints. In the following, we will describe the most common case with $X = V$.

In the QHA, the total free energy (including the static energy at 0 K) is an energy surface as a function of V and T given by

$$F(V,T) = E^{st}(V) + F^{qha}(V,T) \tag{39}$$

For a given temperature (or volume) the corresponding volume (or temperature) is found by minimizing the Helmholtz energy in Eq. 39. Differentiation with respect to volume gives the pressure:

$$P = -\left(\frac{\partial F}{\partial V}\right)_T = -\left(\frac{\partial E^{st}}{\partial V}\right)_T + \frac{1}{V} \sum_{\vec{q},v} \hbar\omega(\vec{q}, v)\gamma(\vec{q}, v)\left[\frac{1}{2} + \frac{1}{e^{-\hbar\omega(\vec{q},v,X)/k_B T} - 1}\right] \tag{40}$$

where γ are the Grüneisen mode parameters

$$\gamma(\vec{q}, v) = -\frac{V}{\omega(\vec{q},v)}\frac{\partial\omega(\vec{q},v)}{\partial V} \tag{41}$$

Other thermodynamic potentials such as the Gibbs energy can be obtained as

$$G(P,T) = F(V,T) + PV \tag{42}$$

Several other thermophysical properties are then derived according to classical thermodynamic laws. For example, the isothermal bulk modulus B_0 is obtained as

$$B_0 = -V \left(\frac{\partial P}{\partial V}\right)_T = V \left(\frac{\partial^2 F}{\partial V^2}\right)_T \tag{43}$$

Or for the volumetric thermal expansion β

$$\beta = \frac{1}{V}\left(\frac{\partial V}{\partial T}\right)_P = -\frac{(\partial P/\partial T)_V}{(\partial P/\partial V)_T} = \frac{1}{B_0}\sum_{\vec{q},v} \hbar\omega(\vec{q},v)\gamma(\vec{q},v)\frac{\partial}{\partial T}\left[\frac{1}{e^{-\hbar\omega(\vec{q},v,X)/k_B T}-1}\right] \tag{44}$$

The isochoric (constant-volume) heat capacity C_V is obtained as in the harmonic approximation for the given volume. We note, however, that in the QHA different C_V can be calculated for different volumes (corresponding to different pressures). The isobaric heat capacity C_P is determined either from its definition $C_P = -T\left(\frac{\partial^2 G}{\partial T^2}\right)_P$, or from the thermodynamic expression

$$C_P - C_V = TV\,\beta^2 B_0 \tag{45}$$

where all other quantities have already been introduced.

Although in principle exact, all the above relationships involve derivatives of the free energy $F(V,T)$ for which an analytical expression is not available. The results are normally sets of discrete points which require a numerical scheme for the evaluation of the derivatives.

In the QHA, several thermophysical properties of materials can be calculated as a function of temperature and volume (pressure). As an example, some results for f.c.c. Ni calculated using the Quantum Espresso code are shown in Fig. 6.

Several experimental data sets are also shown for comparison. We remark that the magnetic contribution, particularly important in the heat capacity, is not included and hence the typical peak at the Curie temperature is not reproduced. At low temperatures, the harmonic approximation reproduces the experimental results well for the heat capacity, although at temperatures higher than approximately room temperature, the discrepancy becomes more and more significant. In the harmonic approximation, the high-T heat capacity goes to the limit of 3R, as expected. By considering the QHA and the electronic contribution, the high-T heat capacity is significantly improved and approaches the experimental data. Other properties are shown only in the QHA, since they cannot be evaluated in the harmonic approximation. We note that the temperature-dependence of these properties is in good agreement with experiment, while there is a nearly constant discrepancy for the lattice parameter and bulk modulus, already present at 0 K. The origin of this discrepancy is the exchange-correlation functional used in DFT

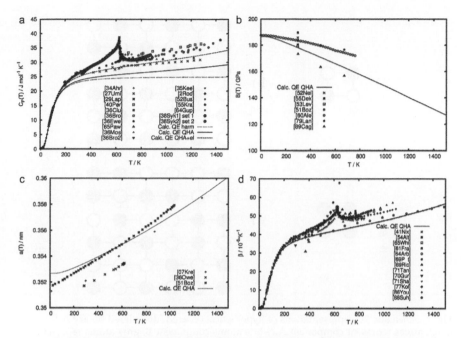

Fig. 6: Thermophysical properties of pure Ni f.c.c. as a function of temperature according to different approximations. (a) Isobaric heat capacity C_p. (b) Lattice parameter a (of f.c.c. unit cell). (c) Isothermal bulk modulus B. (d) Volumetric thermal expansion β. Experimental data from various literature sources are reported for comparison. See ref. (Palumbo et al. 2014) for the full list of references on the experimental data sets.

calculations which, as we have already commented, represents the major source of uncertainties in such results to date.

Ab initio Methods for Disordered Alloys

The methods described up to this point are suitable for application to ordered systems, that is, compounds and alloys where each position in the unit cell is occupied by a given atomic species. In a real alloy, however, some degree of disorder may occur and hence the atoms may more or less randomly occupy the unit cell sites.

Let us consider a binary alloy $A_{0.5}B_{0.5}$ where the subscripts refer to its global composition. Two distinct extreme scenarios can occur; these are represented in two-dimensional schemes in Fig. 7, where we assume that the alloy has a simple square 2D lattice.

The alloy could be completely ordered with A atoms (black) occupying the top-left and bottom-right positions in the square cell in Fig. 7a, while B atoms (white) occupy the other two positions. The fully ordered alloy is

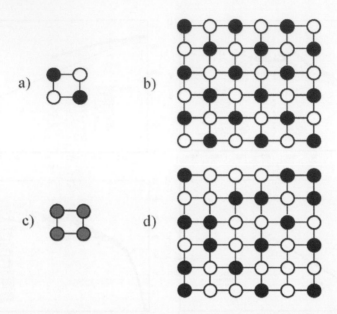

Fig. 7: Schematic 2D representations of hypothetical ordered/disordered binary A-B alloys. Black atoms represent component A, white atoms component B, grey atoms represent the random occurrence of either A or B atoms. (a) Unit cell for a fully ordered alloy A0.5B0.5. (b) A larger portion of the ordered solid obtained from the unit cell in a. (c) Unit cell for a disordered alloy A0.5B0.5. (d) A larger portion of the disordered alloy from the unit cell in c).

obtained from repetitions of the unit cell upon translation and is shown in Fig. 7b. Each A atom is always surrounded by four B atoms as nearest neighbours and vice versa.

The other scenario is that the alloy is fully disordered. In this case, each position in the unit cell is randomly occupied by A or B atoms, that is, there is a 50% probability that it is occupied by A atoms and a 50% probability that it is occupied by B atoms. Such a situation is represented in Fig. 7c and 7d. For the unit cell (Fig. 7c) the grey colour represents the "random" occupancy by either A or B atoms. Each A atom is *on average* surrounded by two A atoms and two B atoms as nearest neighbours and vice versa. However, if we take a particular A atom in the lattice, it may happen by chance that it is surrounded for instance by four B atoms or by four A atoms and all possibilities in between.

As shown in Fig. 7d, in the extended lattice the translational invariance is lost and we cannot infer the properties of the disordered alloy from the properties of a single unit cell. This implies that the *ab initio* methods described in the previous sections cannot be applied to random alloys without proper modifications. A brute-force approach would simply consist of solving the Kohn-Sham equations for a system of the size (ideally) of the real crystal, without applying the Bloch theorem. This is however computationally unfeasible because of the large number of electrons to be considered (of the order of the Avogadro number).

In general, it is of interest to evaluate the alloy properties at several different compositions or even in the full compositional space. A systematic theory has been developed to tackle the "alloy problem" and we present here only a brief introduction of the methods used to treat disordered alloys. The interested reader can find a detailed discussion of the alloy theory in (Ducastelle 1991; De Fontaine 1994; Sanchez et al. 1984; Ruban and Abrikosov 2008).

We note that in a real alloy the true scenario is often in between the fully ordered alloy in Fig. 7b and the fully disordered alloy in Fig. 7d. Order parameters are introduced to characterize the specific degree of ordering. The energy of the system and thus the relative stability of different ordered/disordered alloys depend on the degree of ordering. It is customary to describe the parent lattice for a disordered alloy using pseudo-spin occupation variables σ_i, which for a binary A-B system can assume the values +1 or −1 referring to the A or B component, respectively. Any arrangement of A and B atoms in the alloy lattice can then be described by a vector σ of these occupation variables.

Neglecting, for simplicity, temperature-dependent excitations, the Helmholtz energy of the (disordered) system can now be expressed as:

$$F(T, V, \xi) = E^{st}(V, \xi) - TS^{conf}(V, \xi) \tag{46}$$

where ξ is an order parameter.

The static (disordered) alloy energy $E^{st}(V, \xi)$ can be calculated with the following first-principles[9] methods:

a. Cluster Expansion (CE). This method parameterizes the energy of the alloy as a polynomial in the occupation variables, adding the contributions from different clusters (a set of lattice sites) that are not symmetrically equivalent. The coefficients of this expansion are determined from the total energies of a number of ordered structures using any DFT code. The sum is truncated according to some convergence criteria and it is hence able to describe the energy of the alloy as a function of composition. This is the most thorough method for computing alloy properties and its accuracy can in principle be tailored to each specific problem.

b. Korringa-Kohna-Rostocker Coherent Phase Approximation (KKR-CPA). Within this approach, a one-component effective medium is derived whose scattering properties are the same as the average of

---

[9] Although practically DFT is mostly used, any first-principles method can be employed.

the alloy components. This effective medium is determined according to a mathematical formalism based on Green's functions and requires DFT codes based on the same formalism. Although the accuracy of the method can be questioned for some systems (Ruban and Abrikosov 2008), it has the significant advantage of being faster than CE and it is easy to calculate any arbitrary alloy composition.

c. Special Quasi-Random Structures (SQS). Suitable supercells with a "reasonable" size are constructed which can reproduce the disordered nature of the alloy (Zunger et al. 1990). The size of these supercells is usually in the range 32–256 atoms. Each supercell represents an ordered alloy configuration which can be calculated using any DFT code. Despite its simplicity, it has to be remarked that the construction of an ordered supercell which can reproduce the properties of a disordered alloy is not obvious and caution must be applied in using this approach (Ruban and Abrikosov 2008). Furthermore, the set of disordered alloy compositions which can be reliably calculated is limited by the size of the supercell.

Among the various approaches for calculating the configurational entropy $S^{conf}(V, \xi)$, the Bragg-Williams (BW) approximation (Bragg and Williams 1934) is the most widely employed for metallic alloys. This is mainly because of its mathematical and physical simplicity so that the method can be applied to complex multicomponent alloys. For instance, most commercial software for phase equilibria calculations is based on this approximation. This approach is already described in detail in the Chapter "Computational Thermodynamics" and we will not repeat its derivation here. We note, however, that this approach has the serious shortcoming of being "single-site" in nature, while it is well known that in a typical alloy, atomic interactions are not confined to the nearest-neighbour pairs, but extend more widely over distant pairs. This deficiency is usually overcome in computational thermodynamics approaches such as CALPHAD by introducing some corrective terms in the excess energy (see "Computational Thermodynamics: From Experiments to Applications").

A possible alternative is a more refined treatment for the configurational entropy including long-range atomic correlations. The cluster variation method (CVM) (Kikuchi 1951) offers a viable and more physical way of calculating the configurational entropy of alloys. This method, combined with cluster expansion, has been successfully applied to several binary systems. It has, however, the drawback of being mathematically cumbersome and the extension to multicomponent alloys such as steels and Ni-base superalloys is not realistic at the moment.

Coupling *ab initio* Methods and Semi-Empirical (CALPHAD) Models

Despite the tremendous progress in developing reliable DFT-based methods for calculating thermophysical properties and the continuous increase in available computational power, there are some applications for which *ab initio* methods are still too demanding. For instance, phase stability and equilibria calculations in metallic alloys such as steels and Ni-base superalloys require knowledge of the free energy as a function of temperature and composition for several different ordered and disordered phases in a composition space of up to 15 elements. Furthermore, the present accuracy of DFT calculations may not be sufficient for some applications, mostly due to the unknown exchange-correlation potential. An important example is pure iron, whose b.c.c. to f.c.c. (and then back to b.c.c.) transition with increasing temperature depends on a free energy difference of the order of 1 meV/at, higher than the typical difference between a LDA and GGA calculation.

Traditionally, such systems have been modelled with semiempirical methods such as CALPHAD, in which a number of model parameter values are obtained by fitting experimental data (Saunders and Miodownik 1998; Lukas et al. 2007). Although complex problems can be successfully tackled in this way, relying only on experimental data limits the predictive ability of such models and their applicability to systems where few experimental results are available. Furthermore, the physical soundness of these fitting parameters has been questioned.

A viable compromise that is increasingly becoming popular is to use DFT results coupled with CALPHAD thermodynamic models. The benefits of this approach are several:

- Increase in the predictive ability of CALPHAD models through reduction of the number of parameters that need to be fitted from experiment.
- Enabling the possibility of using complex (and more accurate) thermodynamic models with many parameters since many of them can be estimated using DFT.
- Providing a strong physical base for model parameters obtained using DFT-based methods clearly linked to the originating physical phenomena.
- Extending the applicability of CALPHAD models to systems where few or no experimental values are available.
- Extending the applicability of *ab initio* methods to systems that are too demanding for present-day approaches.

Being a field in continuous evolution, there is no consensus yet on how this coupling must be carried out. Several approaches have been proposed and applied in recent years, with different levels of use of DFT results. A workshop took place in 2013 in the Ringberg castle in Germany where leading experts from both the *ab initio* and CALPHAD communities discussed state-of-the-art methods and how to combine them (Hickel et al. 2014). In the following, we will illustrate some of the current ideas in the field, showing some examples from the work of the author and co-workers. Other references on the topic are for example (Turchi 2005; Liu 2009) and references therein.

An initial and now widely accepted way of incorporating DFT results into thermodynamic modeling is by using them as "experimental" data in order to fit model parameters. Mostly energies of formation at 0 K have been used for this purpose and the DFT results are usually limited to a number of key structures. In this case, DFT values complement different kinds of experimental data and are consistently used within the CALPHAD approach. DFT results can cover regions in the temperature-pressure-composition space where experimental values are scarce and help in the assessment of the reliability of conflicting data sets. DFT results are particularly important for metastable (or unstable) phases, for which experimental determination is difficult or impossible, and in cases where achieving equilibrium experimentally is a long-term process, especially at low temperatures. An example is the reassessment of the Fe-Ni system (Cacciamani et al. 2010). DFT calculations were carried out in this system in order to determine enthalpies of formation, lattice parameters and magnetic moments for several structures of interest. These results were used together with several experimental data sets available in the literature for a revaluation of the phase diagram and thermodynamic properties of this system. As a result, a new set of thermodynamic model parameters were obtained (Cacciamani et al. 2010) and the resulting phase diagram is shown in Fig. 8.

The use of DFT results in thermodynamic models can, however, be more extended (and beneficial). We note here that thermodynamic modeling is mostly based on the Gibbs energy G rather than on the Helmholtz energy F, as is the case for DFT methods. The reason is just practical: the parameter models in CALPHAD being traditionally based on experimental data carried out mostly at constant (ambient) pressure. The analysis of experimental data is for the same reason based on G. For the sake of simplicity, we neglect the pressure dependence in the following and assume $P = 1$ *bar*. In fact, we note that, although in principle possible, most commercial thermodynamic databases do not model the pressure dependence.

Let us take a relatively complex phase, the sigma (σ) phase, as an illustrative example. This phase has a tetragonal unit cell (Pearson symbol tP30, space group n.136 $P4_2/mnm$), with 5 different Wyckoff positions ($2a$, $4f$, $8i_1$, $8i_2$, $8j$) and 30 atoms in total. Its Gibbs energy can be modeled using the Compound Energy Formalism (CEF) or sublattice model (Hillert 2001), which has been introduced in Chapter 1 "Computational Thermodynamics: From Experiments to Applications". According to this model, each phase is considered as being composed of interlocking sublattices, their number being equal to the different Wyckoff positions of the crystal structure of the phase. The CEF Gibbs energy with five sublattices for the sigma phase then reads

$$G_\sigma(T, y_i^k) = H_\sigma^{st}(y_i^k) - TS_\sigma^{conf}(y_i^k) + G_\sigma^{ex}(T, y_i^k) \tag{47}$$

where $i = A, B$ in a binary system with A, B components; we have introduced the site fractions y_i^k, which are related to the composition by $x_i = \Sigma_k y_i^k$. The site fraction y_i^k gives the fraction of the component i in the sublattice k. The first term in the above equation is referred to as the CEF reference surface of energy and for five sublattices it is given by

$$H_\sigma^{st} = \sum_{ijklm} y_i^1 y_j^2 y_k^3 y_l^4 y_m^5 \Delta H_{ijklm}^\sigma \tag{48}$$

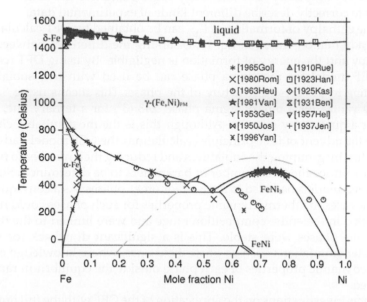

Fig. 8: Critically assessed Fe-Ni phase diagram according to the thermodynamic evaluation in (Cacciamani et al. 2010). The model parameters from which the phase diagram is calculated have been determined including the DFT results together with available experimental data from the literature.

where each index i,j,k,l,m runs over the possible components in the system (A, B in this example) and $\Delta H^{\sigma}_{ijklm}$ is the enthalpy of formation of the (hypothetical) compound with the crystal structure of the σ phase in which atoms i fully occupy the first sublattice sites, atoms j fully occupy the second sublattice sites, etc. These hypothetical compounds are also called "endmembers" and their enthalpies of formation "endmember energies". For instance, $\Delta H^{\sigma}_{AAAAA}$ is the enthalpy of formation of the hypothetical pure A σ phase, where all sublattices are occupied by A atoms. We remark here that some of the $ijklm$ hypothetical compounds in the model may be (and usually are) metastable or even unstable. In a binary system (2 components) and for a phase with 5 sublattices, there is a total of 32 (2^5) possible endmembers. In a ternary system this number rises to 243 (3^5) and so on.

The configurational entropy term is usually modelled according to the BW approximation as

$$S^{conf}_{\sigma}(y^k_i) = -R \sum_s a^s \sum_{i=A,B} y^s_i ln(y^s_i) \tag{49}$$

where R is the gas constant and a^s is the multiplicity of the s sublattice.

The last term, $G^{ex}_{\sigma}(T, y^k_i)$, is called the excess energy and describes what is left out by the first two terms. It may be implemented with differing degrees of complexity and ultimately must guarantee the ability of the model to correctly describe different kinds of experimental data.

The enthalpy of formation $\Delta H^{\sigma}_{ijklm}$ can be obtained by DFT calculations as already demonstrated in Eq. 17, assuming the difference between the enthalpy and the energy of formation is negligible. By using DFT results, the CEF model for the sigma phase can be used with five sublattices according to the crystal structure of the phase. This means using 32 (2^5) energy values for a binary sigma phase, 243 (3^5) for a ternary one, 1024 (4^5) for a quaternary one, etc. Although this is the most suitable choice, before the advent of first-principles calculations the CEF model had to be simplified by grouping the sublattices and reducing their number to two in order to reduce the total number of parameters to be determined. Similar simplified models had also to be used for other phases. As a consequence, in some cases the thermodynamic properties for such phases could not be calculated in the entire composition range and were limited to the region where the phases were stable. This is a significant drawback for some computer simulations such as phase-field ones, where knowledge of the thermodynamic properties of each phase outside the equilibrium range is required.

Some investigations on the application of the CEF with the full number of sublattices for each phase and using DFT energies have been carried out, for example (Palumbo et al. 2011; Palumbo et al. 2014). Figure 9 shows the calculated ternary enthalpy surface in the Cr-Ni-Re system obtained in the entire composition range with the CEF+DFT for the sigma phase.

In these investigations, no excess term has been used $G_\sigma^{ex}(T, y_i^k) = 0$ and hence no parameters were optimized. The results obtained, including phase diagram calculations, are in qualitative agreement with the experimental data, but significant differences still exist. A possible way to further improve these results, without applying more refined DFT methods, consists in optimizing some parameters in the excess term using experimental information. A similar approach has been used for example in (Palumbo et al. 2010; Mathieu et al. 2013).

Alternatively, more refined DFT methods can be used to include further interactions and physical contributions. For example, temperature-dependent DFT results can be included in the above formalism for each endmember compound. Equation 48 would then change into

$$H_\sigma^{st} = \sum_{ijklm} y_i^1 y_j^2 y_k^3 y_l^4 y_m^5 (\Delta H_{ijklm}^\sigma + \Delta G_{ijklm}^{\sigma,el} + \Delta G_{ijklm}^{\sigma,vib} + \Delta G_{ijklm}^{\sigma,magn}) \quad (50)$$

where the three terms corresponding to the different possible excitations are included. For instance, the electronic term is given by

$$\Delta G_{ijklm}^{\sigma,el} = G_{ijklm}^{\sigma,el} - x_A G_A^{\alpha,el} - x_B G_B^{\beta,el} \quad (51)$$

$G_{ijklm}^{\sigma,el}$ being the electronic contribution for the hypothetical endmember $ijklm$, x_A and x_B the mole fractions of A and B for the endmembers, $G_A^{\alpha,el}$ and $G_B^{\beta,el}$ the electronic contributions for pure A and B in the α and β structures, and α and β the stable structures for A and B at 0 K. All terms in

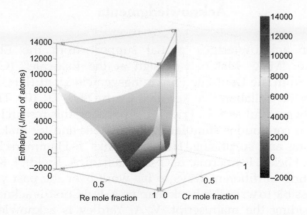

Fig. 9: Calculated enthalpy of the ternary sigma phase in the Cr-Ni-Re system in the entire composition range at T = 373 K. Details on the calculations can be found in (Palumbo et al. 2011).

Eq. 51 are dependent on temperature, and can also depend on the volume (cf. Section 4). We remark that we have used G instead of F in the above equation and assumed their difference is negligible. At 0 K, assuming the structures are fully relaxed and hence $P = 0$, $\Delta H^\sigma_{ijklm} = \Delta E^\sigma_{ijklm}$ (as we already pointed out in Section 3) and $\Delta G^\sigma_{ijklm} = \Delta F^\sigma_{ijklm}$. If the pressure dependence is not modelled, $G_\sigma(T, y^k_i)$ is usually assumed to be at constant $P = 1$ *bar* and the difference $\Delta H^\sigma_{ijklm} (1\ bar) - \Delta H^\sigma_{ijklm} (0\ bar) \cong 0$. In the general case, to derive G from F (and H from E), the relationship $P = P(V)$ must be determined, for instance by fitting DFT data with the equation of state as described in Section 3.1.

Although in principle straightforward, the inclusion of temperature-dependent excitations presents some practical problems which have to be solved in order to incorporate first-principles results in thermodynamic databases. For example, the output of the DFT methods described in Section 4 is a set of discrete data points as a function of temperature and volume. The available thermodynamic software usually deals with analytical representation of the same properties, mostly using polynomials. Different strategies can then be implemented, either fitting the DFT results with analytical functions or directly incorporating the discrete data sets as tables. Alternatively, some properties can be calculated "on the fly" from quantities such as electronic and phonon DOS. Additional work is still necessary to determine which strategies are the most suitable for multicomponent alloy systems. For a more detailed discussion, the reader is referred to (Palumbo et al. 2014).

Acknowledgments

The author acknowledges partial support from the EU Centre of Excellence "MaX—Materials Design at the Exascale" (Grant No. 676598) and from the Deutsche Forschungsgemeinschaft (DFG) through project C6 as the Collaborative Research Centre SFB/TR 103. The author is particularly grateful to S. G. Fries for supporting this research over the past years, for continuous stimulating discussions and for making every day interesting and surprising. He also thanks T. Hammerschmidt, A. Dal Corso, T. Hickel, F. Koermann, G. Cacciamani, A. Breidi, J. Kossmann for contributing in different ways to his work over the past years thus making it possible to write this book chapter. A. Dal Corso is acknowledged for proof reading the manuscript. V. A. Yardley is acknowledged for checking the English of the manuscript.

Keywords: Density Functional Theory (DFT), Thermodynamic modeling, Vibrational properties, Thermophysical properties, Phonons, Quasi-harmonic approximation

References

Asker, C., A.B. Belonoshko, A.S. Mikhaylushkin and I.A. Abrikosov. 2008. First-principle solution to the problem of Mo lattice stability. Phys. Rev. B 77: 220102(R).

Baroni, S., S. De Gironcoli, A. Dal Corso and P. Giannozzi. 2001. Phonons and related crystal properties from density-functional perturbation theory. Rev. Mod. Phys. 73: 515.

Baroni, S., P. Giannozzi and E. Isaev. 2010. Density-functional perturbation theory for quasi-harmonic calculations. Rev. Mineral. Geochem. 71: 39–57.

Baroni, S., P. Giannozzi and A. Testa. 1987. Green's function approach to linear response in solids. Phys. Rev. Lett. 58: 1861–1864.

Becker, C.A., J. Ågren, M. Baricco, Q. Chen, S.A. Decterov, U.R. Kattner et al. 2014. Thermodynamic modeling of liquids: CALPHAD approaches and contributions from statistical physics. Phys. Status Solidi B 251: 33–52.

Born, M. and K. Huang. 1954. Dynamical Theory of Crystal Lattices. Clarendon Press, Oxford.

Bragg, W.L. and E.J. Williams. 1934. The effect of thermal agitation on atomic arrangement in alloys. Proc. Roy. Soc. A145: 699–730.

Cacciamani, G., A. Dinsdale, M. Palumbo and A. Pasturel. 2010. The Fe-Ni system: Thermodynamic modelling assisted by atomistic calculations. Intermetallics 18: 1148–1162.

De Fontaine, D. 1994. Cluster approach to order-disorder transformations in alloys. Solid State Physics 47: 33–176.

Ducastelle, F. 1991. Order and phase stability in alloys. *In*: F.R. de Boer and D.G. Pettifor [eds.]. Cohesion and Structure Series, vol. 3. North-Holland, Amsterdam.

Hickel, T., U. Kattner and S.G. Fries. 2014. Computational thermodynamics: Recent developments and future potential and prospects. Phys. Status Solidi B 251: 9–13.

Hillert, M. 2001. The compound energy formalism. J. Alloy Compd. 320: 161–176.

Hohenberg, P. and W. Kohn. 1964. Inhomogeneous electron gas. Phys. Rev. 136: B864–B871.

Kikuchi, R. 1951. A theory of cooperative phenomena. Phys. Rev. 81: 988–1003.

Körmann, F., A.A.H. Breidi, S.L. Dudarev, N. Dupin, G. Ghosh, T. Hickel et al. 2014. Lambda transitions in materials science: Recent advances in CALPHAD and first-principles modelling. Phys. Status Solidi B 251: 53–80.

Kohn, W. and L.J. Sham. 1965. Self-consistent equations including exchange and correlation effects. Phys. Rev. 140: A1133–A1139.

Kresse, G., J. Furthmuller and J. Hafner. 1995. *Ab initio* force constant approach to phonon dispersion relations of diamond and graphite. Europhys. Lett. 32: 729–735.

Liu, Z.-K. 2009. First-principles calculations and CALPHAD modeling of thermodynamics. J. Phase Equilib. Diff. 30: 517–534.

Lukas, H.L., S.G. Fries and B. Sundman. 2007. Computational Thermodynamics. The Calphad Method. Cambridge University Press, New York.

Martin, R.M. 2004. Electronic Structure: Basic Theory and Practical Methods. Cambridge University Press.

Mathieu, R., N. Dupin, J.-C. Crivello and J.-M. Joubert. 2013. CALPHAD description of the Mo-Re system focused on the sigma phase modeling. CALPHAD 43: 18–31.

Palumbo, M., B. Burton, A. Costa e Silva, B. Fultz, B. Grabowski, G. Grimvalle et al. 2014. Thermodynamic modelling of crystalline unary phases. Phys. Status Solidi B 251: 14–32.

Palumbo, M., T. Abe, S.G. Fries and A. Pasturel. 2011. First-principles approach to phase stability for a ternary σ phase: Application to Cr-Ni-Re. Phys. Rev. B 83: 144109.

Palumbo, M., T. Abe, C. Kocer, H. Murakami and H. Onodera. 2010. *Ab initio* and thermodynamic study of the Cr-Re system. CALPHAD 34: 495–503.

Palumbo, M., S.G. Fries, A. Dal Corso, F. Koermann, T. Hickel and J. Neugebauer. 2014. Reliability evaluation of thermophysical properties fom first-principles calculations. J. Phys.: Condens. Matter 26: 335401.

Palumbo, M., S.G. Fries, T. Hammerschmidt, T. Abe, J.-C. Crivello, A. Breidi et al. 2014. First-principles-based phase diagrams and thermodynamic properties of TCP phases in Re-X systems (X = Ta, V, W). Comput. Mater. Sci. 81: 433–445.

Palumbo, M., S.G. Fries, A. Pasturel and D. Alfe. 2014. Anharmonicity, mechanical instability, and thermodynamic properties of the Cr-Re-σ phase. J. Chem. Phys. 140: 144502.

Parlinski, K., Z.Q. Li and Y. Kawazoe. 1997. First-principles determination of the soft mode in cubic ZrO_2. Phys. Rev. Lett. 78: 4063–4067.

Perdew, J.P., K. Burke and M. Ernzerhof. 1996. Generalized gradient approximation made simple. Phys. Rev. Lett. 77: 3865–3869.

Ruban, A.V. and I.A. Abrikosov. 2008. Configurational thermodynamics of alloys from first-principles: Effective cluster interactions. Rep. Prog. Phys. 71: 046501.

Sampoli, M.P. Benassi, R. Dell'Anna, V. Mazzacurati and G. Ruocco. 1998. Numerical study of the low-frequency atomic dynamics in a Lennard-Jones glass. Phil. Mag. 77: 473–484.

Sanchez, J.M., F. Ducastelle and D. Gratias. 1984. Generalized cluster description of multicomponent systems. Physica A 128: 334–350.

Saunders, N. and A.P. Miodownik. 1998. CALPHAD. Calculation of phase diagrams. *In*: R.W. Cahn [ed.]. Pergamon Materials Series. Pergamon, Oxford.

Turchi, P.E.A., I.A. Abrikosov, B. Burton, S.G. Fries, G. Grimvall, L. Kaufman et al. 2005. Interface between quantum-mechanical-based approaches, experiments and CALPHAD methodology. CALPHAD 31: 4–27.

Zunger, A., S.H. Wei, L.G. Ferreira and J.E. Bernard. 1990. Special quasirandom structures. Phys. Rev. Lett. 65: 353–357.

1.3

The Formation Volume in Rare Earth Intermetallic Systems

A Representation by Means of Atomic Physical Quantities

*M. Pani** and *F. Merlo*

INTRODUCTION

The formation of the intermetallic phases is often accompanied by large volume effects, and several interpretations of this phenomenon have been proposed following different approaches, either from a simply empirical or purely theoretical point of view. In his pioneering work, Biltz (Biltz 1934) employed the experimental density values to list the molar volumes of the elements and of numerous inorganic and organic solids, including intermetallic phases. The systematic analysis of the data allowed the scientist to estimate the effective volumes of the atoms within the compounds, making possible the application of the volume additivity. Machlin studied the effects of electronegativity on energy and volume of formation, assuming that the volume corrections depend on the Gordy electronegativity difference (Machlin 1980). Watson and Bennett obtained

University of Genova, Department of Chemistry and Industrial Chemistry, Via Dodecaneso 31, 16146 Genova, Italy.
* Corresponding author: marcella@chimica.unige.it

a good correlation between the volume effects shown by the phases of the transition elements and a scale resembling the Gordy electronegativity, while a cellular method was applied to estimate the volume changes in phases with alkaline and alkaline earth metals (Watson and Bennett 1982, 1984). Alonso showed that a model of a disordered binary alloy of nontransition metals explains the tendency to a negative deviation from Vegard's law as this lowers the energy of formation (Alonso et al. 1984). In the Miedema model (Miedema and Niessen 1982), the volume contraction in metallic systems can be ascribed to a charge transfer effect, described mainly by the differences both in an electronegativity-like scale (Φ^*) and in the electron density parameter (n_{ws}). Moreover, in systems with atoms of different radius, a further volume contraction may arise from elastic size mismatch energy. The differences in electronegativity are ignored in the Hafner approach, based on the lowest-order pseudopotential perturbation theory (Hafner 1985). This method provides good results for extended solid solutions of homovalent systems (intra-alkaline and intra-alkaline-earth alloys) and for some intermetallics of the cited elements. A phenomenological approach was used to describe the volume effects displayed by the intermetallic compounds formed by alkaline earths (Ca, Sr, Ba) and divalent rare earths (Eu, Yb) (Merlo 1988) and by the trivalent rare earths (Merlo and Fornasini 1993), introducing a charge transfer atomic parameter, correlated with Pauling's electronegativity. More recently, the volume contractions of the binary phases of Ca, Sr, Ba, Eu and Yb were represented by a simple equation containing the electronegativity, the compressibility and the group number (Fornasini and Merlo 2006). The most advanced method is based on the calculation of partial atomic volumes and charges as a function of composition (Baranov et al. 2007).

The present work is dedicated to the binary intermetallic compounds formed by the rare earths with the other elements. The volume effects shown by these systems will be represented following a procedure similar to that proposed in the previous work on the intermetallic phases of divalent elements (Fornasini and Merlo 2006), by a tentative correlation between formation volume and some atomic physical quantities. A possible application of the proposed approach to ternary systems will be also shortly pointed out.

The Volume Effects Representation

The volume of formation of a phase can be predicted by two methods. The original Vegard rule (Pearson 1972) refers to a monodimensional crystallographic parameter, namely supposing a linear variation of the lattice parameter as a function of the composition. This approach, based on the constancy and additivity of the atomic size of the components, can be easily applied to cubic structures and provides positive or negative deviations of the derived cell volume as regards the mean cell volume obtained from the proper combination of the elemental volumes. When the unit cell is not cubic, the use of the Vegard method becomes difficult, owing to the occurrence of independent cell parameters. For this reason, the average atomic volume, namely the elementary cell volume divided by the total number of atoms, is usually plotted as a function of composition. The supposed linear variation, now based on the constancy and additivity of the atomic volume of the elements, the so-called Zen law, is the preferred hypothesis (Zen 1956), which predicts no volume effects ($\Delta V = V_{calc} - V_{obs} = 0$). This simple method, even if erroneously called again as the Vegard law, is applied in the present work, comparing the volumes calculated on the basis of the proper combination of the elemental atomic volumes with the experimental ones.

Let V be the experimental average atomic volume of a given $R_{1-x}X_x$ compound (cell volume divided by the total number of atoms per cell), and V_0 the proper sum of the elemental volumes, V_R^0 and V_X^0, according to the Vegard (or Zen) law

$$V_0 = (1 - x)\, V_R^0 + x V_X^0 \tag{1}$$

The volume effect, expressed by the difference $V_0 - V \equiv \Delta V$, is generally positive, namely a volume contraction is observed, corresponding to a negative deviation from the Vegard law.

In nearly all considered systems the experimental V points give a regular trend with the composition: some examples of systems containing rare earth elements are shown in Fig. 1, using as abscissa the mole fraction of the partner X element.

As previously proposed (Merlo 1988; Merlo and Fornasini 1993; Fornasini and Merlo 2006), the regular distribution of the observed V values suggests that the area comprised between the "ideal" Vegard-like line and the curve connecting the phases can be taken as an average measure of the volume effects of the whole system (grey coloured area in Fig. 2).

This quantity, given by the expression

$$\Delta V_{int} = \int_0^1 (V_0 - V)dx \tag{2}$$

can be defined as "integral volume effect". To evaluate this area for all systems in a similar analytical way, a simple equation can be used:

$$V_0 - V = Kx^n(1 - x)^2 \tag{3}$$

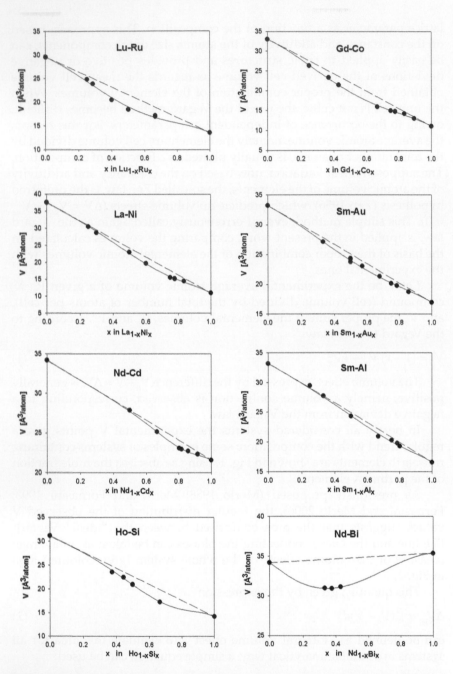

Fig. 1: Average atomic volumes vs. atomic fraction of the partner X element in the Lu-Ru, Gd-Co, La-Ni, Sm-Au, Nd-Cd, Sm-Al, Ho-Si and Nd-Bi systems. Full circles indicate the experimental values of the phases with known structure, broken lines refer to the Vegard-like linear trends and solid lines are calculated by Eq. 3.

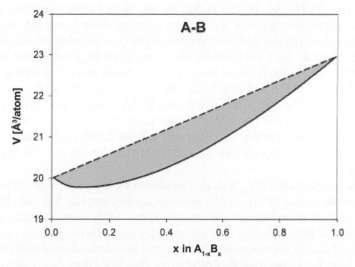

Fig. 2: Typical trend of the experimental volume data (solid line) and corresponding Vegard straight line (dashed) for an intermetallic system.

where K and n are adjustable parameters. The value of n for a given system indicates the composition showing the maximum volume contraction, namely $RX_{n/2}$, which may not correspond to an existing phase. In Fig. 1 the solid lines are calculated by Eq. 3, which is well fitted to the experimental points. Combining Eqs. (2) and (3) one obtains

$$\Delta V_{int} = 2K/[(n + 1)(n + 2)(n + 3)] \tag{4}$$

and the average atomic volume of a given phase can be calculated, in turn, by combining Eqs. (3) and (4):

$$V = V_0 - \frac{1}{2}(n + 1)(n + 2)(n + 3)x^n(1 - x)^2 \Delta V_{int} \tag{5}$$

Volume Effects in Systems Formed by the Trivalent Rare Earths

The experimental volume data of the rare earth binary intermetallic compounds were divided into two groups, which are referred to the normally trivalent rare earths and to the normally divalent lanthanides europium and ytterbium, respectively. For the first group, the examined data belong to the R-X systems where R = La, Ce, Pr, Nd, Sm, Gd, Tb, Dy, Ho, Er, Tm, Lu and X = Fe, Ru, Os, Co, Rh, Ir, Ni, Pd, Pt, Cu, Ag, Au, Mg,

Zn, Cd, Hg, Al, Ga, In, Tl, Si, Ge, Sn, Pb, Sb, Bi, for a total number of 312 binary systems. Some other R-X binary systems containing intermediate phases have been discarded, owing either to the low number of compounds (X = Be, Mn, Tc, Re), or because the atomic arrangements are mostly interpreted on the basis of an iono-covalent bond model showing a low space filling (X = B, C, P, As, S, Se, Te, Po). According to the available data collections (Iandelli and Palenzona 1979; Villars and Calvert 1991, 1997; CRYSTMET 2010; Pearson's Crystal Data 2015/16), the intermediate phases with known crystal structure are at least 2,000, distributed among more than 70 stoichiometries and crystallizing in about 200 different structure types.

Table 1 reports the ΔV_{int} values, calculated by means of Eqs. (3) and (4), for a representative group of 128 systems containing La, Nd, Gd, Ho and Lu. For each X element five values occur, but with some variations. For antimony and bismuth only four systems are considered, with La, Sm, Gd, Ho and La, Nd, Gd, Er, respectively, due to the small number of intermediate phases. For the same reason, not always the five cited rare earths were selected, but a different neighbouring lanthanide: Pr instead of La and Tb instead of Gd in the Fe systems; Sm instead of Nd in the Au, Al, Si and Ge systems; Tm instead of Lu in the Fe, In and Pb systems. The ΔV_{int} values, ranging from 0 $Å^3$/atom (La-Cu and La-Mg systems) to 2.53 $Å^3$/atom (Er-Bi system), were obtained by least squares fitting of Eq. 3 to all literature data, excluding the NaCl-type phases occurring in the Sb and Bi systems (this will be justified subsequently). The values of the elemental volumes V_R^0 and V_X^0 used in the foregoing equations were taken from King (King 1982). For silicon and germanium the atomic volumes of the β-Sn-type form were selected (14.19 and 16.88 $Å^3$/atom, respectively), while for gallium the volume of the In-type form was adopted, namely 17.61 $Å^3$/atom (Villars and Calvert 1991, 1997). These structure types, different from those stable in the normal thermodynamic conditions, were chosen owing to their better space filling, similar to that usually shown by the intermetallic compounds. Table 1 reports also the n values, whose values range from 1 to about 16. Actually, the composition showing the maximum volume contraction goes from the 2:1 stoichiometry (La-Bi system) to the 1:8~ stoichiometry (La-Zn system). To maintain the physical meaning of the calculated V(x) curve, the composition of the phase with the maximum volume contraction must lie within the range of the experimentally observed compounds. So, in 25 systems the n value was constrained to the R-richest composition (RFe$_2$ and RMg phases, La$_2$Co$_3$, La$_2$Bi, Er$_5$Bi$_3$) or to the X-richest composition (NdPd$_3$, La$_{14}$Ag$_{51}$, Nd$_{14}$Ag$_{51}$, LaIn$_3$, LaSi$_2$, SmSi$_2$, GdSi$_2$, HoSi$_2$, LaGe$_2$, SmGe$_2$, GdGe$_2$, HoGe, Lu$_3$Ge$_5$).

As said above, Eq. 5 can be used, by means of the data in Table 1, to calculate the atomic volume of a given intermediate phase. The so obtained values for the examined 631 binary compounds of the trivalent rare earth

Table 1: Values of the Integral Volume Effect (ΔV_{int}) and of the n Parameter in Eqs. (3), (4) and (5), for the 128 Binary Systems Containing Trivalent Rare Earths, Ordered Following a Quasi Periodic Number N.

N	System	n	ΔV_{int} [$Å^3$/atom]
1	Pr-Fe	4	0.82
2	Nd-Fe	4	0.72
3	Tb-Fe	4	0.66
4	Ho-Fe	4	0.66
5	Tm-Fe	4	0.65
6	La-Ru	2.87	1.30
7	Nd-Ru	3.19	1.13
8	Gd-Ru	2.35	1.28
9	Ho-Ru	2.24	1.26
10	Lu-Ru	2.26	1.33
11	La-Os	2.78	1.35
12	Nd-Os	2.89	1.12
13	Gd-Os	2.29	1.40
14	Ho-Os	2.28	1.20
15	Lu-Os	2.20	1.16
16	La-Co	3	1.08
17	Nd-Co	2.19	1.21
18	Gd-Co	2.31	1.29
19	Ho-Co	2.74	1.03
20	Lu-Co	2.29	1.28
21	La-Rh	2.70	1.26
22	Nd-Rh	3.31	1.12
23	Gd-Rh	2.31	1.34
24	Ho-Rh	2.76	1.45
25	Lu-Rh	2.14	1.45
26	La-Ir	2.81	1.23
27	Nd-Ir	2.95	1.06
28	Gd-Ir	2.63	1.43
29	Ho-Ir	2.59	1.44
30	Lu-Ir	2.29	1.48
31	La-Ni	3.70	0.91
32	Nd-Ni	3.70	0.90
33	Gd-Ni	3.51	0.98
34	Ho-Ni	3.81	0.91
35	Lu-Ni	3.53	0.90
36	La-Pd	4.97	0.65
37	Nd-Pd	6	0.72

Table 1 cont....

Table 1 cont...

N	System	n	ΔV_{int} [Å³/atom]
38	Gd-Pd	3.86	1.06
39	Ho-Pd	3.30	1.08
40	Lu-Pd	2.79	1.10
41	La-Pt	3.70	0.96
42	Nd-Pt	5.68	1.09
43	Gd-Pt	3.06	1.25
44	Ho-Pt	2.89	1.15
45	Lu-Pt	2.91	1.16
46	La-Cu	–	0
47	Nd-Cu	8.22	0.05
48	Gd-Cu	3.49	0.45
49	Ho-Cu	3.47	0.46
50	Lu-Cu	2.32	0.58
51	La-Ag	7.3	0.25
52	Nd-Ag	7.3	0.28
53	Gd-Ag	3.70	0.43
54	Ho-Ag	2.64	0.50
55	Lu-Ag	3.71	0.60
56	La-Au	1.89	0.55
57	Sm-Au	2.63	0.57
58	Gd-Au	2.49	0.90
59	Ho-Au	2.92	0.84
60	Lu-Au	2.79	1.02
61	La-Mg	–	0
62	Nd-Mg	2	0.02
63	Gd-Mg	2	0.29
64	Ho-Mg	2	0.34
65	Lu-Mg	2	0.45
66	La-Zn	15.42	0.16
67	Nd-Zn	15.82	0.14
68	Gd-Zn	9.81	0.24
69	Ho-Zn	8.98	0.30
70	Lu-Zn	5.13	0.47
71	La-Cd	9.96	0.17
72	Nd-Cd	7.47	0.28
73	Gd-Cd	5.59	0.51
74	Ho-Cd	4.76	0.53
75	Lu-Cd	3.67	0.68
76	La-Hg	5.38	0.97

Table 1 cont....

Table 1 cont...

N	System	n	ΔV_{int} [Å^3/atom]
77	Nd-Hg	4.62	1.19
78	Gd-Hg	4.12	1.38
79	Ho-Hg	3.40	1.47
80	Lu-Hg	3.38	1.65
81	La-Al	3.68	0.13
82	Sm-Al	4.22	0.33
83	Gd-Al	4.38	0.48
84	Ho-Al	4.42	0.55
85	Lu-Al	4.08	0.70
86	La-Ga	6.32	0.11
87	Nd-Ga	5.78	0.21
88	Gd-Ga	3.73	0.71
89	Ho-Ga	4.01	0.92
90	Lu-Ga	4.47	0.98
91	La-In	6	0.84
92	Nd-In	5.77	1.09
93	Gd-In	4.57	1.36
94	Ho-In	5.08	1.35
95	Tm-In	4.66	1.53
96	La-Tl	3.52	1.43
97	Nd-Tl	3.49	1.67
98	Gd-Tl	3.50	1.95
99	Ho-Tl	3.41	1.97
100	Lu-Tl	3.31	1.95
101	La-Si	4	0.43
102	Sm-Si	4	0.74
103	Gd-Si	4	0.91
104	Ho-Si	4	1.09
105	Lu-Si	2.31	0.90
106	La-Ge	4	0.60
107	Sm-Ge	4	0.85
108	Gd-Ge	4	1.20
109	Ho-Ge	2	0.90
110	Lu-Ge	3.3	1.27
111	La-Sn	3.05	1.34
112	Nd-Sn	5.08	1.10
113	Gd-Sn	3.46	1.47
114	Ho-Sn	3.06	1.63
115	Tm-Sn	2.80	1.60

Table 1 cont....

Table 1 cont...

N	System	n	ΔV_{int} [Å³/atom]
116	La-Pb	5.35	1.00
117	Nd-Pb	4.15	1.25
118	Gd-Pb	3.12	1.72
119	Ho-Pb	2.91	1.79
120	Tm-Pb	2.77	1.79
121	La-Sb	1.85	1.61
122	Sm-Sb	2.14	2.02
123	Gd-Sb	1.82	2.28
124	Ho-Sb	1.96	2.45
125	La-Bi	1	1.86
126	Nd-Bi	1.17	2.06
127	Gd-Bi	1.45	2.52
128	Er-Bi	1.2	2.53

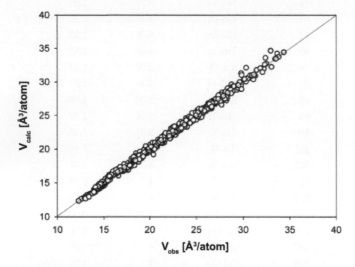

Fig. 3: Average atomic volumes, calculated by Eq. 5, as a function of the observed ones for the intermediate phases of the trivalent rare earths.

are plotted in Fig. 3. The average deviation of the calculated values from the observed ones amounts to 1.15%, and this good agreement confirms both the regularity in the distribution of the experimental points within each system and the ability of the Eq. 3 in the representation of the volume data.

Volume Effects in Systems Formed by Europium and Ytterbium

The second group of examined phases refers to the 54 R-X systems, where R is a divalent rare earth (Eu or Yb), and X is Be, Mn, Re, Fe, Ru, Os, Co, Rh, Ir, Ni, Pd, Pt, Cu, Ag, Au, Be, Mg, Zn, Cd, Hg, Al, Ga, In, Tl, Si, Ge, Sn, Pb, Sb, Bi, with a total number of about 250 intermediate phases. Some systems were not considered in the following calculation, owing to the low number of intermediate phases (Eu-Be, Yb-Be, Yb-Mg, Eu-Re, Yb-Re, Eu-Fe, Yb-Os, Eu-Ir). The main problem is represented by the valence state of Eu and Yb, as in some systems all or part of the intermediate phases show the occurrence of the trivalency of the rare earths. This changes the effective value of their atomic volume, and thus greatly influences the calculation of the integral volume effect. While the elemental volume of ytterbium is 41.25 Å^3/atom in the normal divalent state, the corresponding volume of the ideal trivalent metal can be obtained from an average value between the atomic volumes of thulium and lutetium, namely 29.81 Å^3/atom. Moreover, as reported in the general analysis of the valence anomalies of the rare earths (Iandelli and Palenzona 1979), ytterbium (and rarely europium) can show the phenomenon of the intermediate valence, which can vary with temperature and pressure, as shown by YbAl$_2$. The occurrence of a valence different from two is pointed out by the results of magnetic measurements, or simply by comparing the values of the elementary cell volume of the intermediate phases formed by Eu and Yb with those of the isostructural compounds formed by the other rare earths. It is known (Iandelli and Palenzona 1979) that the lattice parameters of isostructural intermetallic compounds of the lanthanides show a very regular trend (in most cases a linear trend) as a function of the rare earth trivalent ionic radii. The values of the Eu and Yb phases are usually much higher than those of the contiguous rare earths (samarium and gadolinium for europium, thulium and lutetium for ytterbium), and this proves their divalent state. On the other hand, when the Eu or Yb points lie exactly on the trend displayed by the other rare earths, this can be taken as a strong indication of the occurrence of a trivalent state. Some examples of this simple analysis are given in Figs. 4 and 5, where the cubic root of the atomic volume (the cell volume divided by the total number of atoms per cell) ($^3\sqrt{(V/n)}$) is plotted vs. the rare earth trivalent ionic radius for different isostructural series of intermetallic compounds.

As shown in Fig. 4, the large positive deviations of the Eu and Yb points are due to their divalent state in the RCd (CsCl type), RZn$_2$ (CeCu$_2$ type) and RHg$_3$ (Ni$_3$Sn type) phases. On the other hand (see Fig. 5), in the RIr$_2$ (MgCu$_2$ type) phases both Eu and Yb appear trivalent, while in the RSi phases, which crystallize in the closely related CrB or FeB

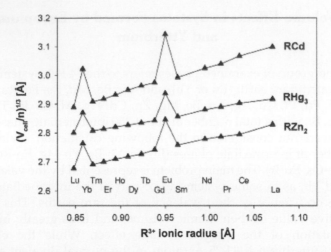

Fig. 4: Cubic root of the atomic volume as a function of the rare earth trivalent ionic radii for the RCd, RZn$_2$ and RHg$_3$ phases.

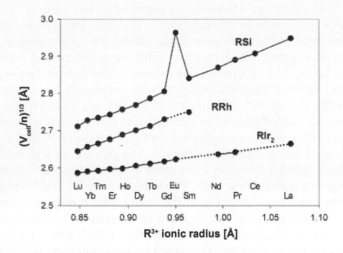

Fig. 5: Cubic root of the atomic volume $\sqrt[3]{(V/n)}$ as a function of the rare earth trivalent ionic radii for the RIr$_2$, RRh and RSi phases.

structure types, EuSi contains divalent Eu and YbSi contains trivalent Yb. Among the RRh compounds (CsCl type), YbRh should contain trivalent Yb, while the non-existence of EuRh can be probably imputed to the low tendency of Eu to acquire the trivalent state (Fig. 5).

A further indication of the valence of Eu and Yb lies in the close chemical and alloying similarity between these rare earths and the alkaline earths calcium, strontium and barium. Actually, when Eu and Yb are

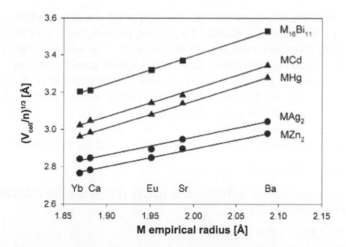

Fig. 6: Cubic root of the atomic volume $^3\sqrt{(V/n)}$ as a function of the M empirical radii for the MZn$_2$, MAg$_2$, MHg, MCd and M$_{16}$Bi$_{11}$ phases, where M = Eu, Yb, Ca, Sr and Ba.

normally divalent, very often they form phases with the same composition and structure as Ca, Sr and Ba. Moreover, for each group of five isotypic compounds a regular dependence of the cell volumes as a function of the size of the divalent metals is usually observed. The following series of empirical radii was obtained by a complete analysis of the available crystallographic data of binary intermetallic compounds: 1.881 Å for Ca, 1.987 Å for Sr, 2.087 Å for Ba, 1.951 Å for Eu, and 1.869 Å for Yb (Fornasini and Merlo 2006). Figure 6 reports the $^3\sqrt{(V/n)}$ values vs. the cited radii for five series of isotypic compounds: MCd and MHg (CsCl type), MZn$_2$ and MAg$_2$ (CeCu$_2$ type), M$_{16}$Bi$_{11}$ (Ca$_{16}$Sb$_{11}$ type); the regular trends shown by the experimental points agree with the occurrence of divalent Eu and Yb, while in some other cases the large negative deviation of the ytterbium point can indicate the tendency of the rare earth to the trivalent state.

On the basis of the foregoing considerations, the following systems were excluded from the volume analysis: Yb-Mn, Yb-Fe, Yb-Ru, Yb-Co, Yb-Rh, Yb-Ir, Yb-Ni, Eu-Pd, Yb-Pd, Yb-Pt, Yb-Au, Yb-Al, Yb-Si, Yb-Sb. Therefore, the remaining 32 systems formed by Eu and Yb with 20 X elements and containing 168 intermediate phases were considered. The calculations of the integral volume effects were initially carried out by means of Eq. 3, but the results were not as good as for the systems of the trivalent rare earths. A much better representation of the observed volumes trend within the Eu and Yb systems was obtained by the following equation, differing from Eq. 3 only in the power exponent of the term $(1 - x)$:

$$V_0 - V = Kx^n(1 - x) \tag{6}$$

This mathematical effect can be imputed to the higher volume contractions shown by the Eu and Yb phases, producing large values of the area comprised between the Vegard-like straight line and the experimental trend. The negative deviations from the Vegard trend are large also at the two ends of the graphs, namely for low concentration of R or X, and Eq. 6 works fairly well in these regions. Analogously to Eq. 3, K and n are adjustable parameters, and again the value of n for a given system indicates the composition showing the maximum volume contraction, namely in this case RX_n. By combining Eqs. (2) and (6) one obtains:

$$\Delta V_{int} = K/[(n + 1)(n + 2)] \tag{7}$$

and the average atomic volume of a given phase can be calculated by coupling Eqs. (6) and (7):

$$V = V_0 - (n + 1)(n + 2)x^n (1 - x)\Delta V_{int} \tag{8}$$

Table 2 reports the values of ΔV_{int} and of n, calculated by means of Eqs. (6) and (7), for the 32 systems containing europium and ytterbium. As for the systems of the trivalent rare earths, in order to maintain the physical meaning of the calculated $V(x)$ curve, the composition of the phase with the maximum volume contraction must lie within the range of the experimentally observed compounds. Therefore, in 10 systems the n value was constrained to the rare earth richest composition ($EuNi_2$, $EuMg$, $YbZn$, $EuCd$, $YbCd$, $EuHg$, $EuAl$, Yb_2Ga, Eu_2Ge, Yb_2Ge).

Also for the systems formed by Eu and Yb, Eq. 8 can be used to obtain the atomic volumes of the 168 binary intermediate phases, and the so calculated values are plotted against the experimental ones in Fig. 7. The average deviation of the calculated values with respect to the observed ones amounts to 1.33%, and this agreement confirms, once again, both the regularity of the distribution of the experimental points within each system and the ability of the Eq. 6 in representing the volume data.

General Considerations on the Observed Volume Effects

The ΔV_{int} values obtained for the systems containing trivalent rare earths are plotted in Fig. 8, as a function of the arbitrary quasi periodic order number N used in Table 1. For each X element, five points are reported, corresponding to the La, Nd, Gd, Ho and Lu systems, except the before

Table 2: Values of the Integral Volume Effect (ΔV_{int}) and of the n Parameter in Eqs. (6), (7) and (8), for 32 Binary Systems Containing the Divalent Rare Earths Europium and Ytterbium, Ordered Following a Quasi Periodic Number M.

M	System	n	ΔV_{int} [Å³/atom]
1	Eu-Rh	0.93	5.80
2	Eu-Ni	2	3.07
3	Eu-Pt	0.84	5.18
4	Eu-Cu	1.62	2.11
5	Yb-Cu	1.22	2.20
6	Eu-Ag	0.77	2.18
7	Yb-Ag	0.70	1.99
8	Eu-Au	0.78	3.13
9	Eu-Mg	1	0.73
10	Eu-Zn	1.14	2.49
11	Yb-Zn	1	2.50
12	Eu-Cd	1	2.21
13	Yb-Cd	1	2.48
14	Eu-Hg	1	3.70
15	Yb-Hg	1.09	3.76
16	Eu-Al	1	2.77
17	Eu-Ga	0.71	3.94
18	Yb-Ga	0.5	3.98
19	Eu-In	0.94	4.03
20	Yb-In	0.91	3.79
21	Eu-Tl	1.12	4.28
22	Yb-Tl	1.04	4.57
23	Eu-Si	0.60	3.88
24	Eu-Ge	0.5	3.98
25	Yb-Ge	0.5	4.36
26	Eu-Sn	0.74	4.48
27	Yb-Sn	0.73	4.66
28	Eu-Pb	0.90	5.09
29	Yb-Pb	0.74	4.73
30	Eu-Sb	0.81	4.68
31	Eu-Bi	1.33	5.02
32	Yb-Bi	1.29	4.83

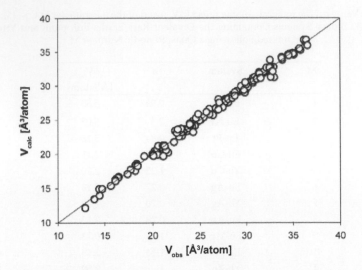

Fig. 7: Average atomic volumes calculated by Eq. 8 as a function of the observed ones for the intermediate phases of europium and ytterbium.

cited cases. Analogously, the ΔV_{int} values obtained for the systems containing europium and ytterbium are plotted in Fig. 9, as a function of the arbitrary quasi periodic order number M used in Table 2.

Some comments can be made:

a) As can be seen in Fig. 8, the ΔV_{int} of the systems formed by the trivalent rare earths shows a roughly regular trend, going from a minimum in the region of the Cu, Ag, Mg and Zn systems, to increasing values towards the group 15 systems on one side, and the transition metals on the other side. The Au, Hg and Tl systems show higher values, while the Fe systems show lower values.

b) For a given X element, the ΔV_{int} values usually increase going from lanthanum to lutetium systems.

c) By comparing Table 1 and Table 2, or Fig. 8 and Fig. 9, it can be noted that the volume contractions shown by the phases of the divalent rare earths are about twice those of the trivalent rare earths. Eu and Yb present similar ΔV_{int} values.

d) Also for the systems containing europium and ytterbium, a periodic trend is visible: a roughly increasing trend is observed going from the Cu to the Bi groups, while the transition metals systems have higher values.

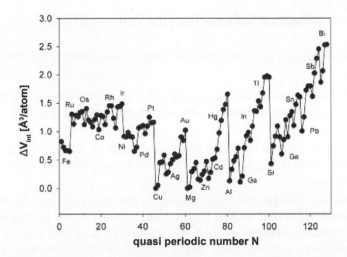

Fig. 8: Integral volume effect (ΔV_{int}) of the 128 R-X studied systems containing trivalent rare earths as a function of the quasi periodic number N. For each X element, five points are reported, corresponding to the La, Nd, Gd, Ho and Lu systems (some exceptions are detailed along the text).

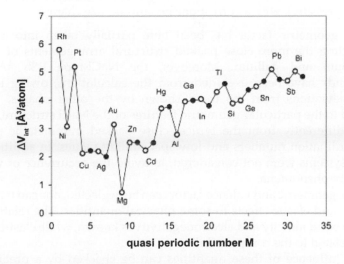

Fig. 9: Integral volume effect (ΔV_{int}) of the 32 R-X studied systems containing europium and ytterbium as a function of the quasi periodic number M. Eu and Yb points are represented by open and full symbols, respectively.

Factors Influencing the Volume Effects

As it is known from the literature, the main factors influencing the volume effects are:

a) ionic effects: the electronegativity difference between different atoms produces a charge transfer with a strengthening and shortening of the chemical bond;

b) elastic effects: the different electron density in the elements gives a change of their Wigner-Seitz cells up to the equilibrium density in the compound;

c) geometric effects: the space filling in the elements can be strongly different from that shown by the phases (silicon and germanium are the classical example, as their diamond-type structure is a very open atomic arrangement, and high volume contractions are expected during the phase formation);

d) valence effects: some rare earths (cerium, samarium, europium, ytterbium) can display valence instabilities, due to chemical (the partner element) and/or thermodynamic conditions (temperature and pressure), with a corresponding atomic size change.

The geometric factor has been here partially taken into account, by selecting the more close packed structural arrangements of silicon, germanium and gallium. Moreover, the NaCl-type RSb and RBi compounds have been excluded from the calculations, owing to their strong deviations from the trend shown by the other phases. This is imputed to the particular geometric features of the NaCl structure, which shows, differently from the typical close packed intermetallic phases, low coordination numbers and low space filling values. In addition, the cerium systems were not considered, to avoid the occurrence of valence instability phenomena.

If the geometric and valence factors can be neglected, one can try to link the ionic and elastic effects to some physical quantities. The relationship between bond ionicity and electronegativity is known, while elastic effects can be related to the compressibility.

The influence of these quantities can be checked by a preliminary comparison of their periodic trends with that of the volume effects.

The factors influencing the elastic effects are initially considered. Figure 10 reports the compressibility (μ_X) of the X elements as a function of the quasi periodic number N. For each X element, five identical values are plotted; this arbitrary choice is adopted for Figs. 10 to 12, in order to allow a better comparison with the experimental ΔV values. The μ_X values ($m^2\,Kg^{-1}\,10^{11}$) are taken from (Gschneidner 1964).

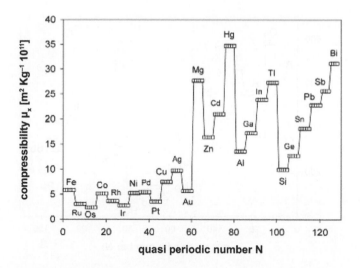

Fig. 10: Compressibility values of the X elements vs. the quasi periodic number N.

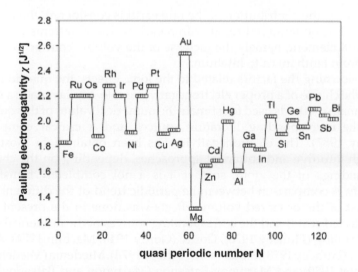

Fig. 11: Pauling electronegativity of the X elements vs. the quasi periodic number N.

Actually, the comparison of the ΔV_{int} and μ_x trends (Fig. 8 and Fig. 10, respectively) provides two qualitative remarks: (a) a roughly similar trend is recognizable between the ΔV_{int} and μ_x values going from zinc to bismuth; (b) the μ_x values for the X metals from iron to gold are nearly constant and sharply smaller than those of the post-transition elements. Therefore, the elastic effects should be important mainly for the post-transition elements.

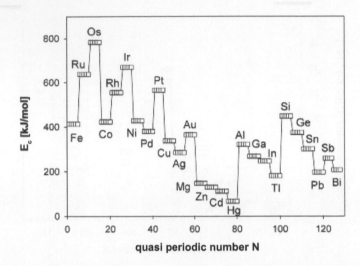

Fig. 12: Cohesive energy E_c of the X elements vs. the quasi periodic number N.

Moreover, if the contribution of the rare earth is considered, the ratio μ_X/μ_R provides the observed trend within each series of systems containing a given X element, namely the increase of the volume contraction values going from lanthanum to lutetium.

Concerning the factors related to the ionic effects, the question arises about the choice of a proper electronegativity scale. Mullay made a review on the main methods used to estimate the numerical values of this quantity, with the corresponding literature sources and a critical comparison (Mullay 1987). The electronegativity has been obtained in most cases through intuitive and empirical approaches, depending on the chemical surroundings of the atom, so that it is a not completely transferable quantity. A comparison between the periodic trend of the different scales and that of the observed volume effects was done in the present work. The selected scales and the corresponding sources of the numerical values were: Pauling (Huheey 1978), Gordy (Gordy 1946; Machlin 1974), Allred-Rochow (Huheey 1978), Sanderson (Huheey 1978), Miedema (Miedema and de Chatel 1980) and Martynov-Batsanov (Martynov and Batsanov 1980). The best result was obtained by Pauling's scale: Fig. 11 reports the corresponding electronegativity as a function of the quasi periodic number N, and a comparative glance to Fig. 8 (ΔV_{int} vs. N) indicates a similar contribution of the ionic effects in all systems. On the other hand, for the transition X metals going from gold group to osmium group the electronegativity values decrease, differently from the corresponding ΔV_{int} trend.

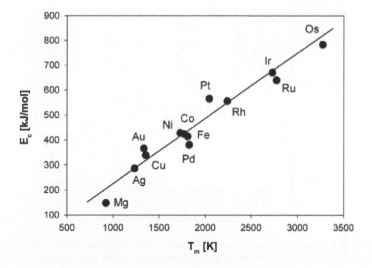

Fig. 13: Cohesive energy E_c of the X elements of the 8 to 11 groups and Mg, vs. the corresponding melting temperatures $T_{m,X}$.

This last observation gives the opportunity to introduce a further factor. In a more general way, if the volume effects are related to the bond strength within the compounds, an atomic index of the bond energy of the elements can be taken into account. The best quantity is the cohesive energy E_c (experimentally measured by the sublimation enthalpy), whose values are taken from Kittel (Kittel 2005) and plotted in Fig. 12 vs. the quasi periodic number N.

It can be seen that transition elements show the known increase going from the gold group to the osmium group. In considering the rare earths a problem arises, because the E_c values of the lanthanides show a not regular, "saw-tooth" trend. This behaviour has been ascribed to an additional energy term due to the different electronic configurations in the gas and solid phases (Beaudry and Gschneidner 1978). The same anomalous trend, according to the Trouton rule, is found also for the boiling point T_b, and again this last quantity cannot be used in the present treatment. On the other hand, it is known that the melting temperature T_m is an approximate measure of the cohesive energy, as shown by Fig. 13, where the T_m values are plotted versus E_c for the transition X elements. If the sum of the melting temperature of the X and R elements ($T_{m,X} + T_{m,R}$) is plotted vs. N for the systems containing Fe to Au (Fig. 14), a decreasing trend is observed from the left to the right side, and an increasing trend is obtained from light to heavy rare earths, in agreement with the ΔV_{int} trend.

From a chemical point of view, the most important factor influencing the bond energy both in the elements and in the compounds is the

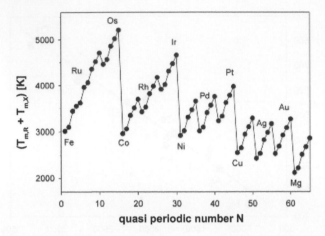

Fig. 14: Sum of the melting temperatures of the rare earth and partner element X ($T_{m,R} + T_{m,X}$) for the systems containing from 8 to 11 group X elements, vs. the quasi periodic number N.

group number, which is related to the number of bond electrons (Brewer 1981). That is usually true for non-transition elements, but the situation becomes complex for transition metals, as shown by the periodic trends of several physical and chemical quantities. Band structure calculations and experimental spectroscopic data can give the basis for evaluating the effective number and type of bond electrons. On the other hand, some physical quantities of the elements related to the bond strength (cohesive energy E_c, boiling point T_b, melting point T_m) show similar trends containing a maximum within the transition series 3d, 4d and 5d. An analogous trend can be hypothized for the effective number of bond electrons, going from the alkaline metals (with indisputable valence one) to the zinc group elements (with indisputable valence two). In the present analysis a quantity G has been used to represent the valence of the X elements: the selected G values come out from the comparison of the periodic behavior of the aforementioned physical quantities with the observed trend of the volume effects.

   In conclusion, the physical quantities used for the theoretical representation of ΔV_{int} are the Pauling electronegativities (χ_X, χ_R), the melting temperatures ($T_{m,X}$, $T_{m,R}$) and the X group number (G) for the systems formed by the Fe group to Cu group elements, while for the Zn group to Sb group elements the factors are again the electronegativities, the X group number and the compressibilities (μ_X, μ_R).

Calculation of the Volume Effects in Systems Containing the Trivalent Rare Earths

In a previous work (Fornasini and Merlo 2006) on the volume effects shown by the binary intermetallics in M-X systems formed by the divalent elements (M = Ca, Sr, Ba, Eu, Yb), the following empirical equation was proposed to describe the integral volume effects:

$$\Delta V_{int} = a_0 + a_1 (\chi_X - \chi_M) + a_2 G(\mu_M + \mu_X) \tag{9}$$

where χ_X and χ_M are the electronegativities on the Pauling scale, μ_M and μ_X are the compressibilities, G is related to the group number of the X element, and a_0, a_1, a_2 are refinable parameters.

Going to the present systems containing the trivalent rare earths, several analytical representations were tried, employing the cited atomic quantities. For the R-X systems where R = trivalent rare earth and X = Fe, Ru, Os, Co, Rh, Ir, Ni, Pd, Pt, Cu, Ag, Au, Mg, the following equation gives acceptable results:

$$\Delta V_{int} = b_0 + b_1 G(\chi_X - \chi_R) + b_2(T_{m,X} + T_{m,R}) \tag{10}$$

where χ_X and χ_R are the electronegativities on the Pauling scale, $T_{m,X}$ and $T_{m,R}$ are the melting temperatures of the two elements, and G is related to the group number of the X element. The refinable parameters b_0, b_1 and b_2 were obtained by the least squares method, with a correlation coefficient $R^2 = 0.97$: $b_0 = -0.64(1)$ [Å^3 atom^{-1}]; $b_1 = 0.14(1)$ [Å^3 atom^{-1} J$^{-1/2}$]; $b_2 = 3.1(2)$ 10^{-4} [Å^3 atom^{-1} K^{-1}]. The best agreement between observed and calculated ΔV_{int} was obtained by assigning the following values to the G quantity: G = 4 for X = Fe, Ru, Os, G = 5 for X = Co, Rh, Ir, G = 4 for X = Ni, Pd, Pt, G = 3 for X = Cu, Ag, Au, G = 2 for X = Mg.

A second equation is proposed for the post-transition elements, with X = Zn, Cd, Hg, Al, Ga, In, Tl, Si, Ge, Sn, Pb, Sb, Bi:

$$\Delta V_{int} = c_0 + c_1 G(\chi_X - \chi_R)^{1/2} + c_2(\mu_X/\mu_R) \tag{11}$$

where χ_X and χ_R are again the Pauling electronegativities, while μ_X and μ_R are the atomic compressibilities (m^2 Kg^{-1} 10^{11}) (Gschneidner 1964). G assumes the usual values of the X group number: G = 2 for X = Zn, Cd, Hg, G = 3 for X = Al, Ga, In, Tl, G = 4 for X = Si, Ge, Sn, Pb, G = 5 for X = Sb, Bi. The fitting parameters were refined by least squares method, with a correlation coefficient $R^2 = 0.98$: $c_0 = -1.17(5)$ [Å^3 atom^{-1}]; $c_1 = 0.41(1)$ [Å^3 atom^{-1}J$^{-1/4}$]; $c_2 = 1.55(5)$ [Å^3 atom^{-1}]. The results obtained via Eqs. (10) and (11) are plotted in Fig. 15 vs. the quasi periodic order number N, together with the observed values reported both in Table 1 and Fig. 8, while Fig. 16 reports the so calculated ΔV_{int} values as a function of the observed ones.

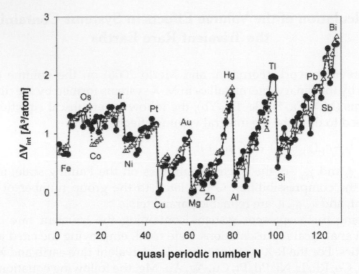

Fig. 15: Integral volume effects ΔV_{int} for the systems containing trivalent rare earths: values calculated by Eqs. (10) and (11) (open triangles and dotted lines) and values observed (full circles and solid lines), vs. the quasi periodic number N.

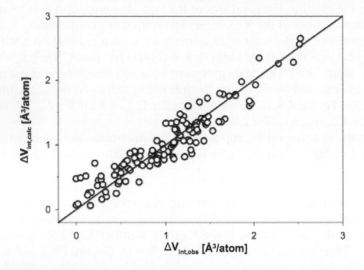

Fig. 16: Integral volume effects ΔV_{int} calculated by Eqs. (10) and (11) as a function of the observed values for the systems containing trivalent rare earths. The solid line represents the equality condition.

Calculation of the Volume Effects in Systems Containing Europium and Ytterbium

Also for the systems containing europium and ytterbium, some analytical representations of the integral volume effects were tried. For the R-X systems with X = Rh, Ni, Pt, Cu, Ag, Au and Mg acceptable results have been obtained by means of Eq. 10, already applied for the systems of the trivalent rare earths, maintaining the same meaning for χ_X and χ_R and for $T_{m,X}$ and $T_{m,R}$. The values of G, related to the group number of the X element, were again 5 for Rh, 4 for Ni, 3 for Cu, Ag and Au and 2 for Mg. The refinable parameters were $b_0 = -3.4(7)$ [$\mathring{A}^3$ atom^{-1}]; $b_1 = 0.53(8)$ [$\mathring{A}^3$ atom^{-1} J$^{-1/2}$]; $b_2 = 1.8(4)$ 10^{-3} [$\mathring{A}^3$ atom^{-1} K^{-1}], as obtained by the least squares method, with a correlation coefficient $R^2 = 0.96$.

For the systems containing the post-transition elements (X = Zn, Cd, Hg, Al, Ga, In, Tl, Si, Ge, Sn, Pb, Sb, Bi), a modified version of Eq. 9 was used:

$$\Delta V_{int} = d_0 + d_1 (\chi_x - \chi_R)^{1/2} + d_2 G(\mu_R + \mu_X) \tag{12}$$

Again, χ_X and χ_R are the electronegativities on the Pauling scale, μ_X and μ_R are the atomic compressibilities (m^2 Kg^{-1} 10^{11}) (Gschneidner 1964), and G assumes the usual values of the X group number: G = 2 for X = Zn, Cd, Hg, G = 3 for X = Al, Ga, In, Tl, G = 4 for X = Si, Ge, Sn, Pb, G = 5 for X = Sb, Bi. The refinement process produced a correlation coefficient $R^2 = 0.88$, and the following values of the fitting parameters: $d_0 = -1.9(4)$ [$\mathring{A}^3$ atom^{-1}]; $d_1 = 5.3(5)$ [$\mathring{A}^3$ atom^{-1}J$^{-1/4}$]; $d_2 = 4.1$ (5) 10^{-3} [$\mathring{A}^3$ atom^{-1} m^{-2} Kg 10^{-11}].

The ΔV_{int} of the systems containing europium and ytterbium, calculated by means of the Eqs. (10) and (12) are plotted in Fig. 17 as a function of the quasi periodic order number M, together with the observed values reported in Table 2. Figure 18 shows the so calculated ΔV_{int} values vs. the observed ones.

Volume Effects in Ternary Systems

The above treatment regards the binary compounds. In order to extend the analysis to polynary phases, the interactions among three or more types of atoms should be studied, and the problem can be complex, owing to the occurrence of numerous stoichiometries and crystal structures. An empirical approach was tested for a limited number of ternary equiatomic RMX compounds, where R = Ca, Sr, Ba, Eu, Yb; M = Cu, Ag, Zn, Cd;

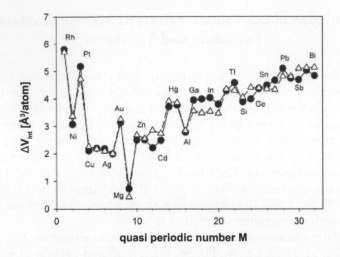

Fig. 17: Integral volume effects ΔV_{int} for the systems containing europium and ytterbium: values calculated by Eqs. (10) and (12) (open triangles and dotted lines) and values observed (full circles and solid lines) vs. the quasi periodic number M.

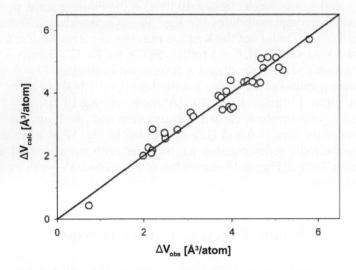

Fig. 18: Integral volume effects ΔV_{int} calculated by Eqs. (10) and (12) as a function of the observed values for the systems containing europium and ytterbium. The solid line represents the equality condition.

X = Si, Ge, Sn, Pb, Sb, Bi (Merlo et al. 1990, 1991). In such an approach, the RMX phases were considered as pseudo binary phases $R(M_{0.5}X_{0.5})_2$, and their volume was calculated following the procedure described for binary compounds (Merlo 1988). The values of the charge transfer atomic parameter were those proposed for the alkaline and rare earth R elements, while the average value of the M and X partner elements was selected. For the 74 RMX phases the mean difference between experimental and the so calculated volumes amounts to 2.3%. Following these positive results, an analogous calculation of the volume effects was done for a wider sample of equiatomic compounds (Fornasini and Merlo 1995): 210 RMX phases where R is either an alkaline earth or the divalent rare earths Eu and Yb, and 330 phases where R is a trivalent rare earth. The simple hypothesis was applied, that the volume of each RMX phase is given by the average volume of the two binary phases RM_2 and RX_2, as shown below:

$$V_{calc}(RMX) = \frac{1}{2}\,[V_{calc}(RM_2) + V_{calc}(RX_2)]$$

The volumes of the binary compounds were obtained by using the equations and the atomic parameter values already proposed (Merlo 1988) and (Merlo and Fornasini 1993), also for not existing or not structurally known phases. The mean deviations from the experimental data were 3.4% or 2.2%, when R is a divalent or trivalent metal, respectively. According to these results, in the possible two-step synthesis

$$R + M + X \rightarrow \frac{1}{2}\,RM_2 + \frac{1}{2}\,RX_2 \rightarrow RMX$$

most of the volume formation effects occur in the first step. In other words, the binary R-M and R-X interactions are mostly responsible for the total volume effects, while the following ternary interactions seem to play a secondary role. Therefore, the representation of the volume effects for the ternary compounds, based on those of the binary compounds, seems a satisfactory procedure. However, the cited works (Merlo et al. 1990, 1991; Fornasini and Merlo 1995) regard only a particular composition, the equiatomic stoichiometry RMX, which is easily connected to the 1:2 formula of the corresponding binary phases.

If an analogous calculation is attempted for a ternary compound with general formula $A_xB_yC_z$, two problems arise: the choice of the binary phases, and the selection of a proper combination of their volumes. It is known that some thermodynamic quantities of ternary phases can be derived from those of binary phases, following four geometrical methods suggested by Colinet, Muggianu, Toop and Kohler, respectively (Lukas et al. 2007). The extension of these methods to the evaluation of the volume effects in ternary intermetallic compounds was proposed (Pani and Merlo 2011), together with a slightly different choice of the binary phases. The graphical representation of four selecting modes is shown in Fig. 19, where the full circles indicate a generic ternary compound and the six (Colinet method), three (Muggianu and Kohler methods) and nine binary phases (Pani and Merlo method) used in a weighted combination to estimate the ternary volume. These methods were tested for eight systems containing 91 ternary phases with known crystal structure. A reasonable agreement between observed and calculated volume values was shown both by the Kohler and by the Pani and Merlo method, the mean deviation approaching 2.5%. This result confirms, as already found (Merlo et al. 1990, 1991; Fornasini and Merlo 1995), that passing from binary to ternary systems the additional ternary interactions may be disregarded. All the 91 examined systems (Pani and Merlo 2011) contained a rare earth or an alkaline earth metal, and the volumes of the selected binary phases were

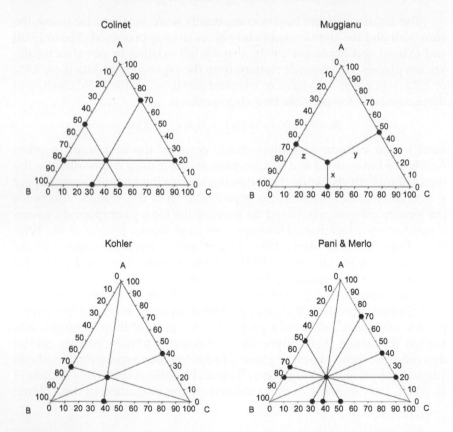

Fig. 19: Methods of selecting compositions of the binary phases to estimate a physical property of a ternary phase.

obtained by fitting to the experimental volume data of the binary systems the same Eqs. (3) and (6) used here.

On the basis of the results of the present work, the estimation of the volume of a rare earth ternary compound can be carried out without considering the observed binary volume values. Actually, the volumes of the binary phases can be obtained by means of the here proposed Eqs. (10), (11) and (12), and either the Kohler or Pani and Merlo procedures can be used.

Conclusions

The overall comparison between observed and calculated volume effects shows a reasonable agreement, but the proposed equations cannot be used for accurate quantitative predictions. It is probable that the equations can be implemented by inserting other physical properties and using more complex analytical expressions, so obtaining a better final result. However, as often happens for any empirical approach, the improvement of the analytical treatment may occur at the expense of the physical meaning of the whole representation. Anyway, some useful information can be extracted also from the present simple approach.

The elastic effects, measured by the compressibility, appear negligible for the systems containing transition metals, while their contribution is high in the post-transition region, mainly in the systems containing X elements of the 12 and 13 groups. The bond energy of the elements, measured by the melting temperature, represents the major contribution (from 60 to 80%) in the systems containing transition metals. The ionic effects, imputed to charge transfer phenomena induced by the usually high electronegativity differences, occur in all systems, particularly for 14 and 15 group X elements.

The periodic character of the volume effects in the phase formation is shown by the fundamental role played by the G parameter, which is related to the effective number of bond electrons.

The relationship between volume effects and energy of formation, proposed for many years and extended to the intermetallic systems by Kubaschewski (Kubaschewski et al. 1967), was tested for the binary phases formed by the divalent rare earth and alkaline earth metals (Fornasini and Merlo 2006), obtaining a rough confirmation of such dependence. For the here studied systems an analogous analysis was not made, but probably this relation is weaker, because of the smaller volume effects in the systems containing the trivalent rare earths.

As last notation, it can be pointed out that general correlations have been proposed between the formula-unit volume of solids, mainly in ionic compounds and minerals, and thermodynamic quantities, like entropy, lattice potential energy, enthalpy and Gibbs energy of formation, isothermal compressibility (Mallouk et al. 1984; Jenkins et al. 2010). Some simple equations allow scientists to estimate with good accuracy physical quantities not available in the literature, so confirming the important role of the volume in the solid state thermodynamics.

Acknowledgement

The Authors are very grateful to Prof. Maria L. Fornasini for her continuous assistance during the preparation of this work and for the accurate reading of the text.

Keywords: Volume effects, Rare earth elements, Intermetallic compounds, Binary systems, Ternary systems

References

Alonso, J.A., D.J. Gonzàles and M.P. Iñiguez. 1984. Role of the excess volume of formation on alloying. Phys. Stat. Sol. 123: 485–489.

Baranov, A., M. Kohout, F.R. Wagner, Y. Grin and W. Bronger. 2007. Spatial chemistry of the aluminium-platinum compounds: a quantum chemical approach. Z. Kristallogr. 222: 527–531.

Beaudry, B.J. and K.A. Gschneidner, Jr. 1978. Preparation and basic properties of the rare earth metals. pp. 173–232. *In*: K.A. Gschneidner, Jr. and L. Eyring [eds.]. Handbook on the Physics and Chemistry of Rare Earths, Vol. 1. North-Holland, New York, NY, USA.

Biltz, W. 1934. Raumchemie der festen Stoffe. Verlag Leopold Voss, Leipzig, Germany.

Brewer, L. 1981. The role and significance of empirical and semiempirical correlations. pp. 155–174. *In*: M. O'Keeffe and A. Navrotsky [eds.]. Structure and Bonding in Crystals, Vol I. Academic Press, New York, NY, USA.

CRYSTMET. 2010. Structure and powder database for metals. Materials Toolkit.

Fornasini, M.L. and F. Merlo. 1995. Equiatomic ternary phases formed by alkaline earths and rare earths. J. Alloys Compd. 219: 63–68.

Fornasini, M.L. and F. Merlo. 2006. Binary intermetallic phases formed by Ca, Sr, Ba, Eu and Yb: similarities and differences. Z. Kristallogr. 221: 382–390.

Gordy, W. 1946. A new method of determining electronegativity from other atomic properties. Phys. Rev. 69: 604–607.

Gschneidner, Jr., K.A. 1964. Physical properties and interrelationships of metallic and semimetallic elements. pp. 275–426. *In*: F. Seitz and D. Turnbull [eds.]. Solid State Physics, Vol. 16. Academic Press, New York, NY, USA.

Hafner, J. 1985. A note on Vegard's and Zen's laws. J. Phys. F: Met. Phys. 15: L43–L48.

Huheey, J.E. 1978. Inorganic Chemistry: Principles of Structure and Reactivity, 2nd Ed. Harper & Row, New York, NY, USA.

Iandelli, A. and A. Palenzona. 1979. Crystal chemistry of intermetallic compounds. pp. 1–54. *In*: K.A. Gschneidner, Jr. and L. Eyring [eds.]. Handbook on the Physics and Chemistry of Rare Earths, Vol. 2. North-Holland, New York, NY, USA.

Jenkins, H.D.B., L. Glasser and J. Lee. 2010. Volume-based thermoelasticity: consequences of the (near) proportionality of isothermal compressibility to formula-unit volume. Inorg. Chem. 49: 9978–9984 and references therein.

King, H.W. 1982. Temperature-dependent allotropic structures of the elements. Bull. Alloy Phase Diag. 2: 577.

Kittel, C. 2005. Introduction to Solid State Physics, 8th Edn. Wiley, New York, NY, USA.

Kubaschewski, O., E.L.L. Evans and C.B. Alcock. 1967. Metallurgical Thermochemistry, 4th Edn. Pergamon Press, London. U.K.

Lukas, H., S.G. Fries and B. Sundman. 2007. Computational Thermodynamics. The Calphad Method. Cambridge University Press, Cambridge, U.K., and references therein.

Machlin, E.S. 1974. Pair potential model of intermetallic phases-I. Acta Metall. 22: 95–108.

Machlin, E.S. 1980. A pair potential analysis of bonding in alloy phases. pp. 127–193. *In*: L.H. Bennett [ed.]. Theory of Alloy Phase Formation. Met. Soc. AIME, Warrendale, Pa., USA, and references therein.

Mallouk, T.E., G.L. Rosenthal, G. Muller, R. Brusasco and N. Bartlett. 1984. Fluoride ion affinities of GeF_4 and BF_3 from thermodynamic and structural data for $(SF_3)_2GeF_6$, ClO_2GeF_5 and ClO_2BF_4. Inorg. Chem. 23: 3167–3173.

Martynov, A.I. and S.S. Batsanov. 1980. A new approach to the determination of the electronegativity of atoms. Russ. J. Inorg. Chem. 25: 1737–1739.

Merlo, F. 1988. Volume effects in the intermetallic compounds formed by Ca, Sr, Ba, Eu and Yb with other elements. J. Phys. F: Met. Phys. 18: 1905–1911.

Merlo, F. and M.L. Fornasini. 1993. Volume effects in rare earth intermetallic compounds. J. Alloys Compd. 197: 213–216.

Merlo, F., M. Pani and M.L. Fornasini. 1990. RMX compounds formed by alkaline earths, europium and ytterbium. I. Ternary phases with M = Cu, Ag, Au; X = Sb, Bi. J. Less-Common Met. 166: 319–327.

Merlo, F., M. Pani and M.L. Fornasini. 1991. RMX compounds formed by alkaline earths, europium and ytterbium. II. Ternary phases with M = Zn, Cd; X = Si, Ge, Sn, Pb. J. Less-Common Met. 171: 329–336.

Miedema, A.R. and P.F. de Châtel. 1980. A semi-empirical approach to the heat of formation problem. pp. 344–389. *In*: L.H. Bennett [ed.]. Theory of Alloy Phase Formation. Met. Soc. AIME, Warrendale, PA, USA.

Miedema, A.R. and A.K. Niessen. 1982. Volume effects upon alloying of two transition metals. Physica B 114: 367–374.

Mullay, J. 1987. Estimation of atomic and group electronegativities. pp. 1–25. Structure and Bonding, Vol. 66. Springer, Berlin, Germany.

Pani, M. and F. Merlo. 2011. A new method to estimate the atomic volume of ternary intermetallic compounds. J. Solid State Chem. 184: 959–964 and references therein.

Pearson, W.B. 1972. The Crystal Chemistry and Physics of Metals and Alloys. Wiley, New York, NY, USA, and references therein.

Pearson's Crystal Data. 2014/2015. Crystal Structure Database for Inorganic Compounds. ASM International, Materials Park, Ohio, USA.

Villars, P. and L.D. Calvert. 1991. Pearson's Handbook of Crystallographic Data for Intermetallic Phases, 2nd Edn. ASM International, Materials Park, Ohio, USA.

Villars, P. and L.D. Calvert. 1997. Pearson's Handbook of Crystallographic Data for Intermetallic Phases, Desk Edition. ASM International, Materials Park, Ohio, USA.

Watson, R.E. and L.H. Bennett. 1982. Volume effects in transition metal alloying. Acta Metall. 30: 1941–1955.

Watson, R.E. and L.H. Bennett. 1984. Model predictions of volume contractions in transition-metal alloys and implications for Laves phase formation-II. Acta Metall. 32: 491–502.

Zen, E-an. 1956. Validity of "Vegard's law". Am. Mineral 41: 523.

PART 2
ROLE OF MODELING ON THE DESIGN OF ALLOYS AND INTERMETALLIC COMPOUNDS

2.1

Metal-Ceramic Interactions in Brazing Ultra High Temperature Diboride Ceramics

Fabrizio Valenza,[1,*] *Cristina Artini,*[1,2] *Sofia Gambaro,*[1]
Maria Luigia Muolo[1] *and Alberto Passerone*[1]

INTRODUCTION

Ultra high temperature ceramics (UHTCs) constitute a class of materials of primary importance for the industry (Fahrenholtz 2014). Typically, UHTCs are borides (TiB_2, HfB_2, TaB_2, ZrB_2), carbides (HfC, ZrC, TiC, NbC) and nitrides (HfN, TaN) of the transition metals in Groups IV–VI; the common feature of these compounds is to have a melting temperature above 3000°C (Wuchina et al. 2007). Although these materials have been studied for a long time especially in 60's and 70's in relation to the space race, the habit to refer to them as UHTCs is relatively recent. In the last 20 years, the interest of the scientific and technological community towards these materials arose peremptorily fuelled by their attractive combinations of properties such as extremely high melting temperature, high hardness and chemical inertness, thermal shock resistance and, for transition-metal

---

[1] National Research Council - Institute of Condensed Matter Chemistry and Technologies for Energy (CNR-ICMATE), Via de Marini 6, 16149, Genova, Italy.
[2] Department of Chemistry and Industrial Chemistry, University of Genova, Via Dodecaneso 31, 16146, Genova, Italy.
* Corresponding author: fabrizio.valenza@ge.icmate.cnr.it

ceramic diborides, also high thermal and electrical conductivity. This work is focused on diborides of the 4th Group TiB_2, ZrB_2 and HfB_2 with reference to the studies on their interactions with liquid metals with the goal of designing brazing procedures.

Relevant examples of applications for diborides not strictly related to their use as thermal shields in aerospace (Fahrenholtz et al. 2007; Levine et al. 2002) are in energy production (combustion chambers of gas turbines, nuclear plants, absorbers in solar plant), in handling of molten metals, in heating elements, in plasma arc electrodes, in cutting tools, etc. (Alfano et al. 2014; Sciti et al. 2013; Levine et al. 2009; Telle et al. 2000; Norasetthekul et al. 1999; Murata et al. 1970).

These materials could be used for either complete or selective replacement of existing components, but in both cases, the difficulties that arise in producing large-scale ceramic-based structures of complex shape or the need to reliably integrate ceramic components into a structure require an effective joining technology. Ceramic/ceramic and ceramic/metal joints must be reliable, robust, and the methods of joining must be able to successfully accommodate different materials combinations in order to create structures of increasing geometric complexity and size.

The strength, reliability, and other performance characteristics of joints are strongly affected by the chemical and physical properties of the interface generated during the bonding/joining process.

In this work, a general background is presented related to the properties of diborides, wetting at high temperature and joining by brazing techniques. Then, a detailed survey of the wetting data of liquid metals on Ti, Zr and Hf diborides is provided, with special reference to the work undertaken by this research group. Moreover, we present a review of the recent findings related to liquid assisted joining of diborides with a discussion about the role of processes and microstructures on the joint final mechanical response.

Background

Transition Metals Diborides Properties

The properties, characteristics and behaviour of transition metal diborides were extensively reviewed in (Justin and Jankowiak 2011; Fahrenholtz et al. 2007; Munro 2000; Opeka et al. 1999; Post et al. 1954); here, only the most peculiar characteristics of these materials are recalled. The following table summarizes the material properties of TiB_2, ZrB_2 and HfB_2 at room temperature.

All these diborides crystallize in the hexagonal structure and besides very high melting points and hardnesses, transition metal diborides exhibit characteristics which are usually pertaining to metals such as good thermal and electrical conductivities because of the partly covalent and partly metallic bonding of the transition element ceramics (Dempsey 1963). As explained in the next sections, this metallic character has consequences for the wetting and surface properties.

Diborides are subjected to oxidation when exposed to air and elevated temperatures ($\approx > 700°C$) and the resulting oxides are very stable ($\Delta G°_f = -626, -766$ and -846 kJ/mol at 1500°C for TiO_2, ZrO_2 and HfO_2 respectively):[1]

$$TiB_2 + 5/2 \, O_2 \rightarrow TiO_2 + B_2O_3, \, \Delta G°(T) = -1950 + 0.9747T$$

$$ZrB_2 + 5/2 \, O_2 \rightarrow ZrO_2 + B_2O_3, \, \Delta G°(T) = -2075 + 1.0056T$$

$$HfB_2 + 5/2 \, O_2 \rightarrow HfO_2 + B_2O_3, \, \Delta G°(T) = -2050 + 0.9757T$$

XO_2 (X = Ti, Zr and Hf) and B_2O_3 form a continuous protective layer until, at high temperatures, B_2O_3 evaporates and the oxide layers remain in porous form. However, the protection against oxidation of this layer starts to fail at temperatures above 1400°C. In order to increase the oxidation resistance in air of these diborides, the addition of secondary phases such as SiC or silicides has been extensively used (Carney et al. 2011; Silvestroni and Sciti 2011; Fahrenholtz et al. 2007; Parthasarathy et al. 2007). A protective glassy layer forms at the boride surface when Si-containing phases are added, providing a protection barrier up to about 1600°C through passive oxidation. These processes lead to the possibility of application of diborides in very aggressive environments at high temperature. Moreover, the addition of secondary phases to borides has been extensively adopted in order to improve the relatively low resistance to thermal shock of single phase materials (Chamberlain et al. 2004). The choice of a secondary phase to enhance the mechanical properties should take into account the matching of the thermal expansions and of the particle sizes (Watts et al. 2011): underestimating this issue could produce stress-induced microcracking on particulate reinforced composites. Silicon carbide is the compound which has been most extensively studied for the enhancement of mechanical and oxidation properties. In the case of ZrB_2-SiC composites additions of up to 30 vol% is reported to increase significantly the flexural strength up to ≈ 1100 MPa and the toughness up to ≈ 5.5 MPa·m$^{0.5}$ with a resulting increase of the thermal shock resistance (Guo 2009; Liu et al. 2009; Monteverde 2007; Chamberlain et al. 2004). The role of SiC, as any other additives, is to deflect cracking along the grain

---

[1] Thermodynamic data from: Chase M., NIST-JANAF Thermochemical Tables, 4th Ed, 1998.

boundaries while for pure diborides intragranular cracking is usually observed (Guo 2009).

Besides SiC, also other compounds, alone or in combination with SiC, are reported to increase the oxidation and mechanical properties of diborides; among them refractory silicides ($MoSi_2$, $HfSi_2$, $ZrSi_2$, Ta_5Si_3), graphite or other diborides are used (Hu et al. 2010; Guo 2009; Sciti et al. 2006). The introduction of small amounts of these compounds is also useful and very often adopted in order to enhance the sintering of diborides. In fact, the densification of XB_2 materials is very difficult because of their strong bonds and low bulk and grain boundary diffusion rates. Therefore, high temperatures and pressures as well as long processing times are required for an effective sintering (Guo et al. 2008; Monteverde et al. 2003) and even at these tough conditions the final microstructures are not satisfactory. To give an order of magnitude, isostatic pressing sintering of ZrB_2 requires 20–30 MPa at 2100°C or extremely high pressures (800 Mpa) at 1800°C (Guo 2009). High-densification of the pure powders could also be achieved by using very fine granulometries due to the higher surface/volume ratio. For example (Chamberlain et al. 2004), a relative density of 99.8% has been obtained after sintering of fine (d < 0.5 µm) ZrB_2 powders at 1900°C and 32 MPa; however, it should be mentioned that the production of fine powders is, in itself, a high energy consuming process due to the high hardness of diborides. Moreover, the presence of oxides in the starting powders inhibits densification and promotes grain growth. The use of sintering aids, may help to overcome these issues; besides silicides, also nitrides, carbides or pure metals were successfully used. These additives may control the microstructure and decrease the temperature required to achieve full density by reducing the oxygen content (e.g., B_4C reacts with XO_2 to give XB_2 and volatile CO and B_2O_3), by inhibiting grain growth or by forming intergranular liquid phases (e.g., when pure metals like Ni of $HfSi_2$ are used). However, it is worth reminding that the presence of grain boundary phases deriving from sintering aids could be detrimental for the high temperature properties especially when liquid phases can form.

In addition to the traditional and predominant hot-pressing technique, alternative processing routes have been recently investigated to improve the purity and densification behaviour of UHTCs or to reduce processing parameters such as temperature, time or pressure. A review on these techniques could be found in (Guo 2009); among them, spark plasma sintering, reactive hot pressing, self-propagating high-temperature synthesis, and polymer precursors pyrolysis routes (Xie et al. 2012) are reported to be feasible.

Basics of Wetting at High Temperature

Wetting could be defined as the tendency of a liquid to adhere to a solid surface. Wetting phenomena are omnipresent and have a fundamental role in many processes in nature, biology, industry, and, in general, whenever a liquid comes in contact with a solid (de Gennes et al. 2004).

Making reference to the sessile drop method (Fig. 1), which constitutes a common and relatively simple way to display and assess wettability, the equilibrium of interfacial forces at the triple line, that is the site where the three phases (solid, liquid, vapour) are in common contact, is described by the well known Young and Young–Dupré equations (Eqs. 1 and 2):

$$\frac{\sigma_{SV} - \sigma_{LS}}{\sigma_{LV}} = cos\theta \tag{1}$$

$$W_a = \sigma_{LV}\,(1 + cos\theta) \tag{2}$$

where σ_{SV}, σ_{SL}, σ_{LV} represent the solid/vapour (S/V), solid/liquid (S/L), and liquid/vapour (L/V) interfacial tensions respectively, and θ the contact angle at the triple line; Wa is the thermodynamic work of adhesion. It should be noted that Eq. 1 is strictly valid only if the solid surface can be considered rigid, or if the atomic mobility is not sufficient to rearrange the surface structure.

When speaking of wetting of liquid metals (or alloys, glasses) on solid substrates (refractory metals or ceramics), phenomena which are specific of the aggressive environment related to high temperatures come into play (Eustathopoulos et al. 1999; Passerone et al. 2012a). This topic is pertinent to many technological and scientific areas such as metallurgy (casting, coatings) (Valenza et al. 2009), glass industry, microelectronics (Matsumoto and Nogi 2008), joining of ceramics by brazing (Nicholas 2006), production of composites by pressureless infiltration (Aghajanian et al. 1991), stabilization of liquid metal foams by solid particles (Kaptay 2003), etc.

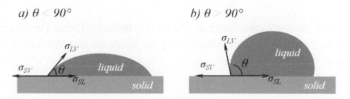

Fig. 1: Liquid drops in contact with solid surfaces in (a) wetting and (b) non wetting situations. The three interfacial forces on which contact angles depend are indicated.

At variance with processes involving water or organic liquids characterized by room temperature environments, processes involving liquid metals are characterized by high temperatures, which means that the atomic mobility is high, diffusion processes are very active and the reactivity between the different phases results in a large amount of chemical energy. Thus, even if capillary phenomena still play an important role, the final equilibrium configuration of the system cannot be foreseen only on the basis of the Young and Young–Dupré equations.

When a liquid phase is brought into contact with a solid, in a well defined atmosphere, the interfacial interactions that could happen at the interfaces between the three phases can be summarized as follow:

- Solid/Liquid: dissolution of the ceramic phase, infiltration of the liquid into the solid, formation of interfacial compounds and adsorption of active elements (e.g., Ti, Cr).
- Liquid/Vapor: evaporation, adsorption of active metals and gaseous compounds (e.g., oxygen) and oxidation.
- Solid/Vapor: oxidation-deoxidation of the surface, selective evaporation (e.g., SiO) and surface restructuring.

Depending on interfacial interactions, these wetting typologies can be recognized:

- Non reactive: in this ideal and rare situation; no atoms exchange occurs between the solid and liquid phases. In this case, in metal-ceramic systems the contact angles are usually very high.
- Adsorptive: adsorption of gaseous species at the L/V interface (e.g., oxygen) or active elements at the S/L interface (e.g., active metals like Ti, Cr, Zr, etc.) could decrease σ_{LV} and σ_{SL} modifying the wetting of liquid metals and, thus, the final contact angle.
- Dissolutive: the liquid dissolves part of the ceramic phase until the chemical potentials of the various components in the solid and in the liquid are the same. In this case, σ_{LV} can change dramatically and the S/L interface does not lie anymore on a plane invalidating Young's law.
- Reactive: new species form at the S/L interface. These products could form continuous layers which may be more or less wettable than the starting surface.

Basics of Ceramic Joining by Liquid State Processes

Metal/ceramic or ceramic/ceramic joining processes are necessary in order to integrate ceramic components into metallic structures, to obtain a final product with enhanced characteristics with respect to the starting materials or to assemble ceramics in complex shapes whenever, as in the case of diborides, sintering is difficult.

Within this field, when the target is to obtain a product for high temperature use, several processes are possible as, for example, mechanical joining (by rivets, bolts, fastening), diffusion bonding (solid state processes), friction welding, brazing, partial transient liquid phase bonding, etc. The last two methods involve a liquid phase acting as a medium for the contact between the surfaces to be joined. The use of a liquid for joining is often preferable to limit surface preparation, to avoid high pressures or when the adjoining surfaces are not flat or smooth.

Brazing relies on capillarity to distribute the liquid alloy along the adjoining interfaces. Making reference to the following picture (Fig. 2), it could be demonstrated that spontaneous liquid infiltration occurs when the sum of the two contact angles of the brazing alloy on the adjoining materials (for example one metal and one ceramic) is less than 180°.

Thus, the knowledge of wettability processes, of their kinetics (i.e., contact angles or contact area and dimensions as a function of time) and of how processing parameters (e.g., atmosphere, thermal cycling) and surface properties (e.g., roughness) affect them is fundamental. Moreover, the evolution of the metal/ceramic interfaces and the resulting chemical compositions and microstructures must be known and analyzed. This information, used to define and refine the choice of brazing alloys, can be also foreseen by effective modeling such as CALPHAD, Density Functional Theory (DFT), Molecular Dynamics (MD), etc. The initial characteristics of the materials to be joined combined to the ones of the resulting interfaces affect in a serious way the final mechanical response. In fact, during cooling, stresses arise at the interface between the two different neighbouring materials due to the thermal expansion mismatch when it

Fig. 2: Sketch of a metal-ceramic couple joined via liquid metal capillary infiltration. The process is spontaneous when $\alpha + \beta < 180°$.

is not accommodated by elastic or plastic deformation. This stress could be roughly estimated, taking into account only the elastic deformation, by the formula (Ning et al. 1989):

$$\sigma = \frac{E_1 E_2}{E_1 + E_2} (\alpha_1 - \alpha_2)(T_j - T_0),$$ (3)

where E indicates the elastic modulus, α the coefficients of thermal expansion (CTE) of the two materials, T_j the joining temperature and T_0 the room temperature; normally, in metal-ceramic joints, the metal surface at the interface with the ceramic is in tension, while, symmetrically, the ceramic surface is compressed.

Finally, mechanical tests at room and high temperature provide a means to assess whether the joints meet the strength requirements and have reproducible properties. Currently, the lack of a common and standardized testing method for the shear of joined samples makes the comparison of different joining solutions from different laboratories difficult (Ferraris and Casalegno 2015). Nevertheless, the study of the failure behavior, crack propagation and comparison of strength data obtained under the same testing conditions, may give useful hints for the assessment of the reliability of joined samples. Failures that occur in the ceramic rather than at the joint are an indication of a joint strength exceeding that of the ceramic, and thus the joint region is not the strength-limiting feature. Special attention paid to obtain good wetting and an acceptable CTE mismatch at the metal-ceramic interfaces should avoid poorly designed systems which encounter failures in the interfacial zone at low stresses.

Wetting of Transition Metal Diborides by Liquid Metals

In principle, due to their metallic character, TiB_2, ZrB_2 and HfB_2 should behave as metals when contacted by liquid metals. For liquid metals on inert solid metallic surfaces, adhesion is usually quite high as metallic bonds are established at the liquid-solid interface; consequently, contact angles are very low and the spreading, if the process is governed only by viscous forces, is finished in a time interval of less than a second (Dezellus and Eustathopoulos 2010; Saiz et al. 2010). In practice, as detailed later on, different phenomena were experimentally observed for transition metal diborides which lead to very complex phenomena and to different typologies of wetting. The pioneering and extensive work which has been made in the 70's about high temperature wettability and reactivity of transition metals diborides has been recently reviewed (Passerone et al. 2012b). In 80's and early 90's only a few papers appeared about this topic while in the last two decades, the renewed concern towards these materials arose an interest on their wetting properties too. In the Table 2, a review is presented of wetting results obtained for diborides of Ti, Zr and Hf in the last 15 years. Results obtained on composites are presented as well whenever the content in diboride is higher than 50 vol%. The nature of interfacial phenomena is also indicated as non reactive, dissolutive (i.e., the liquid dissolved part of the ceramic substrate) or forming new interfacial products. In principle, due to the high reactivity associated with high temperatures, for these systems a real *non reactive* system is hard to find: here, the term is used for systems which macroscopically showed neither dissolution of the ceramic phase nor formation of interfacial compounds.

As one can see from Table 2, most of the work was devoted to study the interactions of borides with Al, Ag, Cu, Au and Ni and their alloys.

The investigations on the system Al/TiB_2 demonstrated good wettability accompanied with low reactivity and solubility making this material combination especially suitable for casting Al alloys or for the use of TiB_2 as evaporation boats for Al. For these systems, the contact angles decrease when temperature increases due to the deoxidation of the droplet surface.

For the pure metals of the Ib Group (Cu, Ag, Au), which may constitute the basis for brazing alloys in the medium temperature range, a non-reactive behaviour was generally observed while a significant scatter of contact angles has been found in the literature. The interpretation of the wetting results for these metals should be made taking into account the chemistry of the ceramic surface which in turn depends on its composition. In fact, as mentioned before, diborides are often combined to SiC to form

Table 1: Material properties for TiB_2, ZrB_2 and HfB_2.

Material property	TiB_2	ZrB_2	HfB_2
Melting (°C)	3225	3245	3380
Density (g/cm³)	4.52	6.08	10.5
Young modulus (GPa)	530–565	420–489	480–530
Flexural strength (MPa)	400	568	440–480
Hardness (GPa)	25–35	21–23	28
Toughness (MPa·m⁰·⁵)	6.2	3.4–4.2	7.5
CTE (°C⁻¹)	$4.8 \cdot 10^{-6}$	$5.9–6.2 \cdot 10^{-6}$	$6.3–6.6 \cdot 10^{-6}$
Electrical conductivity (S/m)	$5 \cdot 10^{6}$	$1.0 \cdot 10^{7}$	$9.1 \cdot 10^{6}$
Thermal conductivity (W/(m·K))	25	60	104

Table 2: Contact angles of metals or alloys on XB_2 ($X = Ti$, Zr, Hf) based ceramics.

Ceramic	Metal/alloy	Temperature	Contact angle	Interfacial phenomena	Reference
TiB_2 – AlN 35 vol%	Ag	960–1000	158°	Non reactive	(Mattia et al. 2005)
TiB_2 – AlN 35 vol%	Ag-Cu eut.	?	140°	Non reactive	(Mattia et al. 2005)
TiB_2 – AlN 35 vol%	Ag-Cu eut. + 3 wt% Ti	770–850	60°	Interfacial product	(Mattia et al. 2005)
TiB_2	Al	800	77°	Non reactive	(Xi et al. 2015)
TiB_2	Al	1030–1200	12°	Non reactive	(Xi et al. 2015)
TiB_2	AlNi	1650–1670	11°	Non reactive	(Umanskyi et al. 2015)
TiB_2	Au	1150	15°	Non reactive	(Aizenshtein et al. 2012)
TiB_2	Au-2 at% B	1150	10°	Non reactive	(Aizenshtein et al. 2012)
TiB_2	Cr	1600–1900	10°–73°	Dissolutive	(Zhunkovskii et al. 2011)
TiB_2 – AlN 35 vol%	Cu	1100	131°	Non reactive	(Mattia et al. 2005)
TiB_2	Cu	1150	91°	Non reactive	(Passerone et al. 2006)
TiB_2	Cu	1150	50°	Non reactive	(Aizenshtein et al. 2012)
TiB_2	Cu-8 at% B	1150	10°	Non reactive	(Aizenshtein et al. 2012)
TiB_2 – AlN 35 vol%	Ni	1460	-	Dissolutive	(Mattia et al. 2005)
ZrB_2	Ag	1050	> 140°	Non reactive	(Muolo et al. 2005)
ZrB_2	Ag	1023	150°	Non reactive	(Passerone et al. 2007)
ZrB_2	Ag	1023	117	Non reactive	(Valenza et al. 2012a)
ZrB_2- SiC 20 vol%	Ag	1023	97	Non reactive	(Valenza et al. 2012a)
ZrB_2	Ag-Cu eut.	1050	141°	Non reactive	(Muolo et al. 2005)

Material	Alloy	Temp	Angle	Interaction	Reference
ZrB_2	Ag-Cu eut. + 2.4 at% Zr	1050	52°	Interfacial product	(Muolo et al. 2005)
ZrB_2	Ag-2.4 at% Ti	1050	27°	Interfacial product	(Muolo et al. 2005)
ZrB_2- SiC 20 vol%	Ag-9 at% Ti	1050	< 10°	Interfacial product	(Valenza et al. 2012a)
ZrB_2	Ag-2.4 at% Zr	1050	21°	Interfacial product	(Muolo et al. 2005)
ZrB_2	Ag-2.4 at% Hf	1050	75°	Interfacial product	(Muolo et al. 2005)
ZrB_2	Ag-2.4 at% Hf	1150	53°	Interfacial product	(Muolo et al. 2005)
ZrB_2	AlNi	1650–1670	20°	Non reactive	(Umanskyi et al. 2015)
ZrB_2	Au	1130	34°	Non reactive	(Passerone et al. 2007)
ZrB_2	Au-40 at% Ni	980	< 10°	Interfacial product	(Voytovych et al. 2007)
ZrB_2	Au-40 at% Ni	1170	25°	Dissolutive	(Voytovych et al. 2007)
ZrB_2	Cr	1800–1950	45°–90°	Dissolutive	(Zhunkovskii et al. 2011)
ZrB_2	Cu	1150	> 140°	Non reactive	(Muolo et al. 2005)
ZrB_2	Cu	1150	80°	Non reactive	(Passerone et al. 2007)
ZrB_2	Cu	1150	72°	Non reactive	(Valenza et al. 2012a)
ZrB_2- SiC 20 vol%	Cu	1150	80°	Non reactive	(Valenza et al. 2012a)
ZrB_2- SiC 20 vol%	Cu-22 at% Ti	1150	19°	Interfacial product	(Valenza et al. 2012a)
ZrB_2 – Si_3N_4 5 vol%	Ni	1500	< 10°	Dissolutive	(Valenza et al. 2012b)
ZrB_2 – Si_3N_4 5 vol%	Ni-17 at% B	1130	< 10°	Dissolutive & liquid-solid-liquid transition	(Artini et al. 2013)
ZrB_2 – Si_3N_4 5 vol%	Ni-17 at% B	1150	22°	Dissolutive	(Artini et al. 2013)
ZrB_2 – Si_3N_4 5 vol%	Ni-17 at% B	1200	21°	Dissolutive	(Valenza et al. 2012b)

Table 2 cont....

Table 2 cont....

Ceramic	Metal/alloy	Temperature	Contact angle	Interfacial phenomena	Reference
$ZrB_2 - Si_3N_4$ 5 vol%	Ni-17 at% B	1500	< 10°	Dissolutive	(Valenza et al. 2012b)
$ZrB_2 - Si_3N_4$ 5 vol%	Ni-50 at% B	1200	58	Non reactive	(Valenza et al. 2012b)
$ZrB_2 - Si_3N_4$ 5 vol%	Ni-50 at% B	1500	38	Non reactive	(Valenza et al. 2012b)
HfB_2	Ag	1023	130°	Non reactive	(Passerone et al. 2006)
HfB_2	Au	1130	98°	Non reactive	(Passerone et al. 2006)
HfB_2	Cu	1150	47°	Non reactive	(Passerone et al. 2006)
$HfB_2 + HfSi_2$ 5 vol%	Ni	1520	18°	Dissolutive	(Passerone et al. 2009a)
HfB_2	Ni	1500	< 10°	Dissolutive	(Sobczak et al. 2010)
$HfB_2 + HfSi_2$ 5 vol%	Ni-17 at% B	1520	< 10°	Dissolutive	(Passerone et al. 2009a)
$HfB_2 + B_4C$ 7 vol%	Ni-17 at% B	1200	37°	Dissolutive	(Passerone et al. 2010)
$HfB_2 + B_4C$ 7 vol%	Ni-17 at% B	1130	-	Dissolutive & solidification at testing temperature	(Passerone et al. 2010)
$HfB_2 + B_4C$ 7 vol%	Ni-50 at% B	1520	< 10°	Non reactive	(Passerone et al. 2010)
HfB_2	Ni-45 at% Pd	1250	21°	Dissolutive	(This Group, unpublished data)
$HfB_2 + HfSi_2$ 5 vol%	Ni-3 at% Ti	1520	20°	Dissolutive	(Passerone et al. 2009a)
$HfB_2 + HfSi_2$ 5 vol%	Ni-5 at% Ti	1520	< 10°	Dissolutive	(Passerone et al. 2009a)
$HfB_2 + HfSi_2$ 5 vol%	Ni-50 at% Ti	1520	< 10°	Dissolutive	(Passerone et al. 2009a)

composites and sintering aids are used to enhance densification during production. Moreover, the stability of XB_2 depends on the environmental conditions (i.e., atmosphere) and on the temperatures which are used during experiments. Speaking of pure diborides, below 700°C the surface oxidation in air is negligible, while up to 1100°C, the formation of a continuous layer of XO_2 and of B_2O_3 provides a passive protection which may hinder the wettability. At temperatures higher than 1100°C, the vapour pressure of the liquid B_2O_3 becomes relevant: specifically for the temperatures in the medium range, the vapour pressure goes from $3 \cdot 10^{-2}$ Pa at 1000°C to $4 \cdot 10^{-1}$ Pa at 1150°C (Hildebrand et al. 1963). Even though in most laboratory tests the oxygen partial pressure is kept to a minimum, so that strong oxidation of the substrate is not expected, a native oxide layer on the surface always exists and its evaporation cleans gradually the surface, making it more metallic and thus, more wettable by liquid metals.

The effect of the evaporation of B_2O_3 at high temperature applies only for the pure XB_2. In fact, in the case of composites with SiC or when Si-containing sintering enhancers are used, B_2O_3 and SiO_2 (from the oxidation of SiC) form a borosilicate glass which provides protection against oxidation (Opeka et al. 1999) above 1100°C.

These cleaning mechanisms depend strongly on temperature and environmental conditions: poor wettability is due most probably to uneffective deoxidation because of low temperatures, as in the case of experiments with Ag, or to the use of a protective atmosphere instead of vacuum which facilitates the evaporation of B_2O_3. The small dissolution of the ceramic phase into the molten metal should be also considered. In fact, even though SEM images have demonstrated that, at the macroscopic scale, the metal-ceramic interface remains planar after interaction with liquid Cu, Ag or Au, the EDS analysis showed that traces of Ti, Zr or Hf, B and, when present, Si could be found in the solidified drop. Ti, Zr or Hf may act as interfacial active element while boron may constitute a deoxidizing agent by reacting with B_2O_3 to form B_2O_2 which is a gaseous compound at the experimental temperatures. Moreover, both B and Si lower the surface tension of the liquid metal (Adachi et al. 2010; Passerone et al. 2009b). In any case, when alloys are used made of Ag, Cu and Au with additions of active metals such as Ti, Zr, and Hf, a striking improvement of the wetting is reported. These elements segregate and react with the oxide barriers at the metal/ceramic interface with the effect of a noticeable decrease of the liquid-solid interfacial tension lowering the contact angle, despite the concurrent increase in liquid surface tension. In some cases, SEM observations showed that continuous layers of new interfacial compounds, rich in the active elements, were formed which are more wettable by the liquid phase (Fig. 3).

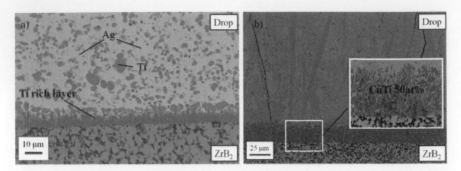

Fig. 3: SEM (scanning electron microscope) images (back scattered electrons – BSE) of the cross sectioned Ag-Ti 9 at% (a) and Cu-Ti 22 at% (b) drops on ZrB$_2$ substrates showing the formation of Ti-rich layers at the metal-ceramic interface.

Nickel is a metal of great importance from the technological point of view; moreover, its high melting point makes it a good candidate for brazing alloys aimed at high temperature applications. Therefore, many efforts were made, especially by Dr. A. Passerone and collaborators, in order to investigate the high temperature interactions of this metal and its alloys with ZrB$_2$ and HfB$_2$ by means of wetting experiments (Passerone et al. 2009a; Passerone et al. 2010; Valenza et al. 2012b) and their interpretation by using multicomponent phase diagrams calculated for these systems by the CALPHAD method (Artini et al. 2014; Cacciamani et al. 2011; Kaufman et al. 2010; Passerone et al. 2010).

When molten Ni is brought into contact with HfB$_2$ at 1520°C (Passerone et al. 2009a), fast and good wetting was observed (Fig. 4). However, strong dissolution of the ceramic phase occurred as observable from cross-sectioned samples showing an extensive damage of the ceramic phase. The dissolution and the following spreading of the saturated alloy led to the characteristic sigmoidal shape of the metal/ceramic interface (Fig. 5) which is due to different concurrent phenomena as schematized in Fig. 6: (a) melting of the droplet and initial spreading, (b) dissolution of the ceramic, and (c) further spreading. Moreover, the dissolution caused the introduction of B and Zr into the liquid alloy and, upon cooling, reprecipitation of HfB$_2$ crystals and formation of the Hf$_2$Ni$_{21}$B$_6$ compound; in addition, an infiltrated layer formed (Fig. 7). The phase diagram for the B-Ni-Hf system (Fig. 8), evaluated by the CALPHAD method (Kaufman et al. 2010), predicts that, at this temperature, the dissolution of the boride is greatly reduced and even suppressed if a Ni–B alloy, instead of pure Ni, is put in contact with HfB$_2$. Experimental results showed that, keeping the contact angles well below 90° (Fig. 4), the addition of B leads to a decrease (NiB 17 at% alloy) and, finally, to almost the disappearance of the substrate dissolution (NiB 50 at% alloy,

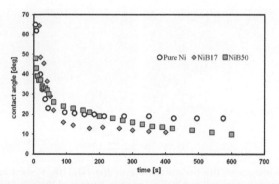

Fig. 4: Contact angles vs. time for Ni-B alloys on HfB_2 at 1520°C.

Fig. 5: SEM pictures of the cross section of the Ni/HfB_2, NiB 17 at%/HfB_2 and NiB 50 at%/ HfB_2 samples after test at 1520°C for 10 min showing the decreasing of the dissolution of the ceramic phase with increasing boron content.

Fig. 6: Sketch of the evolution of the metal-ceramic interphase during Ni-HfB_2 contact.

Fig. 5). The behaviour of Ni-B alloys in contact with ZrB_2 surfaces is similar from both the wetting and interfacial reactivity points of view (Fig. 9). As predicted by phase diagrams (Valenza et al. 2012b), during sessile drop tests at 1500°C pure Ni dissolved an important quantity of ZrB_2, while, when adding boron the dissolution decreased and disappeared when a NiB 50 at% alloy is used. In any case the wetting of the liquid is good, with angles below 20° for any Ni-B composition tested, and fast, as most of the process happens in the first minutes from melting.

The use of Ni alloyed with B, beside the effect of reducing the interfacial reactivity, enables the use at lower temperatures making the NiB alloys appealing from the point of view of brazing processes provided

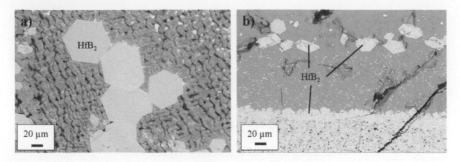

Fig. 7: SEM-BSE images of the microstructure of the Ni/HfB$_2$ system tested at 1520°C showing (a) hexagonal recrystallized HfB$_2$ at the top surface of the solidified drop, and (b) cross sectioned sample with HfB$_2$ recrystallization at the interface and in the bulk drop.

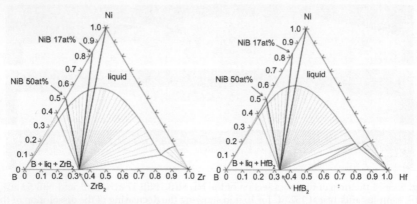

Fig. 8: Isothermal section of the B-Ni-Zr and B-Ni-Hf phase diagrams at 1500°C.

that wetting and interfacial phenomena are studied and understood. Wetting experiments conducted at 1200°C showed a decrease in wettability (however well acceptable for brazing processes) for both the NiB 17 and NiB 50 at% alloys being the contact angles 21° and 58° respectively in the case of ZrB$_2$. At the same time, cross-sectioned samples exhibited a metal-ceramic interface macroscopically planar, indicating that, as foreseen by phase diagrams, the dissolution took place only to a very limited extent.

The use of the NiB 17 at% in contact with both ZrB$_2$ and HfB$_2$ at lower temperatures led also to the interesting observation of solidification of the drop with further re-melting at the testing temperature (Artini et al. 2013; Passerone et al. 2010). This liquid-solid-liquid transition has also been used in order to validate phase diagrams calculations. Looking at the spreading kinetics shown in Fig. 10, at 1130°C the wetting of NiB 17 at% on ZrB$_2$ starts with a rapid decrease of the contact angle associated to a reduction of the drop height and to an increase of the drop diameter (part a), but, after a while, a change in the drop shape occurs caused by a partial solidification of the alloy (part b). A few minutes later (at 1130°C), a further melting takes place and the drop remains in the liquid state until the end of the test (part c).

This phenomenon can be successfully interpreted on the basis of the ternary computed phase diagrams. The isothermal section at 1130°C is reported in Fig. 11: when the NiB 17 at% alloy is put in contact with ZrB_2, due to the small dissolution of ZrB_2, the composition of the liquid drop changes with time along the dotted line until it reaches saturation (point B). In the A region, the reprecipitation of $Zr_2Ni_{21}B_6$ occurs, until the global composition lies inside the (liquid + $Zr_2Ni_{21}B_6$) two-phase field. The complete remelting of the drop explains why no $Zr_2Ni_{21}B_6$ could be found at the metal/substrate interface or in the bulk of the solidified drop.

Interactions of ZrB_2 with Ni have been also assessed at relatively low temperatures by (Voytovych et al. 2007) using the Au-Ni 40 at% azeotrope with melting point of 955°C. This alloy forms an interfacial reaction

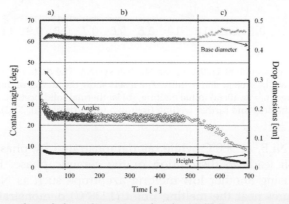

Fig. 9: Contact angle and drop dimensions for the Ni/ZrB_2 sample at 1500°C. Dotted lines delimit the steps of the wetting kinetics: (a) initial spreading, (b) stationary stage and dissolution of the ceramic, (c) further spreading (redrawn from Valenza et al. 2012b).

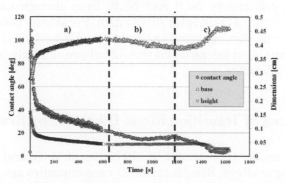

Fig. 10: NiB 17 at%/ZrB_2 spreading kinetics at 1130°C (redrawn from Artini et al. 2013).

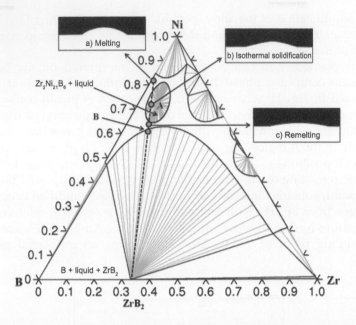

Fig. 11: Isothermal section at 1130°C of the Ni-Zr-B system. The dotted line joining NiB 17 at% to ZrB$_2$ shows the system under study. The insets show the images of the sessile drop at the different stages of Fig. 10: (a) initial melting, (b) isothermal solidification, (c) remelting of the drop and further wetting (redrawn from Artini et al. 2013).

compound, Ni$_2$B, at 980°C while, as demonstrated by phase diagrams, at 1170°C Ni leads to dissolution of the ceramic phase. Experiments performed using another low-melting azeotrope such as Ni-Pd 45 at% at temperatures near the melting point (1238°C) demonstrated a strong interfacial reactivity with dissolution of the ceramic phase (Valenza et al. unpublished data). A complex microstructure has been observed (Fig. 12) mainly constituted, beside the starting alloy, of HfPd$_3$ and Ni-B compounds, presumably Ni$_2$B and Ni$_3$B. Even though no quaternary phase diagrams are available for this system, the affinity between Ni and Pd and the similarities between binary Pd-B and Pd-Hf to Ni-B and Ni-Hf phase diagrams let us infer that Pd should behave similarly to Ni when in contact with diborides.

Joining of Transition Metal Diborides by Brazing

In Table 3, a review is presented for diborides of Ti, Zr and Hf joined to themselves or to alloys. Results obtained on composites are presented as well if the content in diboride is higher than 50 vol%. When available,

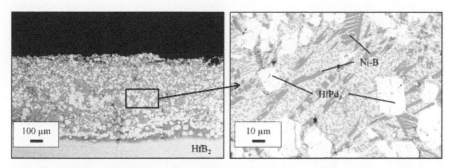

Fig. 12: SEM-BSE image of the cross section of a Ni-Pd 45 at%/HfB$_2$ wetting couple after sessile drop test at 1250°C showing the complex microstructure due to high interfacial reactivity.

the values of shear strength are presented as well as the indication of the rupture mode (e.g., in the ceramic phase, in the interlayer, mixed mode, etc.); in this context, caution should be paid when comparing shear strength values obtained under different conditions. When strength data were not provided by the authors, the information on the joint characteristics as detected by microstructural analysis (e.g., defect-free, cracked, etc.) is reported.

Most of the studies summarized in Table 3 regard the joining of composite materials rather than pure ceramics and, in particular, the ZrB$_2$-SiC composite which represents, among the transition metal diborides family, the most attractive and feasible combination for aerospace applications due to its acceptable density.

Brazing is the most commonly used option for joining diborides; an extensive work on this has been performed by Asthana and Singh which reported brazing of ZrB$_2$-SiC (fibres, particles) to themselves or to alloys such as Ti, Ti6Al4V and Inconel-625 using several brazing fillers (review in Asthana and Singh 2013). In literature it has been found that, when brazing processes are pursued using Ag or Cu-based alloys, commercially available Ag-Cu-Ti alloys included, defect-free joints are usually obtained with satisfactory mechanical performances. These joining options may be sufficient for extreme operating conditions for short times. For example, when a 20-mm thick ZrB$_2$ slab undergoes high surface thermal fluxes (T > 1800°C) for about 6 min, the underlying structure is reported to remain below 600°C (Muolo et al. 2003).

Higher service temperatures could be obtained by using fillers with higher melting temperatures such as glasses or Ni-based alloys. In the previous session we discussed the dissolution and reaction phenomena associated to the high temperature contact between Ni and diborides. Joints made using commercial Pd-based alloys (Singh and Asthana 2010) such as Palco and Palni exhibited great interaction with the ceramic phase

Table 3: Joining combinations for diborides and composites joined to themselves or to metals.

Material 1	Material 2	Filler	Temperature [°C]	Shear strength [MPa]	Crack propagation/ Joint characteristics	Reference
TiB_2 – TiC 20 vol%	Al_2O_3	TiB_2-Ni-Al_2O_3 graded interlayers	1470	350 (flex. strength)	-	(Gam et al. 1999)
TiB_2 – TiC 20 vol%	TiB_2 – TiC 20 vol%	No filler (Plasma arc welding)	NA	-	Large reaction area and pores	(King et al. 2014)
ZrB_2 – SiC 20 vol%	Ti6Al4V	Ag-9 at% Ti Foil, 60 µm	1050	79	Ceramic	(Valenza et al. 2012a)
ZrB_2 – Si_3N_4 5 vol%	Ti6Al4V	Ag-2.4 at% Zr 100 µm	1050	-	Defect-free	(Muolo et al. 2003)
ZrB_2 – SiC 20 vol% – C 15 vol%	GH99 (Ni-based superalloy)	Ag-Cu-Ti/Cu mesh composite foils	840–920	121	Ceramic	(Tian et al. 2015)
ZrB_2 – SiC 20 vol%	Ti6Al4V	Ag-Cu Eut. 100 µm	880	141	-	(Yang et al. 2013a)
ZrB_2 – SiC 20 vol%	ZrB_2 – SiC 20 vol%	Ag-Cu Eut. 100 µm + Ti powder 15 wt%	900	146	Interlayer, ductile	(Yang et al. 2012)
ZrB_2 – SiC 14 vol% – C 30 vol%	Cu-clad-Mo	Ag-35.3Cu-1.75Ti (wt%) Cusil-ABA	830–835	-	Defect-free	(Singh and Asthana 2009)
ZrB_2 – SiC 14 vol% – C 30 vol%	Cu-clad-Mo	Ag-26.7Cu-4.5Ti (wt%) Ticusil	915–920	-	Defect-free	(Singh and Asthana 2009)
ZrB_2 – SiC 20 vol%	Ti6Al4V	Cu Foil, 40 µm	1050	< 20 MPa	Brazing seam	(Valenza et al. 2012a)

ZrB$_2$ – SiC 20 vol%	Ti6Al4V	Cu (20 μm) Ti (30 μm) Foils	900–960	41–91	-	(Yang et al. 2015)
ZrB$_2$ – SiC 14 vol% - C 30 vol%	ZrB$_2$ – SiC 14 vol% - C 30 vol%	Pd-Ni 40 wt% (Palni) 50 μm	1253–1258	-	Defected	(Asthana and Singh 2009)
ZrB$_2$ – SiC 14 vol% - C 30 vol%	ZrB$_2$ – SiC 14 vol% - C 30 vol%	Pd-Co 35 wt% (Palco) 50 μm	1234–1239	-	Defect-free	(Asthana and Singh 2009)
ZrB$_2$ – SiC 20 vol%	ZrB$_2$ – SiC 20 vol%	Pd-Ni 40 wt% (Palni) 50 μm	1253–1258	-	Defected	(Asthana and Singh 2009)
ZrB$_2$ – SiC 20 vol%	ZrB$_2$ – SiC 20 vol%	Pd-Co 35 wt% (Palco) 50 μm	1234–1239	-	Defect-free	(Asthana and Singh 2009)
ZrB$_2$	ZrB$_2$	Ni Powder < 75 μm	1400	33–60	-	(Yuan and Zhang 2011)
ZrB$_2$ – SiC 20 vol%	ZrB$_2$ – SiC 20 vol%	Ni Powder < 75 μm	1400	24–43	-	(Yuan and Zhang 2011)
ZrB$_2$ – SiC 20 vol%	Ti6Al4V	Ni-50 at% B Paste	1100	74	Ceramic	(Valenza et al. 2014)
ZrB$_2$ – SiC 20 vol%	Nb	Ni foam 4 mm	750	156	Interlayer, ductile	(Yang et al. 2013b)
ZrB$_2$ – SiC 14 vol% - C 30 vol%	Ti	MBF-20 – 50 μm Ni-6.48Cr-3.13Fe-4.38Si-3.13B (wt%)	1039–1044	-	Defect-free	(Singh and Asthana 2007)

Table 3 cont....

Table 3 cont....

Material 1	Material 2	Filler	Temperature [°C]	Shear strength [MPa]	Crack propagation/ Joint characteristics	Reference
ZrB_2 – SiC 14 vol% – C 30 vol%	Ti	MBF-30 – 50 μm Ni-4.61Si-2.8B (wt%)	1069–1074	-	Defected (ceramic)	(Singh and Asthana 2007)
ZrB_2 – SiC 20 vol%	Ti	MBF-20 – 50 μm Ni-6.48Cr-3.13Fe-4.38Si-3.13B (wt%)	1039–1044	-	Defect-free	(Singh and Asthana 2007)
ZrB_2 – SiC 20 vol%	Ti	MBF-30 – 50 μm Ni-4.61Si-2.8B (wt%)	1069–1074	-	Defected (ceramic)	(Singh and Asthana 2007)
ZrB_2 – SiC 14 vol% – C 30 vol%	ZrB_2 – SiC 14 vol% - C 30 vol%	MBF-20 – 50 μm Ni-6.48Cr-3.13Fe-4.38Si-3.13B (wt%)	1039–1044	-	Interfacial microcracking	(Singh and Asthana 2007)
ZrB_2 – SiC 14 vol% – C 30 vol%	Ti	Pd-Ni 40 wt% (Palni) 50 μm	1253–1258	-	Defect-free Interfacial reactivity	(Singh and Asthana 2010)
ZrB_2 – SiC 14 vol% – C 30 vol%	Ti	Pd-Co 35 wt% (Palco) 50 μm	1234–1239	-	Defect-free Interfacial reactivity	(Singh and Asthana 2010)
ZrB_2 – SiC 20 vol%	Ti	Pd-Ni 40 wt% (Palni) 50 μm	1253–1258	-	Defected (ceramic)	(Singh and Asthana 2010)
ZrB_2 – SiC 20 vol%	Ti	Pd-Co 35 wt% (Palco) 50 μm	1234–1239	-	Defected (ceramic)	(Singh and Asthana 2010)

ZrB_2 – SiC 14 vol% – C 30 vol%	Inconel-625	Pd-Ni 40 wt% (Palni) 50 μm	1253–1258	-	Unreliable: strong interfacial reactivity	(Singh and Asthana 2010)
ZrB_2 – SiC 14 vol% – C 30 vol%	Inconel-625	Pd-Co 35 wt% (Palco) 50 μm	1234–1239	-	Defect-free Interfacial reactivity	(Singh and Asthana 2010)
ZrB_2 – SiC 14 vol% – C 30 vol%	Cu-clad-Mo	Pd-Co 35 wt% (Palco) 50 μm	1253–1258	-	Defected (Cu-clad-Mo/Palni interface)	(Singh and Asthana 2009)
ZrB_2 – SiC 20 vol%	Inconel-625	Pd-Co 35 wt% (Palco) 50 μm	1234–1239	-	Defect-free Interfacial reactivity	(Singh and Asthana 2010)
ZrB_2 – SiC 20 vol%	Cu-clad-Mo	Pd-Co 35 wt% (Palco) 50 μm	1253–1258	-	Defect-free Interfacial reactivity	(Singh and Asthana 2009)
ZrB_2 – SiC 14 vol% – C 30 vol%	Inconel-625	Pd-Ni 40 wt% (Palni) 50 μm	1253–1258	-	Unreliable: strong interfacial reactivity	(Singh and Asthana 2010)
ZrB_2 – SiC 20 vol%	Cu-clad-Mo	Pd-Ni 40 wt% (Palni) 50 μm	1234–1239	-	Defect-free Interfacial reactivity	(Singh and Asthana 2009)
ZrB_2 – SiC 20 vol%	Cu-clad-Mo	Pd-Ni 40 wt% (Palni) 50 μm	1234–1239	-	Defect-free Interfacial reactivity	(Singh and Asthana 2009)
ZrB_2 – SiC 20 vol%	ZrB_2 – SiC 20 vol%	SiO_2-Al_2O_3-CaO glass	1440	277 (RT) – 195 (@800°C) – 88 (@1000°C) (three point)	-	(Esposito and Bellosi 2005)

Table 3 cont....

Table 3 cont....

Material 1	Material 2	Filler	Temperature [°C]	Shear strength [MPa]	Crack propagation/Joint characteristics	Reference
ZrB_2–SiC 20 vol%	ZrB_2–SiC 20 vol%	SiO_2-Al_2O_3-Y_2O_3 glass	1440	-	Incomplete bonding	(Esposito and Bellosi 2005)
ZrB_2–SiC 14 vol%–C 30 vol%	Ti	Ti-375 (Ti-37.5Zr-15Cu-10Ni wt%)	858–863	-	Defect-free	(Asthana and Singh 2014)
ZrB_2–SiC 14 vol%–C 30 vol%	Inconel-625	Ti-375 (Ti-37.5Zr-15Cu-10Ni wt%)	858–863	-	Defected (ceramic)	(Asthana and Singh 2014)
ZrB_2–SiC 20 vol%	Ti	Ti-375 (Ti-37.5Zr-15Cu-10Ni wt%)	858–863	-	Defect-free	(Asthana and Singh 2014)
ZrB_2–SiC 20 vol%	Ti	Ti-120 (Ti-12Zr-22Cu-12Ni-1.5Be-0.8V wt%)	830–835	-	Defect-free	(Asthana and Singh 2014)
ZrB_2–SiC 20 vol%	Inconel-625	Ti-375 (Ti-37.5Zr-15Cu-10Ni wt%)	858–863	-	Defected (ceramic)	(Asthana and Singh 2014)
ZrB_2–SiC 20 vol%	Inconel-625	Ti-120 (Ti-12Zr-22Cu-12Ni-1.5Be-0.8V wt%)	830–835	-	Defected (ceramic)	(Asthana and Singh 2014)
ZrB_2–SiC 20 vol%	ZrB_2–SiC 20 vol%	Ti-375 (Ti-37.5Zr-15Cu-10Ni wt%)	858–863	-	Defect-free	(Asthana and Singh 2014)

ZrB₂–SiC 20 vol%	Nb	Ti	1200	158	Interfacial TiB	(He et al. 2012)
ZrB$_2$–SiC 20 vol%	Nb	Ti	1200	158	Interfacial TiB	(He et al. 2012)
ZrB$_2$–SiC 30 vol%	ZrB$_2$–SiC 30 vol%	Zr-B (Spark plasma joining)	1800	311 (@RT) 284 (@1350°C)	Mixed mode	(Pinc et al. 2011)
ZrB$_2$–ZrC 20 vol%	ZrB$_2$–ZrC 20 vol%	No filler (Plasma arc welding)	NA	140–250 (4-point bending)	-	(King et al. 2015)
HfB$_2$–MoSi$_2$ 10 vol%	HfB$_2$–MoSi$_2$ 10 vol%	Ni 2 μm – Nb 125 μm – Ni 2 μm	1500	-	Defect-free	(Saito et al. 2012)
HfB$_2$	HfB$_2$	Ni-50 at% B Capillary infiltration	1300	146	Ceramic	(Muolo et al. 2010)
HfB$_2$	Ta	Ni Capillary infiltration	1500	44	Mixed mode	(Sobczak et al. 2010)

although, in some cases, defect-free joints were produced. On the basis of the insights given by wetting experiments and thermodynamic calculations, diborides were joined using Ni-B alloys. HfB_2 specimens were joined to themselves at 1300°C using a NiB 50 at% as filler (Muolo et al. 2010). The alloy was allowed to flow by capillarity into the space between the two ceramic pieces. The metallographic section (Fig. 13) shows a very good penetration without any voids and any dissolution of the solid ceramic phase. This evidence and the fact that the CTE value for NiB 50 at% is not far from that of HfB_2 ($\alpha_{pure\,HfB2} = 6.3 \cdot 10^{-6}\,C^{-1}$; $\alpha_{NiB50} = 7.1 \cdot 10^{-6}\,C^{-1}$) led to a very good mechanical performance. An average shear strength of 146 MPa was measured with a brittle fracture occurring into the ceramic body and not along the brazing seam. The NiB 50 at% alloy was used as a brazing filler also for joining ZrB_2-SiC to Ti6Al4V system (Valenza et al. 2014). In this case, diffusion of Ti led to a complex multilayer interfacial structure (Fig. 14), with a thickness of about 100 µm. Interfacial layers were reputed to have an important role in accommodating the thermal mismatch between the diboride composite and the Ti alloys and the shear strength, averaged over three samples, was 74 ± 16 MPa.

Even though brazing constitutes a relatively easy, fast, reliable and flexible way to obtain joints, it is generally limited by the fact that the service temperature is below the one adopted for the joining process so that, in some cases, other joining options should be sought. The possibility to obtain joined interfaces having a service temperature higher than that of the process is offered by the transient liquid phase bonding (TLPB) technique. Saito et al. reported on the possibility to join HfB_2 + $MoSi_2$ composites at 1500°C by the TLPB technique using Ni/Nb/Ni interlayers. Even though Ni-Nb alloys were reported to wet but to dissolve the ceramic phase, the use of very thin (2 µm) layers of Ni assured wetting of the adjoining surfaces with limited dissolution. The results have shown that the interlayer and composite were well-bonded, with no cracks in the interfacial region and, most important, with a re-melting temperature which is in principle that of pure Nb (2477°C).

A similar approach has been attempted (Valenza et al. 2012; Yang et al. 2015) for joining ZrB_2 composites to Ti6Al4V. Using Cu as interlayer allows contact melting to occur at the interface between Cu and the Ti alloy because of the presence of eutectics in the binary Cu-Ti phase diagram. The subsequent diffusion of Ti into the liquid phase can lead to the formation of high-melting phases and their subsequent isothermal solidification. The toughening mechanism by whiskers of TiB embedded in ductile metallic phases and the fact that TiB has an intermediate CTE with respect to that of ZrB_2 and Ti6Al4V contribute to the final mechanical joint strength (Yang et al. 2012).

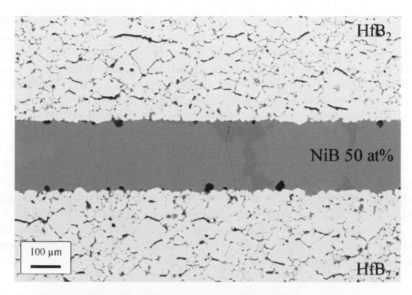

Fig. 13: SEM-BSE image of the cross section of a HfB₂-NiB-HfB₂ joint obtained at 1300°C by capillary infiltration.

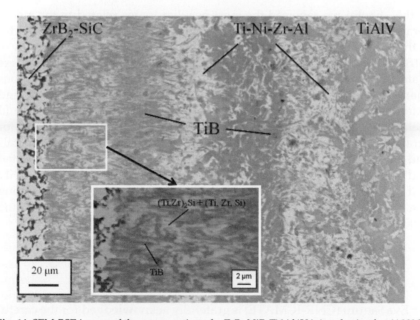

Fig. 14: SEM-BSE image of the cross section of a ZrB₂-NiB-Ti6Al4V joint obtained at 1100°C.

The high electrical conductivity of the diborides offers the possibility to exploit current assisted methods for their joining. To this end, very recently, attempts have been reported about the feasibility of direct joining of TiB_2 or ZrB_2 composites by plasma arc welding (King et al. 2015; King et al. 2014a,b). Even though the control of the reaction area or the formation of large pores at the interface between the adjoining materials still represents an issue, this joining technique is promising especially from the point of view of the final service temperature. In fact, the final interfaces are in principle formed of the same phases constituting the starting material. A similar feature has been reported for spark plasma joining using Zr-B fillers (Pinc et al. 2011); joints obtained by this technique exhibit interfaces undistinguishable from the surrounding material with resulting strength similar to that of the base material. In this case, the issue is constituted by the fact that large joined pieces cannot be realized due to the relatively small dimensions of SPS furnaces; moreover, the joining of curved surfaces may be also difficult.

Conclusions

In this chapter we reported the most peculiar issues related to the metal-ceramic interactions encountered when joining transition metal diborides of the 4th group (TiB_2, ZrB_2 and HfB_2) which belong to the broader family of ultra high temperature ceramics. The most peculiar characteristics of these materials were recalled especially in relation to environmental conditions characterized by extremely high temperatures, mechanical loads and oxidation phenomena. Basics about high temperature wetting of liquid metals on ceramics and their importance in view of brazing processes for ceramics were presented. Recent data about the wetting of liquid metals and alloys on TiB_2, ZrB_2 and HfB_2 and their composites has been collected and discussed. Contact angle data, information on interfacial reactivity and final microstructures, and their interpretation through modeling (e.g., by CALPHAD) constitute the basis for the correct design of joining processes in view of high demanding applications.

Acknowledgments

Dr. Eva Santini (CNR-ICMATE) is gratefully acknowledged for the careful and critical reading of the manuscript.
    The authors would also like to thank all the colleagues who contributed to this work in the last years, and, in particular, Prof. G. Cacciamani (DCCI, University of Genova) for the extensive work in phase diagrams calculations.

Keywords: Borides, Wetting, Joining, Brazing, Metal-ceramic interfaces, Ultra high temperature ceramics, Interfacial reactivity, Phase diagrams, Liquid metals

References

Adachi, M., M. Schick, J. Brillo, I. Egry and M. Watanabe. 2010. Surface tension and density measurement of liquid Si–Cu binary alloys. J. Mater. Sci. 45: 2002–2008.

Aghajanian, M.K., M.A. Rocazella, J.T. Burke and S.D. Keck. 1991. The fabrication of metal matrix composites by a pressureless infiltration technique. J. Mater. Sci. 26: 447–454.

Aizenshtein, M., N. Froumin and N. Frage. 2012. The nature of TiB_2 wetting by Cu and Au. J. Mater. Eng. Perform. 21: 655–659.

Alfano, D., R. Gardi, L. Scatteia and A. del Vecchio. 2014. UHTC-based hot structures. pp. 416–436. *In*: W.G. Fahrenholtz, E.J. Wuchina, W.E. Lee and Y. Zhou [eds.]. Ultra-High Temperature Ceramics: Materials for Extreme Environment Applications. Wiley-American Ceramic Society.

Artini, C., M.L. Muolo, A. Passerone, G. Cacciamani and F. Valenza. 2013. Isothermal solid–liquid transitions in the $(Ni,B)/ZrB_2$ system as revealed by sessile drop experiments. J. Mater. Sci. 48: 5029–5035.

Artini, C., M.L. Muolo, A. Passerone, F. Valenza, P. Manfrinetti and G. Cacciamani. 2014. Experimental investigations and thermodynamic modeling in the ZrB_2/Ni section of the BNiZr system. J. All. Comp. 592: 115–120.

Asthana, R. and M. Singh. 2009. Joining of ZrB_2-based ultra-high-temperature ceramic composites using Pd-based braze alloys. Scripta Mater. 61: 257–260.

Asthana, R. and M. Singh. 2013. Active metal brazing of advanced ceramic composites to metallic systems. pp. 323–360. *In*: D. Sekulic [ed.]. Advances in Brazing. Science, Technology and Applications, Woodhead Publishing Series, Cambridge, UK.

Asthana, R. and M. Singh. 2014. Evaluation of amorphous Ti brazes to join zirconium diboride-based ultra-high-temperature ceramics to metallic systems. Int. J. Appl. Ceram. Tech. 11: 502–512.

Cacciamani, G., P. Riani and F. Valenza. 2011. Equilibrium between MB2 (M = Ti, Zr, Hf) UHTC and Ni: A thermodynamic database for the B–Hf–Ni–Ti–Zr system. Calphad 35: 601–619.

Carney, C.M., T.A. Parthasarathy and M.K. Cinibulk. 2011. Oxidation resistance of hafnium diboride ceramics with additions of silicon carbide and tungsten boride or tungsten carbide. J. Am. Cer. Soc. 94: 2600–2607.

Chamberlain, A.L., W.G. Fahrenholtz, G.E. Hilmas and D.T. Ellerby. 2004. High-strength zirconium diboride-based ceramics. J. Am. Cer. Soc. 87: 1170–1172.

De Gennes, P.G., F. Brochard-Wyart and D. Quéré. 2004. Capillarity and Wetting Phenomena. Springer, New York, USA.

Dempsey, E. 1963. Bonding in the refractory hard-metals. Phil. Mag. 8: 285–299.

Dezellus, O. and N. Eustathopoulos. 2010. Fundamental issues of reactive wetting by liquid metals. J. Mater. Sci. 45: 4256–4264.

Esposito, L. and A. Bellosi. 2005. Joining ZrB_2-SiC composites using glass interlayers. J. Mater. Sci. 40: 4445–4453.

Eustathopoulos, N., M.G. Nicholas and B. Drevet. 1999. Wettability at High Temperatures. *In*: R.W. Cahn [ed.]. (Vol. 3) Pergamon Materials Series, Pergamon.

Fahrenholtz, W.G., G.E. Hilmas, I.G. Talmy and J.A. Zaykoski. 2007. Refractory diborides of zirconium and hafnium. J. Am. Cer. Soc. 90: 1347–1364.

Fahrenholtz, W.G. 2014. A historical perspective on research related to ultra-high temperature ceramics. pp. 6–32. *In*: W.G. Fahrenholtz, E.J. Wuchina, W.E. Lee and Y. Zhou [eds.].

Ultra-High Temperature Ceramics: Materials for Extreme Environment Applications. Wiley-American Ceramic Society.

Ferraris, M. and V. Casalegno. 2015. Integration and joining of ceramic matrix composites. pp. 551–546. *In*: N.P. Bansal and J. Lamon [eds.]. Ceramic Matrix Composites: Materials, Modeling and Technology. John Wiley & Sons, Inc., Hoboken, NJ, USA.

Gam, J.S., K.S. Han, S.S. Park and H.C. Park. 1999. Joining of TiB_2-Al_2O_3 using compositionally graded interlayers. Mater. Manuf. Proc. 14: 537–546.

Guo, S.Q., J.M. Yang, H. Tanaka and Y. Kagawa. 2008. Effect of thermal exposure on strength of ZrB_2-based composites with nano-sized SiC particles. Comp. Sci. Tech. 68: 3033–3040.

Guo, S.Q. 2009. Densification of ZrB_2-based composites and their mechanical and physical properties: A review. J. Eur. Ceram. Soc. 29: 995–1011.

He, P., W. Yang, T. Lin, D. Jia, J. Feng and Y. Liu. 2012. Diffusion bonding of ZrB_2–SiC/Nb with *in situ* synthesized TiB whiskers array. J. Eur. Ceram. Soc. 32: 4447–4454.

Hildebrand, D.L., W.F. Hall and N.D. Potter. 1963. Thermodynamics of vaporization of lithium oxide, boric oxide, and lithium metaborate. J. Chem. Phys. 39: 296–301.

Hu, P., X.H. Zhang, J.C. Han, X.G. Luo and S.-Y. Du. 2010. Effect of various additives on the oxidation behavior of ZrB_2-based ultra-high-temperature ceramics at 1800°C. J. Am. Cer. Soc. 93: 345–349.

Justin, J.F. and A. Jankowiak. 2011. Ultra high temperature ceramics: Densification, properties and thermal stability. Aerospace Lab J. 8: 1–11.

Kaptay, G. 2003. Interfacial criteria for stabilization of liquid foams by solid particles. Coll. Surf. A: Physicochemical and Engineering Aspects 230: 67–80.

Kaufman, L., G. Cacciamani, M.L. Muolo, F. Valenza and A. Passerone. 2010. Wettability of HfB_2 by molten Ni(B) alloys interpreted by CALPHAD methods, Part 1: Definition of the B–Hf–Ni system. Calphad 34: 2–5.

King, D.S., G.E. Hilmas and W.G. Fahrenholtz. 2014a. Plasma arc welding of TiB_2-20 vol% TiC. J. Am. Cer. Soc. 97: 56–59.

King, D.S., G.E. Hilmas and W.G. Fahrenholtz. 2014b. Plasma arc welding of ZrB_2–20 vol% ZrC ceramics. J. Eur. Ceram. Soc. 34: 3549–3557.

King, D.S., G. Hilmas and W. Fahrenholtz. 2015. Mechanical behavior and applications of plasma arc welded ceramics. Int. J. Appl. Ceram. Technol. doi:10.1111/ijac.12402.

Levine, S.R., E.J. Opila, M.C. Halbig, J.D. Kiser, M. Singh and J.A. Salem. 2002. Evaluation of ultra-high temperature ceramics for aeropropulsion use. J. Eur. Ceram. Soc. 22: 2757–2767.

Levine, J.B., S.H. Tolbert and R.B. Kaner. 2009. Advancements in the search for superhard ultra-incompressible metal borides. Adv. Funct. Mater. 19: 3519–3533.

Liu, Q., W. Han, X. Zhang, S. Wang and J. Han. 2009. Microstructure and mechanical properties of ZrB_2-SiC composites. Materials Letters 63: 1323–1325.

Matsumoto, T. and K. Nogi. 2008. Wetting in soldering and microelectronics. Ann. Rev. of Mater. Res. 38: 251–273.

Mattia, D., M. Desmaison-Brut, D. Tétard and J. Desmaison. 2005. Wetting of HIP AlN-TiB_2 ceramic composites by liquid metals and alloys. J. Eur. Cer. Soc. 25: 1797–1803.

Monteverde, F., S. Guicciardi and A. Bellosi. 2003. Advances in microstructure and mechanical properties of zirconium diboride based ceramics. Mater. Sci. Eng. A 346: 310–319.

Monteverde, F. 2007. Hot pressing of hafnium diboride aided by different sinter additives. J. Mater. Sci. 43: 1002–1007.

Munro, R.G. 2000. Material properties of titanium diboride. J. Res. National Inst. Stand. Technol. 105: 709.

Muolo, M.L., E. Ferrera, L. Morbelli, C. Zanotti and A. Passerone. 2003. Joining of zirconium boride based refractory ceramics to Ti6Al4V. Materials in a Space Environment ESA SP 540: 467–472.

Muolo, M.L., E. Ferrera and A. Passerone. 2005. Wetting and spreading of liquid metals on ZrB_2-based ceramics. 40: 2295–2300.

Muolo, M.L., F. Valenza, N. Sobczak and A. Passerone. 2010. Overview on wetting and joining in transition metals diborides. Adv. Sci. Technol. 64: 98–107.

Murata, Y. 1970. Cutting Tool Tips and Ceramics Containing Hafnium Nitride and Zirconium Diboride. U.S. Patent #3487594.

Nicholas, M.G. 1998. Joining Processes. Kluwer Academic Publishers, Dordrecht, The Netherlands.

Ning, X., T. Okamoto and Y. Miyamoto. 1989. Effect of oxide additive in silicon nitride on interfacial structure and strength of silicon nitride joints brazed with aluminium. J. Mater. Sci. 24: 2865–2870.

Norasetthekul, S., P.T. Eubank, W.L. Bradley, B. Bozkurt and B. Stucker. 1999. Use of zirconium diboride copper as an electrode in plasma applications. J. Mater. Sci. 34: 1261–1270.

Opeka, M.M., I.G. Talmy, E.J. Wuchina, A. Zaykoski and S.J. Causey. 1999. Mechanical, Thermal, and Oxidation Properties of Refractory Hafnium and Zirconium Compounds. J. Eur. Ceram. Soc. 19: 2405–2414.

Parthasarathy, T.A.A., R.A.A. Rapp, M. Opeka and R.J.J. Kerans. 2007. A model for the oxidation of ZrB_2, HfB_2 and TiB_2. Acta Mater. 55: 5999–6010.

Passerone, A., M.L. Muolo and D. Passerone. 2006. Wetting of Group IV diborides by liquid metals. J. Mater. Sci. 41: 5088–5098.

Passerone, A., M.L. Muolo, R. Novakovic and D. Passerone. 2007. Liquid metal/ceramic interactions in the (Cu, Ag, Au)/ZrB_2 systems. J. Eur. Ceram. Soc. 27: 3277–3285.

Passerone, A., M.L. Muolo, F. Valenza, F. Monteverde and N. Sobczak. 2009a. Wetting and interfacial phenomena in Ni-HfB_2 systems. Acta Mater. 57: 356–364.

Passerone, A., M.L. Muolo, F. Valenza and R. Novakovic. 2009b. Thermodynamics and surface properties of liquid Cu-B alloys. Surface Science 603: 2725–2733.

Passerone, A., M.L. Muolo, F. Valenza and L. Kaufman. 2010. Wettability of HfB_2 by molten Ni(B) alloys interpreted by CALPHAD methods, Part 2: Wetting and interfacial reactivity. CALPHAD 34: 6–14.

Passerone, A., F. Valenza and M.L. Muolo. 2012a. Wetting at high temperature. pp. 299–334. *In*: M. Ferrari, L. Liggieri and R. Miller [eds.]. Drops and Bubbles in Contact with Solid Surfaces, Series: Progress in Colloid and Interface Science. CRC Press, Taylor and Francis Group.

Passerone, A., F. Valenza and M.L. Muolo. 2012b. A review of transition metals diborides: from wettability studies to joining. J. Mater. Sci. 47: 8275–8289.

Pinc, W.R., M. Di Prima, L.S. Walker, Z.N. Wing and E.L. Corral. 2011. Spark plasma joining of ZrB_2-SiC composites using zirconium-boron reactive filler layers. J. Am. Cer. Soc. (W. Fahrenholtz, ed.) 94: 3825–3832.

Post, B., F.W. Glaser and D. Moskowitz. 1954. Transition metal diborides. Acta Metall. 2: 20–25.

Saito, N., H. Ikeda, Y. Yamaoka, A.M. Glaeser and K. Nakashima. 2012. Wettability and transient liquid phase bonding of hafnium diboride composite with Ni–Nb alloys. J. Mater. Sci. 47: 8454–8463.

Saiz, E., M. Benhassine, J. De Coninck and A.P. Tomsia. 2010. Early stages of dissolutive spreading. Scripta Mater. 62: 934–938.

Sciti, D., S. Guicciardi, A. Bellosi and G. Pezzotti. 2006. Properties of a pressureless-sintered ZrB_2-$MoSi_2$ ceramic composite. J. Am. Cer. Soc. 89: 2320–2322.

Sciti, D., L. Silvestroni, L. Mercatelli, J.L. Sans and E. Sani. 2013. Suitability of ultra-refractory diboride ceramics as absorbers for solar energy applications. Solar Energy Mater. Solar Cells 109: 8–16.

Silvestroni, L. and D. Sciti. 2011. Oxidation of ZrB_2 ceramics containing SiC as particles, whiskers, or short fibers. J. Am. Cer. Soc. (M. Cinibulk, ed.) 94: 2796–2799.

Singh, M. and R. Asthana. 2007. Joining of zirconium diboride-based ultra high-temperature ceramic composites using metallic glass interlayers. Mater. Sci. Eng. A 460-461: 153–162.

Singh, M. and R. Asthana. 2009. Joining of ZrB_2-based ultra-high-temperature ceramic composites to Cu-Clad-Molybdenum for advanced aerospace applications. Int. J. Appl. Ceram. Technol. 6: 113–133.

Singh, M. and R. Asthana. 2010. Joining and integration of ZrB_2-based ultra-high temperature ceramic composites using advanced brazing technology. J. Mater. Sci. 45: 4308–4320.

Sobczak, N., R. Nowak, A. Passerone, F. Valenza, M.L. Muolo, L. Jaworska, F. Barberis and M. Capurro. 2010. Wetting and joining of HfB_2 and Ta with Ni. Trans. Found. Res. Inst. 5–14.

Telle, R., L.S. Sigl and K. Takagi. 2000. Boride-based hard materials. pp. 802–945. *In*: R. Riedel [ed.]. Handbook of Ceramic Hard Materials. Wiley-VCH, Weinheim, Germany.

Tian, X.Y., J.C. Feng, J.M. Shi, H.W. Li. and L.X. Zhang. 2015. Brazing of ZrB_2–SiC–C ceramic and GH99 superalloy to form reticular seam with low residual stress. Ceram. Intern. 41: 145–153.

Umanskyi, O., O. Poliarus, M. Ukrainets and M. Antonov. 2015. Physical-chemical interaction in $NiAl$-MeB_2 systems intended for tribological applications. Welding Journal 94(July): 225–230.

Valenza, F., M.L. Muolo and A. Passerone. 2009. Wetting and interactions of Ni- and Co-based superalloys with different ceramic materials. J. Mater. Sci. 45: 2071–2079.

Valenza, F., C. Artini, A. Passerone and M.L. Muolo. 2012a. ZrB_2–SiC/Ti6Al4V joints: wettability studies using Ag- and Cu-based braze alloys. J. Mater. Sci. 47: 8439–8449.

Valenza, F., M.L. Muolo, A. Passerone, G. Cacciamani and C. Artini. 2012b. Control of interfacial reactivity between ZrB_2 and Ni-based brazing alloys. J. Mater. Eng. Perf. 21: 660–666.

Valenza, F., C. Artini, A. Passerone, P. Cirillo and M.L. Muolo. 2014. Joining of ZrB_2 Ceramics to Ti6Al4V by Ni-based interlayers. J. Mater. Eng. Perf. 23: 1555–1560.

Voytovych, R., A. Koltsov, F. Hodaj and N. Eustathopoulos. 2007. Reactive vs. non-reactive wetting of ZrB_2 by azeotropic Au–Ni. Acta Mater. 55: 6316–6321.

Watts, J., G. Hilmas and W.G. Fahrenholtz. 2011. Mechanical characterization of ZrB_2-SiC composites with varying SiC particle sizes. J. Am. Cer. Soc. 94: 4410–4418.

Weirauch, D.A., W.J. Krafick, G. Ackart and P.D. Ownby. 2005. The wettability of titanium diboride by molten aluminum drops. J. Mater. Sci. 40: 2301–2306.

Wuchina, E., E. Opila, M. Opeka, W. Fahrenholtz and I. Talmy. 2007. UHTCs: Ultra-high temperature ceramic materials for extreme environment applications. Electrochem. Soc. Interf. Winter 2007: 30–36.

Xi, I., I. Kaban, R. Nowak, B. Korpała, G. Bruzda, N. Sobczak, N. Mattern and J. Eckert. 2015. High-temperature wetting and interfacial interaction between liquid Al and TiB_2 ceramic. J. Mater. Sci. 50: 2682–2690.

Xie, C., M. Chen, X. Wei, M. Ge and W. Zhang. 2012. Synthesis and microstructure of zirconium diboride formed from polymeric precursor pyrolysis. J. Am. Cer. Soc. 95: 866–869.

Yang, W., T. Lin, P. He, Y. Huang and J. Feng. 2012. Microstructural investigation of *in situ* TiB whiskers array reinforced ZrB_2-SiC joint. J. All. Comp. 527: 117–121.

Yang, W., T. Lin, P. He, M. Zhu, C. Song, D. Jia and J. Feng. 2013a. Microstructural evolution and growth behavior of *In situ* TiB whisker array in ZrB_2-SiC/Ti6Al4V brazing joints. J. Am. Cer. Soc. 96: 3712–3719.

Yang, W., P. He, T. Lin, C. Song, R. Li and D. Jia. 2013b. Diffusion bonding of ZrB_2-SiC and Nb using dynamic compressed Ni foam interlayer. Mater. Sci. Eng. A 573: 1–6.

Yang, W., P. He, L. Xing and T. Lin. 2015. Microstructural evolution and mechanical properties of ZrB_2-SiC/Cu/Ti/Ti6Al4V brazing joints. Adv. Eng. Mater. 17: 1556–1561.

Yuan, B. and G.J. Zhang. 2011. Microstructure and shear strength of self-joined ZrB_2 and ZrB_2-SiC with pure Ni. Scripta Mater. 64: 17–20.

Zhunkovskii, G.L., T.M. Evtushok, O.N. Grigor'Ev, V.A. Kotenko and P.V. Mazur. 2011. Activated sintering of refractory borides. Powd. Metal. Met. Ceram. 50: 212–216.

Metal Surfaces in Medicine
Current Knowledge of Properties, Modeling and Biological Response

S. Spriano,* S. Ferraris, C. Balagna and *M. Cazzola*

INTRODUCTION

Metal alloys (Ti, Fe and Co based) are widely used in medicine, mainly in contact with bone in dental, trauma and orthopaedic applications, such as dental implants, prostheses (hip, knee, shoulder, elbow, wrist, ankle), fracture fixation (nails, plates, intramedullary nails) and intervertebral discs for spine surgery. Moreover, Nitinol based on NiTi intermetallic compound is used in several cardiovascular devices.

Metal implants must meet mechanical, biological and surgical requirements. As first, adequate mechanical strength, under static or cyclic loading (fatigue), is requested and it currently is an almost met demand (0.5% of people experiencing hip prosthesis breakage). Concerning the surface properties, stable anchorage in the bone (osseointegration), high corrosion resistance and minimization of the friction forces and wear on the articular surfaces are the main needs. Solutions must avoid introducing further complications due to, for instance, excessive surface roughness, stress concentration factors or material changing due to added thermal

Politecnico di Torino, Corso Duca degli Abruzzi, 24 10129 Torino, Italy.
* Corresponding author: silvia.spriano@polito.it

treatments. The biological requirements are related to biocompatibility of the materials and eventual wear debris, bioactive behaviour, restitution of motility, ensuring long-term stability.

Current Knowledge of Properties

Osseointegration Ability

Rapid bone remodeling and osseointegration of a dental or orthopaedic implant are needed, in order to avoid micro-motion and to allow rapid physiological healing and functional recovery, mainly in the case of dental implants with immediate loading. A bad bone quality and/or quantity require specific stimulation for a fast new bone generation, the problem is of interest in the dental implant field in particular in patients with a severely resorbed ridge, but it is important also for the orthopaedic implants (Bauer et al. 2013; Geetha et al. 2009; Dohan Ehrenfest et al. 2010). Fast healing and new bone formation are due to the cooperative action of several phenomena: bioactivity of the surface and apatite precipitation on it (mineralization), protein adsorption, fast adhesion and proliferation of the osteoblastic cells, a high degree of cell differentiation, as well as the absence of infections. The problems of slow osseointegration and fast bacteria proliferation are interconnected because rapid osseointegration helps to low the infection risk, in fact, a sort of competition between the cells and bacteria ("race for the surface") has been described for the implanted biomaterials (Gristina 1987).

Surface composition, energy, charge and roughness/topography of a material affect cell, adhesion, proliferation and metabolism on it (Boyan et al. 2001).

As far as surface topography is concerned, the dimension of the surface features can affect different aspects of cell behaviour. In particular, the surface macro-topography (features > 100 microns) affects primary fixation and the surface micro-topography (between 1 and 100 microns) influences cell recruitment, adhesion, orientation and morphology and even gene expression. The sub-micron and nano features influence the focal contacts and cytoskeletal arrangement, the orientation of the cells and their communication, as well as adsorption and conformation of the proteins and biomolecules on the surface (Boyan et al. 2001; Hayes and Richards 2010; Hayes et al. 2010). The cellular response to the nanoscale features is currently under investigation, however, it can be inferred that the cells effectively sense nanostructures, because they interact *in vivo* with the Extracellular Matrix proteins (ECM), which are characterized by nanometric collagen fibrils; moreover, both the filopodia and lamellopodia are structures on the nano-scale, as well as the integrins (Anselme et al. 2010). Cellular morphology after adhesion to a substrate strongly depends on the surface roughness and on the cellular type: the osteoblasts (as rough-philic cells) attach better to a rough titanium surface while the fibroblasts and epithelial cells (rough-phobic cells) prefer a smooth one (Boyan et al. 1996). The osteoblasts are well spread on the smooth titanium surfaces (Ra values lower than 0.5 microns or 0.2 microns), while they show a more rounded morphology, with extended pseudopodia, and higher differentiation on the rough ones (Ra values in the 0.2–2 micron or 4–7 micron ranges) (Hayes and Richards 2010; Hayes et al. 2010; Boyan et al. 1999). The strength of adhesion is also affected by the surface roughness; in fact, cell detachment from the substrate is easier for the smooth surfaces while the protein production is higher for the rougher ones. Finally, the cells exhibit an enhanced differentiation and reduced proliferation on micro-textured titanium (Hayes and Richards 2010; Hayes et al. 2010; Zinger et al. 2005).

Furthermore, the cells show more filopodia and a more differentiated phenotype on the surfaces with multi-scale topographies, suggesting a synergistic effect of different roughness on different scales (Zhao et al. 2007). A combination of micro-pits (0.5–1.5 micron) and TiO_2 nano-nodules (200 nm) on the Ti-cp surface (Hori et al. 2010) increases both the osteoblast proliferation and differentiation, overcoming the above-cited situation, that foresees reduced proliferation when differentiation is favoured. An opposite effect has been observed in the case of the fibroblasts.

The chemical characteristics of a surface significantly affect cell behaviour too. Sandblasted and acid etched surfaces (mod-SLA), presenting a low surface contamination, can significantly increase fibronectin adsorption and osteoblast differentiation (Zhao et al. 2007). Treatments able to enhance the formation of hydroxyl groups on the metal surface are of interest in order to stimulate the cell differentiation and adhesion (Feng et al. 2012; Zhao et al. 2005; Schwarz et al. 2009; Lu et al. 2010; Zinelis et al. 2012). *In vitro* studies have pointed out that the specific properties noted for hydroxylated titanium surfaces have a significant influence on cell differentiation and growth factor production. A surface enrichment in calcium and phosphate groups has been observed, after immersion in simulated body fluids of the treated samples. It is widely reported in the scientific literature (Kokubo and Takadama 2006) that this effect is an evidence of the material's ability to stimulate, after implantation, the growth of the bone mineral component (hydroxyapatite). It is of interest to get bioactive behaviour of the surface by introducing a thin and well adherent layer of titanium oxide, avoiding the use of thick coatings of a foreigner material. In fact, it is widely reported in the literature that bioactive apatite coatings led to numerous failures in the dental applications, both because of delamination and bio-resorption (Jarcho 1981; Hench 1991; Kodama et al. 2009; Yamada et al. 1997).

Different coatings and surface treatments have been proposed in order to improve the bone-bonding ability of titanium and its alloys. Mechanical, chemical, physical and even biological surface modifications have been considered in the scientific literature (Le Guehennec et al. 2007; Spriano et al. 2010; Hanawa 2010; Bhola et al. 2011; Morra 2006).

Despite wide research, a fast and physiological bone integration of the metallic surfaces remains a challenge. The main issue is to get a multifunctional surface, able to simultaneously give inorganic, as well as biological signals to the surrounding tissues, in order to induce fast and physiological integration of implants and tissue healing. In this context, the authors deeply investigated the surface modification of commercially pure titanium and Ti6Al4V alloy and developed a surface modification treatment able to produce a multiscale surface topography (micro and nanotexture), high density of hydroxyl groups (Ferraris et al. 2011a) and optionally biological molecules (Ferraris et al. 2011b; Ferraris et al. 2012) or antibacterial agents (e.g., silver) on the titanium surface (Ferraris et al. 2014).

The great importance of surface roughness for bone integration is strongly reflected by the development of numerous typologies of commercial implants (both in the dental and orthopaedic fields) that differ for their surface finishing. Three main classes can be described: smooth (machined), rough and porous surfaces.

Machined surfaces, almost smooth (roughness 0.3–1 μm) represent the first generation of implants and guaranteed a good clinical performance (Ballo et al. 2011; Buser 2001).

Rough surfaces have been developed on the basis of the documented improvement of bone integration for rough titanium (Dohan Ehrenfest et al. 2010; Spriano et al. 2010; Ballo et al. 2011; Buser 2001; Lausmaa 2001; Xiao et al. 2001; Junker et al. 2009; Wennerberg and Albrektsson 2009) with the aim to induce bone on-growth on the implant surface. The main surface modification process currently used (Dohan Ehrenfest et al. 2010; Ballo et al. 2011; Buser 2001; Junker et al. 2009) are: (1) sandblasted surfaces (final roughness of 0.5–2.0 μm) obtained by surface blasting with particles of various sizes (25 μm–250 μm) and materials (Al_2O_3, TiO_2, ZrO_2); (2) acid etched surfaces (final roughness of 0.3–1.0 μm); (3) sandblasted and acid-etched surfaces (SLA, final roughness 1–2 μm), microtextured surfaces obtained by various acid etching performed on sandblasted surfaces; (4) laser modified surfaces with micro and nanostructures; (5) anodized surfaces, porous titanium oxide layers, obtained by electrochemical oxidation in various electrolytes enriched with different ions (e.g., Ca, P); (6) Bioactive coatings of hydroxyapatite, calcium phosphates and even bioactive glasses (Dohan Ehrenfest et al. 2010; Spriano et al. 2010; Ballo et al. 2011; Buser 2001; Lausmaa 2001; Xiao et al. 2001; Junker et al. 2009; Wennerberg and Albrektsson 2009; De Jonge et al. 2008; Rizzi et al. 2004) have been obtained by different techniques (plasma spray, electrochemical techniques, RF magnetron sputtering, electrospray, pulsed laser deposition, Ion Beam Assisted deposition, sol-gel deposition, biomimetic depositions, electrophoretic deposition, micro-arc oxidation, ultrasonic spray pyrolysis, and also the discrete crystalline deposition of calcium phosphate nanocrystals).

Among these types of implant surfaces, SLA became a golden standard for the dental implants. Clinical outcome of the implants with SLA or similar surfaces (Cochran et al. 2002) reports a considerably reduced failure rate if a short term response is monitored. On the other side, it was recently reported (Albrektsson et al. 2014) that a not negligible long-term failure rate still occurs in the dental implants, as well as infections and periimplantitis. Concerning coatings, the main drawback of the coatings is creation of an interface that is potentially critical for delamination, both during the surgical implantation and physiological loading (Lausmaa 2001). Moreover, the crystallinity and stoichiometry of the deposited apatite cannot be strictly controlled; bio-resorption of the amorphous apatite coating can occur, leaving back a wide gap between the implant and the bone and it can promote the mobilization of the implant. Lastly, bacteria are more likely to grow onto or next to the hydroxyapatite coated implants than on the titanium implants (Vogely 2000). Concerning micro rough surfaces a deep investigation of inflammation response is needed.

Porous surfaces for bone in-growth can be obtained by several techniques presenting different features. The main characteristics affecting their performance are not so related to the pore shape, but mainly to the size of the interconnecting pores. Pore sizes between 100 and 400 microns are necessary in order to optimise mineralised bone in-growth. An increasing tendency for the formation of the fibrous tissue seems to appear when the pore size is increased beyond 1 mm. The porosity of most implants is usually determined to compromise between maintaining the mechanical strength of the implant while still providing adequate pore size for tissue in-growth.

Traditional techniques for the porous surfaces are sintering (sintered beads, cancellous structured, fiber meshed) and plasma spray (TPS) (Amigò et al. 2003; Smith and Nepew website; Dall'Oca et al. 2004; Mayman et al. 2007; Biswal and Brighton 2010; Zimmer website; Costa santos et al. 2006). It is known from clinical practice that they have some disadvantages, mainly due to a lower fatigue resistance. Moreover, some substrate microstructural changes are possible, such as precipitation of carbides (Co alloys) or enlargement of the grains during sintering. Furthermore, there is a weak mechanical link between the substrate and the coating and not infrequently it displaces. In this case, some particles can become lodged in the articulating parts and they act as a third abrasive body. Alternative techniques are investment casting, welded meshes (mesh structured), metallic foams, rapid manufacturing or micro-fabrication.

Despite the huge research and clinical experience, it is not well known what is crucial for a long-term response of the bone to a foreign body. A high level of marginal bone contact, at short implantation time, was used for a while as the main parameter, in order to select the best implant surface. Actually, it is not known the ideal percentage of bone to implant contact at short times, in order to get an effective long-term integration. A stronger initial bone response could be coupled to a stronger foreign body reaction and inflammatory response.

Soft Tissues

Contact between the soft tissues and the metal implants is one of the key current matters concerning the dental implants, because a stable gum seal around the implant collar is needed in order to avoid peri-implantitis (bacteria infiltration) and down-growing of the junctional epithelium: a sealing gum tissue around the implant can prevent series of events often leading to bone resorption and the removal of the implant.

Currently, collars are usually as smooth as possible in order to avoid biofilm formation, but different engineered surfaces are under investigation. In fact, in the presence of a dental implant (without a specific three-dimensional geometry), the fibres of the connective tissue are oriented parallel to the implant; the consequence is unwanted down migration of the junctional epithelium and the loss of the crestal bone. Moreover, some areas of loose connective tissue (1–3 micron wide) are often developed around a dental implant, with a poor fluid recirculation and difficulties for the macrophages to reach the tissues.

The introduction of surface topographies with a predominant direction, such as grooves, both at the microscale (depth/width of the grooves 1–90 micron) and nanoscale (depth/width less than 1 micron) can be used in order to guide cell-surface interaction. Contact guidance is the response of many types of cells to these topographies: change of their adhesion and proliferation rates, and/or by alignment, elongation and movement toward the main axis of the grooves (Biela et al. 2009). A possible explanation is a mechanical stress induced by the surface topography on the cells, which causes rupture and formation of new cellular fibrous components, affecting cell spreading and causing alignment.

Furthermore, the surface tension can reorganize the extracellular matrix (ECM) into structures which, in turn, affect behaviour of the individual cells (Walboomers et al. 2000; Schupbach and Glauser 2007; Yoshinari et al. 2003). The surface of the experimental collars consists of alternating grooves and ridges and a balance between cell contact guidance and no increased bacterial colonization of the surface must be achieved (Spriano and Ferraris 2014).

As reported in the previous paragraph, fibroblasts and epithelial cells are more sensitive to phenomena of contact guidance than osteoblasts, but it is reported that microgrooves with Ra exceeding 0.35 micron allow osteoblasts to earlier form focal points (Eisenbarth et al. 2002). Moreover, there are even some differences related to the shape of the grooves and their scale (Eisenbarth et al. 2002; den Braber et al. 1996). The main aspect is the rate and quality of the formation of the focal adhesions: it has been reported that those formed along the grooves are more stable than those formed in the perpendicular direction (Anselme et al. 2010).

It must be remembered that a micro or nano-topography also induces changes in surface chemistry and energy because of oxidation or contaminations. For example, regular micro-structured surfaces show an increase in hydrophobicity and this phenomenon is even more pronounced on structures at the nanoscale.

The dental implants are an interesting model in order to investigate the clinical effects of the contact guidance strategy because they involve contact of the implant with both soft (gum) and hard (bone) tissues; also, dental screws are prone to bacteria contamination.

In order to obtain micro- and nanopatterns on the titanium surfaces and implants, various techniques have been proposed (Lu and Leng 2005; Liu and Jay Webster 2007; Etheridge et al. 2013; Madore et al. 1999): Photolithography (such as LiGA), Ion Milling, Chemical Etching, Micromachining, Jet Electrochemical Machining (JET-ECM), Focused Ion Beam (FIB), Electron Beam Lithography, Laser Structuring with pulsed lasers (fs-Laser; Excimer laser), Plasma Etching, Ion etching, Printing. These techniques allow whoever to obtain grooves and ridges with different dimensions, but also with different surface roughness within the pattern, different geometry of the edges and chemical alterations of the surface (oxidation, contamination/ion implantation). The processes resulting in rounded edges, low slope walls, flat and smooth bottom of the micro or nano grooves and low surface contamination are preferred.

The type of cell used for the *in vitro* tests is an important variable to be considered: cancer cell lines are usually faster in their response than primary cells; fibroblasts are usually able to move faster within the microgrooves and only in the direction of the main axis of the microgroove, while epithelial cells show slower movements, but in multiple directions, even in the cross direction.

Numerous studies have focused on microgrooves with dimensions greater than or equal to 10 microns: these works are mainly focused on behaviour of fibroblasts, keratinocytes and epithelial cells, but it must be kept in mind that such high roughness could be critical for bacterial adhesion in the case of a real dental implant.

If the ridges are less than 2 microns large, the cells within the different microgrooves easily create cross-links between themselves, while, if the ridges are larger, the cells remain isolated for longer culture times (Brunette 2001). This point is particularly important considering behaviour of a whole tissue instead of that of a single cell.

Nevins et al. (Nevins 2012) evidenced the relevance of microgrooves, in fact, micro-channels (width/depth 8–12 micron), made by laser on the abutment, alter the behaviour of the epithelial cells and fibroblasts, inducing an orientation of the connective tissue which became an anatomical barrier to down migration of the junctional epithelium. The micro-channels at the abutment-implant connection allow the growth of the supra-crestal connective tissue, oriented perpendicularly to the surface of the implant. This result reduces the negative influence of the micro-gap usually present at the abutment-implant interface and it prevents down migration of the junctional epithelium. The result was a minimum loss of the crestal bone and a layer of soft tissue 3 mm wide is created as in natural teeth (Nevins 2012).

Concerning morphological response of fibroblasts to a nanoscale pattern (Loesberg et al. 2007), two threshold values, below which no contact guidance occurs, were identified: 35 nm for the depth of the grooves and 100 nm for the width. The explanation of this could be that some tens of nanometers are the magnitude of the diameter of the collagen fibrils: the decisive factor is the topographic feature of the native environment of the cells.

In addition, a comparison among the endothelial and smooth muscle cells showed that the fibroblasts are the most sensitive to nanoscale surface topography.

The importance of the geometric shape of the nano features implemented on the surfaces must be also highlighted. Features that develop mainly in height, as holes or pillar (the tested range is around 300 nm high), reduce considerably adhesion of the fibroblasts, as these cells are essentially rough-phobic (Bauer et al. 2013) while surface features with a major axis parallel to the surface, such as nanogrooves and nanofibers arranged in an orderly way, increase adhesion and proliferation rates of the fibroblasts.

Wear Resistance

The lifespan of the Total Hip Prosthesis (THP) and Total Knee Prosthesis (TKP) is limited by wear: it is lower in the patients with higher levels of physical activity.

Metal-on-Polymer (MoP) is the most economic and commonly used solution for THP (about 60% of the total number of implanted prostheses in EU), with the longest surgical experience and follow-up (< 40 years), but it presents low durability and need of too frequent revisions (after about 15 years). In the case of TKP, the Metal-on-Polymer solution is the only one currently used. MoP solution has high wear rates (0.1 mm/year) and wear debris can cause aseptic loosening of the prosthesis.

Metal stands for cobalt-chromium-molybdenum (CoCrMo) cast or wrought alloys; the use of stainless steel (316L) is very limited. Ti alloys are widely used for the stem of THP and tibial component of TKP, but they are not used for the femoral head due to low wear resistance and high friction. The excellent mechanical properties, high corrosion and wear resistance permit the use of CoCrMo alloys, but severe constraints due to toxicity of the constituent elements must be considered (Marti 2000; Niinomi 2002). Wear and mechanical properties of the Co-based alloys strongly depend on their crystalline structure (Niinomi 2002): the co-existing face-centred cubic (FCC) and the hexagonal closed packed (HCP) crystalline structures. Typically, the FCC phase is the predominant one at room temperature, but the FCC → HCP transformation can be isothermally or strain-induced (Chiba et al. 2007; Saldivar Garcia et al. 1999; Balagna et al. 2012a). The other main feature of the cobalt-based alloys is the presence of carbon forming carbides whose distribution and size are influenced by the manufacturing process; the carbides act as a protective barrier against the matrix delamination because of their coherency with the surrounding matrix. The main factors that increase wear resistance of the Co-based alloys are high carbon amount, homogeneously distributed carbides and presence of the HCP crystal structure (Yan et al. 2007; Long 2005; Cawley et al. 2003). However, it must be considered that there are threshold values because the HCP crystalline structure is very brittle and abrasive; wear damage can be due to pull-out of the hard carbides fractured from the matrix (Kodama et al. 2009). A thermal treatment performed at 970°C decreases hardness and bending ductility into CoCrMo alloys, even if the tribological properties remain unchanged.

Polymer stands for Ultra High Molecular Weight Polyethylene (UHMWPE). Polyethylene debris is the principal cause of osteolysis (Ingham and Fisher 2005) and implants failure for aseptic loosening because it is produced in the size range of biological activation. Highly crosslinked UHMWPE is recently used in order to overcome this constraint: wear rate is reduced by a factor of 10, but long-term clinical data are still missing and its efficacy in TKP is less encouraging.

Ceramics (Al_2O_3 or Al_2O_3-ZrO_2 composites) are used in Ceramic-on-Polymer (CoP), Ceramic-on-Ceramic (CoC) and Ceramic-on-Metal (CoM) coupling. The use of CoM coupling is very recent and limited, with poor clinical outcome. Ceramic-on-Ceramic CoC coupling has the most favourable tribological properties (0.003 mm/year), but it is expensive and not suitable for some patients (high elastic modulus). The risk of catastrophic fragile fracture is unavoidable, even if the clinical occurrence is currently low (0.004–0.02%).

During the last fifteen years, particular attention was paid to Metal on Metal (MoM) coupling for arthroprosthesis, because of the lower volumetric wear compared to the MoP coupling, as well as higher durability than the CoC coupling (Shetty and Villar 2006; Cobb and Schmalzreid 2006). Concerning MoM implants, severe issues emerged over the past ten years: pseudotumours, inflammatory lesions in periprosthetic soft tissues, elevated concentrations of Co/Cr ions in blood and subsequent revision surgery. The MoM designs with unusually high short-term revision rates have been removed from the market. All the other MoM designs are under close observation.

Volumetric wear is less in MoM coupling, as a result of the small-sized debris generated; however, it can release up to 500 times more particles than those released in MoP articulations (Cobb and Schmalzreid 2006). Metal debris, especially from CoCrMo alloy, have large surface area and marked tendency to pure and tribologically enhanced corrosion, which have been linked to metal hypersensitivity, local and systemic adverse effects (Ingham and Fisher 2005; Cobb and Schmalzreid 2006; Huber et al. 2009; Granchi et al. 2008). The patients with MoM hip prostheses have blood with higher Co and Cr concentrations than the unexposed individuals; in the case of the well-functioning metal hip prostheses, these concentrations are not usually toxic. However, the patients with poorly functioning metal hip prostheses might have Co and Cr concentrations so elevated, in the joint synovial fluid and blood, that several adverse biological effects can occur (Campbell and Estey 2013). Hence, even if Co and Cr are present in human body tissue, an increase of their concentration can cause hypersensitivity and drastic inflammatory reactions (Cobb and Schmalzreid 2006; Granchi et al. 2008; Granchi et al. 1999). The potential toxic behaviour of Co and Cr ions, such as Co^{2+}, Cr^{3+} and Cr^{6+}, is well recognized, causing chromosome breakage and DNA damage, cell apoptosis and subsequently necrosis (Cobb and Schmalzreid 2006; Bagchi et al. 2001).

The most typical wear mechanisms that can occur in an arthroprosthesis, during the implant's lifetime, are abrasion, fatigue, adhesion and tribochemical reactions (Schmalzried and Callaghan 1999; Buford and Goswami 2004; La Berge 1998; Affatato et al. 2008). Abrasion occurs when a harder surface (ceramic or metal) rubs against a softer one (UHMWPE), resulting in removal and loss of the latter. Wear fatigue involves failure, occurring after a number of loading cycles when the local surface stresses exceed the fatigue resistance of the material. Adhesion arises from the chemical bonding between two surfaces, when they are pressed together under load, with the transfer of material from a surface to the other one. Tribochemical reactions occur as an interaction between the surfaces and environment, with a chemical reaction, such as oxidation in the presence of air (Schmalzried and Callaghan 1999; La Berge 1998).

UHMWPE in a Metal-on-Polymer (MoP) is predominantly subject to adhesion and abrasion, with the contribution of pitting, delamination and fatigue. Moreover, the roughness increase of the femoral counter-face accelerates abrasive wear of polyethylene. An increment of roughness from 0.01 µm to 0.1 µm results in approximately a 50 fold increase in wear rate (Wang et al. 1998).

MoM joints are mainly subject to abrasion, fatigue and tribochemical reactions. As in MoP, abrasion is the most relevant mechanism. The presence of a stable native oxide Cr_2O_3 layer film (passivation film), formed on the metal surface, avoids the direct contact between the two metallic alloys (Wimmer et al. 1998); however, this native oxide layer is not permanent, since it can be scratched and removed during the surfaces movement, facilitating corrosion. The kinetic of the *in situ* formation of a new oxide layer, during the wear mechanism itself, is a key parameter to keep in mind to avoid an extensive corrosion of the joint (Mathew et al. 2014).

In the natural human joints, the direct contact between the bearing surfaces is avoided by the synovial fluid. However, this liquid is corrosive for implanted materials, enhancing the risk of degradation and adhesion between surfaces. The lubrication regime (boundary, mixed or fluid film) is not completely known in the artificial prostheses, even if it is supposed to be a mixed regime in the MoM and MoP couplings (Ward et al. 2010). *In vitro* tribological tests are usually performed by using lubricants, containing different quantities of proteins and hyaluronic acid, simulating the *in vivo* environment effects on lubrication. The serum proteins act as a solid lubricant system, affecting the wear mechanism in different ways, mainly depending on the bearing surface materials (Wang et al. 1998; Gispert et al. 2006). It is reported that the friction coefficient ranges from 0.054 up to values of 0.14 when the conditions change from lubricated (bovine serum) to dry, for the tribological pairing Al_2O_3/UHMWPE

(Damm et al. 2013). The same authors report that the peak values of the coefficient of friction were always determined during the flexion phase, in an *in vivo* study on hip joint prostheses; the synovia is squeezed out of the intra-articular joint space and the lubrication regime at this step certainly changed into a dry phase.

After the dramatic failure of MoM hip prostheses, there is a high medical need for alternatives to Co alloys. Solutions and improvements in design, manufacturing processes and surface treatment of the artificial joints are under investigation, in order to minimize the material wear. Multilayer (TiN/CrN) coatings, as well as diamond-like carbon (DLC) coatings were investigated, but they resulted in poor clinical outcome despite very low friction coefficient when compared to high carbon cobalt alloys (Bursuc et al. 2013; Ortega-Saenz et al. 2013).

Silicon nitride has recently been investigated for application and introduced as a biomaterial, both as bulk material and coating, because of its excellent biocompatibility, low wear rate, relatively high fracture toughness and strength (Petterson et al. 2013). Moreover, silicon nitride particles dissolve in aqueous media and this, in turn, suggests that wear particles can dissolve *in vivo*, which may reduce the negative body response to the debris. Further research is needed for determining behaviour of these solutions under the complex lubrication regime in the total hip and knee prostheses.

Considering the excellent properties of tantalum in terms of corrosion resistance (Robin and Rosa 2000), tribological behaviour and biocompatibility (Stiehler et al. 2007), Ta-based coatings of pure metal or compounds as nitrides, oxides, carbides could be also deposited through different techniques on several surfaces (Rahmati et al. 2016; Vuong-Hung Pham et al. 2013; Balagna et al. 2011). A well adherent multilayer tantalum carbides coating is deposited on CoCrMo alloy by means of thermal treatment in molten salts (Spriano and Bugliosi 2006; Spriano et al. 2005) with good results in terms of wear resistance improvement and hardness increase (Balagna et al. 2012b; Balagna et al. 2014). However, the use of Ta as a biomaterial is extremely limited because of relatively high cost and difficulties of fabrication.

Hemocompatibility and Cardiovascular Devices

The nickel-titanium (Ni-Ti) alloys, especially the near-equiatomic ones, are characterized by excellent mechanical and functional properties, such as shape memory effect (SME), super-elasticity (SE), low elastic modulus, high corrosion resistance and biocompatibility (Duerig et al. 1999). Pseudoelasticity refers to the property of the Ni-Ti alloy to recover its shape from deformation much larger than conventional elastic deformation in metallic materials, in an isothermal condition. In general, the elastic strain of a metallic material is around 0.2%, and higher loading will result in plastic deformation, due to slip or twin boundary motion. On the other hand, the elastic strain could be up to 10% in the case of the pseudoelastic Ni-Ti alloys. The near-equiatomic Ni-Ti alloys are the only pseudoelastic materials nowadays available for industrial applications. The pseudoelastic behaviour can be obtained at human body temperature, making the material an optimal choice for a self-expanding stent and other cardiovascular devices (Duerig et al. 1999; Duerig et al. 2000); a limited use of NiTi is made as orthopaedic implants and orthodontics wires.

Ni is considered a hazardous and toxic element for human health, but Ni-Ti alloys are characterized by good biocompatibility. In fact, a dense passive layer of Ti_2O continuously forms on the surface of Ni-Ti alloy and a homogeneous surface; if the surface area is limited, roughness is low and in the absence of second phases like Ni-rich particles or coarse carbides and nitrides, the surface oxide layer provides a very high corrosion resistance. These features limit and minimize the Ni ion release, allowing the large use of Nitinol in biomedical field (Khalil-Allafi et al. 2010). The Nitinol and metal devices to be put in contact with the blood are generally coated by a ceramic or polymer layer because the metal surface is not enough hemocompatible. The surface in contact with the blood must be highly hydrophilic, with low friction coefficient and a slightly negative surface charge (in order to not attract platelets and not denature proteins).

Several forms of Nitinol such as wire, tube, ribbon, sheet and bar are currently available into the market, produced by means of conventional melting and sintering processes, usually followed by hot and cold working and by heat treatments to achieve suitable final properties (Elahinia et al. 2012). The Nitinol stents usually derive from wire, sheet and tube-based design (Elahinia et al. 2012; Russell 2000). The third one is the most widely used in the commercial devices and it is produced by laser cutting of the desired patterns on Ni-Ti seamless tubes, with diameter from 1 to 10 mm and thin wall thickness (0.2–0.6 mm). However, the use of Ni-Ti alloys is limited by unstable properties because of composition fluctuation and processing conditions, poor workability and machinability, and inclusions of oxides or carbide particles which often cause non-compliance of the products during quality control.

Other ways to produce the NiTi alloy as semi-finished and near-net-shape final components have been evaluated in the literature in order to reduce grain growth, to control the microstructure, limiting inclusions of carbides, and to minimize the final machining procedures. A rapidly solidified NiTi alloy can be obtained with an amorphous or nanocrystalline structure, by the melt spinning technique or planar flow casting (Mehrabi et al. 2012). The limit of this method is the realization of very thin ribbons or strips, not suitable for applications under high load and stress, because of the brittle nature of the products. Anyway, a study in the literature reported the production of conical shaped samples by means of casting into the water-cooled copper mould in a lower regime of the rapid solidification with respect to the melt-spinning, but achieving higher cooling rates than that in conventional casting (Pan et al. 2014a; Pan et al. 2014b). Interesting results are obtained in terms of thermal behaviour and microstructure, especially in the quicker solidified portion of the samples. Further studies are needed in order to move from lab casting to a process on the industrial scale.

Powder metallurgy is another potential route to realizing products with lower cost and energy requirements, shorter time and higher production rate, with elevated reproducibility (Elahinia et al. 2012). The poor sintering ability of NiTi powders due to TiC and oxide contamination on the surface and low free surface energy of the powder particles is recently bypassed using an innovative powder metallurgical process called Electro-Sinter-Forging (ESF) (Balagna et al. 2016). This technique is based on the powder sintering in a very limited time (less than 1s), combining a single short impulse of intense electric current with a mechanical pulse in a properly designed mould (Fais 2010). The highest density samples obtained by ESF result austenitic at room temperature with a low amount of micro and nano-sized Ti-rich precipitates, high hardness. Low Ni ions release and verified biocompatibility towards fibroblasts make the material suitable for biomedical applications. Anyway, further studies in terms of thermal treatment and mechanical tests are needed.

Risk of Infection

Bacterial contamination can be defined as almost ubiquitous, but when it interests the surfaces for medical application and in particular the implantable surfaces, it becomes critical.

Infections represent one of the main complications related to prostheses and often lead to prolonged antibiotic therapies and revision surgery, with consequent increase of hospitalization time and costs.

Prosthetic infections interest 1% of total joint replacement and this percentage increases in the case of revision surgery (Moran et al. 2010; Papagelopoulos et al. 2006; Campoccia et al. 2006).

An incidence up to 2.5% for primary hip and knee arthroplasties and to 10% for revisions has recently been reported (Romanò et al. 2015).

The prosthetic infections interest 1% of the hip and shoulder prostheses, 2% of the knee prostheses, 9% of the elbow prostheses and there is an increase of up to 40% in case of revision. The most responsible bacteria of these infections are *Staphylococcus aureus* and Coagulase-negative staphylococci (Trampuz and Zimmerli 2005).

The increasing development of bacterial resistance to antibiotic, which has been recently defined as "global threat" (Stephens 2014) makes the problem even more serious and highlight the need for developing effective countermeasures.

Numerous solutions have been proposed in the scientific literature, and recently reviewed, in order to contrast the problem of bacterial contamination of the medical implants (Ferraris and Spriano 2016), but the optimal solution is still far from the clinical application.

Several strategies can be applied for the reduction of bacterial contamination. A first distinction must be made between the passive surfaces, which obstacle bacterial adhesion without exerting any antibacterial action, and the active surfaces that contrast bacteria by the release of antibacterial substances. The first class includes mainly anti-adhesive surfaces that can be obtained by topographical strategies (e.g., mirror polishing) or physical/chemical ones such as UV-irradiation of titanium oxide and polymeric coatings (e.g., polyethylene glycol). On the other hand, the active surfaces present an "active element" that can be released in the physiological fluids in order to perform an antibacterial action. Organic substances (antibiotic or antibacterial compounds), as well as inorganic ones (e.g., metallic ions as silver copper or zinc), can be considered as antibacterial agents for surface doping. Active substances can be directly introduced on the metallic surface or incorporated into a coating or surface oxide layer.

In the specific case of the medical implants, there is a sort of "race for the surface" between tissue cells and bacteria after implantation (Gristina 1987); an ideal surface should improve cellular adhesion and reduce the bacterial one.

Considering this point, a further distinction should be made considering whether the final surface is only antibacterial or also able to bond to tissue (bioactive). In order to obtain antibacterial and bioactive surfaces, antibacterial agents have been incorporated in some surface coatings/modified surface layers able to induce bioactive behaviour (e.g., hydroxyapatite coatings or bioactive titanium oxide coatings).

The starting point of the prosthetic infections is bacterial contamination of the implanted surfaces through a process of bacterial adhesion, aggregation and biofilm formation (Arciola et al. 2012). The bacterial biofilm can be defined as a structured community of bacteria protected in a self-produced protein matrix. It is extremely difficult to remove the biofilm from the synthetic surfaces and, in addition, bacteria within the biofilm result protected from the conventional systemic therapies often leading to the need for the implant removal.

As regards the oral environment, the bacteria form specific combinations of biofilm, known as plaque (Watnick and Kolter 2000). Bacterial colonization of dental implants can lead to peri-implantitis, infection characterized by inflammatory phenomena localized in the tissues surrounding the implant leading to bone resorption, which represent a common cause of failure of the oral implants. The bacteria that cause peri-implantitis are the same ones responsible for periodontitis (*Prevotella intermedia, Porfiromans gengivalis, Actinobacillus actinomicetemcomitans, Prevotella nigrescens, Treponema denticola, Bacteriodes forsythus* (Watnick and Kolter 2000)). The inflammatory processes start from the gum bonded to the implant and then penetrate to the bone, causing bone resorption with consequent implant mobility and failure.

The above cited subgingival plaque, formed in the gingival sulcus and responsible for various types of periodontal disease, is strongly dependent on the surface features of the dental implant and in particular on surface roughness. Since the bacterial adhesion is characterized by an initial phase of weak and reversible bond to the surface, the role of roughness can be explained by preferential reversible binding of the bacteria on negative irregularities (cavities, holes, pits, grooves with a depth > 0.4 micron), where they are protected from the mechanical stress of removal (Quirynen et al. 1993). A rough surface exposes a greater surface, easily accessible for bacteria, and it also reduces the shear stresses acting on the bacteria, causing stripping during the first stage of adhesion. An increase in bacteria contamination on different metallic and ceramic materials (titanium and zirconia) with a roughness > 0.3 microns has been reported (Al-Ahmad et al. 2013). A threshold value of roughness (Ra) below which a change in surface roughness does not result in a change in the amount of accumulated plaque has been individuated as 0.2 microns (Feng et al. 2012). This value can be explained considering the vertical dimension of bacteria (approximately 0.5–5 micron), so they are never protected by the mechanical forces of removal if Ra is less than 0.2 microns.

Metals as Substrates for Functionalized Surfaces

Biological functionalization allows scientists to greatly expand the action of the implant surfaces by the delivery of specific signals directly from the surface to the biological environment and to obtain fully multi-functional surfaces. A deliberate interaction with the tissues can be obtained by providing biological cues that trigger specific responses (Morra 2006). The biomolecules can be coupled to the artificial materials mainly in three ways: adsorption, covalent grafting and inclusion into a resorbable carrier (Hildebrand et al. 2006). Various biomolecules can be grafted to the material surface, proteins from the extracellular matrix (ECM), growth factors and drugs can be cited as typical examples in order to mimic the physiological environment, to improve cellular adhesion and to perform localized therapy (Morra 2006; Bauer et al. 2013; Lee and Shin 2007). Moreover, natural molecules (pure or as a mixture of natural extracts) are attracting increasing interest in the scientific community (Varoni et al. 2012; Ahmad and Viljoen 2015). Over the past two decades, the osteotropic biomolecules (such as the growth factors) are increasingly considered as a source of inspiration for the bio-inspired design of the active implant surfaces (Morra 2006). They have been shown to stimulate the formation of bone tissue around the implants with promising results in enhancing bone formation. Overdoses, as well as systemic therapies, must be avoided, that is why their grafting on the surface, in a small and controlled amount, is of great interest, in order to get a limited local action. Various strategies have been reported in the scientific literature in order to graft biologically active molecules onto the metallic substrates (especially titanium and its alloys), such as silanization, self-assembled monolayers (SAMs) and surface activation with organic chlorides can be cited (Spriano et al. 2010; Hanawa 2010; Bhola et al. 2011; Sprianoo et al. 2015; Ferraris et al. 2016). At the same time, numerous molecules have been considered for the functionalization such as RGD-containing peptides, fibronectin, collagen, alkaline phosphatase, Bone Morphogenetic Proteins (BMPs) and hyaluronan.

The need to have a stable grafting and to promote a gradual release of the grafted agent is still an open debate (Kashiwagi et al. 2009). Few prosthetic implants functionalized with biomolecules have been patented, mainly produced using electrolytic techniques (Ellingsen and Lyngstadaas 2007; Bhatnagar 2006). Despite the variety of solutions proposed and recently reviewed (Hanawa 2010; Bagno et al. 2007; Nanci et al. 1998; Hayakawa et al. 2004; Morra et al. 2003; Puleo et al. 2002), the bio-functionalized metallic surfaces are still far from the market. The main problems that hamper the clinical and commercial diffusion of the biologically functionalized implants can be connected with the production and classification processes. In particular, sterilization and storage of the materials carrying biological molecules are critical passages that can alter the functionalities of the grafted biomolecules (Ferraris et al. 2012). Moreover, an implant with a biologically active molecule grafted on it belongs to class III, increasing the CE marking time and costs compared to the traditional metallic implants. The change in classification with a consequent increase in complexity and costs is particularly critical for the dental implants.

Modeling and Biological Response

Modeling and mathematical simulations of the subsequent stages in the interaction of a metal (titanium) implant with the biological environment where tempted, starting from water and protein adsorption, up to cell adhesion and proliferation (Bousnaki and Koidis 2014).

The first question that has to be answered is whether water adsorbs molecularly or dissociatively on titanium; the presence of an adsorbed carbon layer, as well as different thickness and crystalline structures of titanium oxide on the surface, have an influence on this stage. Results can be summarised as follows: bridging oxygen vacancies in the oxide layer result in water dissociation, topographical negative patterns result in water molecularly adsorption, oxygen molecules reduction results in water molecules clustering. Moreover, Ca ions are attracted by surfaces that have a negative charge and oxygen atoms and thus apatite formation in induced (bioactivity).

Concerning protein adsorption, the three-dimensional structure of proteins is determined by intramolecular interactions and intermolecular interactions of the protein with water molecules (Bousnaki and Koidis 2014).

Simulation models on fibronectin adsorption on Ti model indicate that proteins interact with the surface by the carbonyl anions interactions with the assistance of hydrogen bonds from side chains and that the presence of surface defects increases surface energy and promotes a more stable adsorption (Raffaini and Ganazzoli 2012). Interestingly, a too strong interaction may cause conformational deformation resulting in loss of protein bioactivity. Water density alterations, as a result of surface nano-features, have a direct influence on peptide's adsorption geometry, indicating a strong association between water nanostructure and adhesion forces on the Ti surface. In conclusion, surfaces that lead to increased hydration levels enhance protein layer formation and hydrophilic surfaces can potentially result in a greater bone to implant contact. Nano-topographical surface alterations, such as grooves and step edges (less than 50 nm), increase surface energy and enable a more stable protein orientation.

Concerning cell attachment, it has been proposed a mathematical model to predict the strength of cell adhesion to a nano-structured substrate. According to this model, three regimes were identified: for low surface energy (hydrophobic surfaces) each increase in roughness is harmful to cell adhesion; for high surface energies (highly hydrophilic surfaces) there is an optimal roughness that maximizes the adhesion; for surface energies in between the roughness has less influence on cell adhesion (Anselme et al. 2010).

Keywords: Osseointegration, Soft tissue adhesion, Wear, Hemocompatibility, Infections, Functionalization, Modeling

References

Affatato, S., M. Spinelli, M. Zavalloni, C. Mazzega-Fabbro and M. Viceconti. 2008. Tribology and total hip joint replacement: Current concepts in mechanical simulation. Med. Eng. Phys. 30: 1305–1317.

Ahmad, A. and A. Viljoen. 2015. The *in vitro* antimicrobial activity of Cymbopogon essential oil (lemongrass) and its interaction with silver ions. Phytomedicine 22: 657–665.

Al-Ahmad, A., M. Wiedmann-Al-Ahmad, A. Fackler, M. Follo, E. Hellwig, C. Hannig et al. 2013. *In vivo* study of the initial basterial adhesion on different implant materials. Archives of Oral Biology 58: 1139–1147.

Albrektsson, T., C. Dahlin, T. Jemt, L. Sennerby, A. Turri and A. Wennerberg. 2014. Is marginal bone loss around oral implants the result of a provoked foreign body reaction? Clin. Impl. Dent Related Res. 16: 155–165.

Amigó, V., M.D. Salvador, F. Romero, C. Solves and J.F. Moreno. 2003. Microstructural evolution of Ti–6Al–4V during the sintering of microspheres of Ti for orthopedic implants. J. Mat. Proc. Tech. 141: 117–122.

Anselme, K., P. Davidson, A.M. Popa, M. Giazzon, M. Liley and L. Ploux. 2010. The interaction of cells and bacteria with surfaces structured at the nanometre scale. Acta Biomat. 6: 3824–3846.

Arciola, C.R., D. Campoccia, P. Speziale, L. Montanaro and J.W. Costerton. 2012. Biofilm formation in Staphylococcus implant infections. A review of molecular mechanisms and implications for biofilm resistant materials. Biomat. 33: 5967–5982.

Bagchi, D., M. Bagchi and S.J. Stohs. 2001. Chromium (VI)-induced oxidative stress, apoptotic cell death and modulation of p53 tumor suppressor gene. Molecular and Cellular Biochemistry 222: 149–158.

Bagno, A., A. Piovan, M. Dettin, A. Chiarion, P. Brun, R. Gambaretto et al. 2007. Human hosteoblas-like cell adhesion on titanium substrates covalently functionalized with synthetic peptides. Bone 40: 693–699.

Balagna, C., S. Spriano and M.G. Faga. 2011. Tantalum-based thin film coatings for wear resistant arthroprostheses. J. Nanosci. Nanotech. 11: 1–9.

Balagna, C., S. Spriano and M.G. Faga. 2012a. Characterization of Co–Cr–Mo alloys after a thermal treatment for high wear resistance. Mat. Sci. Eng. C 32: 1868–1877.

Balagna, C., M.G. Faga and S. Spriano. 2012b. Tantalum-based multilayer coating on cobalt alloys in total hip and knee replacement. Mat. Sci. Eng. C 32: 887–895.

Balagna, C., M.G. Faga and S. Spriano. 2014. Tribological behavior of a Ta-based coating on a Co–Cr–Mo alloy. Surf. Coatings Tech. 258: 1159–1170.

Balagna, C., A. Fais, K. Brunelli, L. Peruzzo, M. Horynova, L. Celko et al. 2016. Electro-sinterforged Ni-Ti alloy. Intermetallics 68: 31–41.

Ballo, A.M., O. Omar, W. Xia and A. Palmquist. 2011. Dental implant surfaces—Physicochemical properties, biological performance, and trends. pp. 19–56. *In*: I. Turkyilmaz [ed.]. Implant Dentistry—A Rapidly Evolving, Practice, InTech.

Bauer, S., P. Schmuki, K. von de Mark and J. Park. 2013. Engineering biocompatible implant surfaces: Part I materials and surfaces. Prog. Mater. Sci. 58: 261–326.

Bhatnagar, R.S. Methods for immobilizing molecules on surfaces. Patent WO2006/133027A2.

Bhola, R., F. Su and C.E. Krull. 2011. Functionalization of titanium based metallic biomaterials for implant applications. J. Mater. Sci.: Mater. Med. 22: 1147–1159.

Biela, S.A., Y.S. Joachim, P. Spatz and R. Kemkemer. 2009. Different sensitivity of human endothelial cells, smooth muscle cells and fibroblast to topography in the nano-micro range. Acta Biomat. 5: 2460–2466.

Biswal, S. and R.W. Brighton. 2010. Results of unicompartmental knee arthroplasty with cemented, fixed-bearing prosthesis using minimally invasive surgery. J. Arthroplasty 25: 721–727.

Bousnaki, M. and P. Koidis. 2014. Advances on biomedical titanium surface interactions. J. Biomimetics, Biomaterials & Tissue Engineering 19: 43–64.

Boyan, B.D., T.W. Hummert, D.D. Dean and Z. Schwartz. 1996. Role of material surfaces in regulating bone and cartilage cell response. Biomat. 17: 137–146.

Boyan, B.D., V.L. Sylvia, Y. Liu, R. Sagun, D.L. Cochran, C.H. Lohmann, D.D. Dean and Z. Schwartz. 1999. Surface roughness mediates its effects on osteoblasts via protein kinase A and phospholipase A2. Biomat. 20: 2305–2310.

Boyan, B.D., D.D. Dean, C.H. Lohmann, D.L. Cochran, V.L. Sylvia and Z. Schwartz. 2001. The titanium-bone cell interface *In Vitro*: The role of the surface in promoting osteointegration. pp. 562–585. *In*: D.M. Brunette, P. Tengvall, M. Textor, P. Thomsen [eds.]. Titanium in Medicine. Springer-Verlag, Berlin Heidelberg.

Buford, A. and T. Goswami. 2004. Review of wear mechanisms in hip implants: paper I-general. Mater. Design 25: 385–393.

Bursuc, D.C., L. Capitanu and V. Florescu. 2013. A solution to improving seizure resistence in MOM total hip prostheses with self–directed rolling bodies. Am. J. Mater. Sci. 3: 205–216.

Buser, D. 2001. Titanium for dental applications (II): Implants with roughened surfaces. pp. 875–888. *In*: D.M. Brunette, P. Tengvall, M. Textor and P. Thomsen [eds.]. Titanium in Medicine, Springer–Verlag, Berlin, Heidelberg, New York.

Campbell, J.R. and M.P. Estey. 2013. Metal release from hip prostheses: cobalt and chromium toxicity and the role of the clinical laboratory. Clin. Chem. Lab. Med. 51: 213–220.

Campoccia, D., L. Montanaro and C.R. Arciola. 2006. The significance of infection related to orthopedic devices and issues of antibiotic resistance. Biomat. 27: 2331–2339.

Cawley, J., J.E.P. Metcalf, A.H. Jones, T.J. Band and D.S. Skupien. 2003. A tribological study of cobalt-chromium molybdenum alloys used in metal-on-metal resurfacing hip arthroplasty. Wear 255: 999–1006.

Chiba, A., K. Kumagai, N. Nomura and A.S. Miyakaw. 2007. Pin-on-disk wear behavior in a like-on-like configuration in a biological environment of high carbon cast and low carbon forged Co29Cr6Mo alloys. Acta Mater. 55: 1309–1318.

Cobb, A.G. and T.P. Schmalzreid. 2006. The clinical significance of metal ion release from cobalt-chromium metal-on-metal hip joint arthroplasty. P. I. Mech. Eng. H 220: 385–398.

Cochran, D.L., D. Buser, C.M.T. Bruggenkate, D. Weingart, T.M. Taylor, J.P. Bernard et al. 2002. The use of reduced healing times on ITI® implants with a sandblasted and acid-etched (SLA) surface. Clin. Oral Impl. Res. 13: 144–153.

Costa Santos, E., M. Shiomi, K. Osakada and T. Laoui. 2006. Rapid manufacturing of metal components by laser forming. Int. J. Mach. Tools Man. 46: 1459–1468.

Dall'Oca, C., A. Zambito and R. Aldegheri. 2004. Cancellous screws fixation: preliminary results analysis. G.I.O.T. 30: 73–79.

Damm, P., J. Dymke, R. Ackermann, A. Bender, F. Graichen, A. Halder et al. 2013. Friction in total hip joint prosthesis measured *in vivo* during walking. PLoS One 8: 78373.

De Jonge, L.T., S.C.G. Leeuwenburgh, J.G.C. Wolke and J.A. Jansen. 2008. Organic-inorganic surface modifications for titanium implant surfaces. Pharm. Res. 25: 2357–69.

Den Braber, E.T., J.E. de Ruijter, H.T.J. Smits, L.A. Ginsel, A.F. von Recum and J.A. Jansen. 1996. Quantitative analysis of cell proliferation and orientation on Substrata with uniform parallel surface micro-grooves. Biomat. 17: 1093–1099.

Dohan Ehrenfest, D.M., P.G. Coelho, B.-S. Kang, Y.T. Sul and T. Albrektsson. 2010. Classification of osseointegrated implant surfaces: materials, chemistry and topography. Trends Biotechnol. 28: 198–206.

Duerig, T., A. Pelton and D. Stockel. 1999. An overview of nitinol medical applications. Mat. Sci. Eng. A – Struct. 273: 149–60.

Duerig, T., D.E. Tolomeo and M. Wholey. 2000. An overview of superelastic stent design. Minim. Invasiv. Ther. 9: 235–46.

Eisenbarth, E., P. Linez, V. Biehl, D. Velten, J. Breme and H.F. Hildebrand. 2002. Cell orientation and cytoskeleton organisation on ground titanium surfaces. Biomol. Eng. 19: 233–237.

Elahinia, M.H., M. Hashemi, M. Tabesh and S.B. Bhaduri. 2012. Manufacturing and processing of NiTi implants: A review. Prog. Mater. Sci. 57: 911–946.

Ellingsen, J. and S. Lyngstadaas. 2007. Medical prosthetic devices and implants having improved biocompatibility. Patent US2007/077346.

Etheridge, M.L., S.A. Campbell, A.G. Erdman, C.L. Haynes, S.M. Wolf and J. McCullough. 2013. The big picture on nanomedicine: the state of investigational and approved nanomedicine products. Nanomed-Nanotechnol. 9: 1–14.

Fais, A. 2010. Sintering Process and Corresponding Sintering System, EP 2198993B1.

Feng, B., J. Weng, B.C. Yang, S.X. Qu and X.D. Zhang. 2012. The properties of bioactive TiO2 coatings on Ti-based implants. Surf. Coat Tech. 209: 177–183.

Ferraris, S., S. Spriano, G. Pan, A. Venturello, C.L. Bianchi, R. Chiesa et al. 2011a. Surface modification of Ti-6Al-4V alloy for biomineralization and specific biological response: Part I, inorganic modification. J. Mat. Sci. Mat. Med. 22: 533–545.

Ferraris, S., S. Spriano, C.L. Bianchi, C. Cassinelli and E. Vernè. 2011b. Surface modification of Ti-6Al-4V alloy for biomineralization and specific biological response: Part II, Alkaline phosphatase grafting. J. Mat. Sci. Mat. Med. 22: 1835–1842.

Ferraris, S., G. Pan, C. Cassinelli, L. Mazzucco, E. Vernè and S. Spriano. 2012. Effects of sterilization and storage on the properties of ALP-grafted biomaterials for prosthetic and bone tissue engineering applications. Biomedical Mat. 7: 13.

Ferraris, S., A. Venturello, M. Miola, A. Cochis, L. Rimondini and S. Spriano. 2014. Antibacterial and bioactive nanostructured titanium surfaces for bone integration. Appl. Surf. Sci. 311: 279–291.

Ferraris, S., A. Vitale, E. Bertone, S. Guastella, C. Cassinelli, J. Pan et al. 2016. Multifunctional commercially pure titanium for the improvement of bone integration: multiscale topography, wettability, corrosion resistance and biological functionalization. Mat. Sci. Eng. C 60: 384–393.

Ferraris, S. and S. Spriano. 2016. Antibacterial titanium surfaces for medical implants. Materials Science and Engineering C 61: 965–978.

Geetha, M., A.K. Singh, R. Asokamani and A.K. Gogia. 2009. Ti based biomaterials, the ultimate choice for orthopaedic implants—A review. Prog. Mater. Sci. 54: 397–425.

Gispert, M.P., A.P. Serro, R. Colaco and B. Saramago. 2006. Friction and wear mechanisms in hip prosthesis: comparison of joint materials behaviour in several lubricants. Wear 260: 149–158.

Granchi, D., E. Cenni, D. Tigani, G. Trisolino, N. Baldini and A. Giunti. 2008. Sensitivity to implant materials in patients with total knee arthroplasties. Biomat. 29: 1494–1500.

Granchi, D., G. Ciapetti, S. Stea, L. Savarino, F. Filippini, A. Sudanese et al. 1999. Cytokine release in mononuclear cells of patients with Co-Cr hip prosthesis. Biomat. 20: 1079–1086.

Gristina, A. 1987. Biomaterial-centered infection: microbial adhesion versus tissue integration. Science 237: 1588–1595.

Hanawa, T. 2010. Biofunctionalization of titanium for dental implant. Jpn. Dent. Sci. Rev. 46: 93–101.

Hayakawa, T., M. Nagai, M. Yoshinari, M. Makimura and K. Nemoto. 2004. Cell-adhesive protein immobilization using tresyl chloride-activation technique for the enhancement of initial cell attachment. J. Oral Tissue Engin. 2: 14–24.

Hayes, J.S. and R.G. Richards. 2010. Surfaces to control tissue adhesion for osteosynthesis with metal implants: *in vitro* and *in vivo* studies to bring solutions to the patients. Expert Rev. Med. Devices 7: 131–142.

Hayes, J.S., I.M. Khan, C.W. Archer and R.G. Richards. 2010. The role of surface microtopography in the modulation of osteoblasts differentiation. Eur. Cells Mater. 20: 98–108.

Hench, L. 1991. Bioceramics: from concept to clinic. J. Am. Ceram. Soc. 74: 1487–510.

Hildebrand, H.F., N. Blanchemain, G. Mayer, F. Chai, M. Lefebvre and F. Boschin. 2006. Surface coatings for biological activation and functionalization of medical devices. Surf. Coat. Tech. 200: 6318–6324.

Hori, N., F. Iwasa, T. Ueno, K. Takeuchi, N. Tsukimura, M. Yamada et al. 2010. Selective cell affinity of biomimetic micro-nano-hybrid structured TiO2 overcomes the biological dilemma of osteoblasts. Dent. Mater. 26: 275–287.

Huber, M., G. Reinisch, G. Trettenhahn, K. Zweymuller and F. Lintner. 2009. Presence of corrosion products and hypersensitivity-associated reactions in periprosthetic tissue after aseptic loosening of total hip replacements with metal bearing surfaces. Acta Biomat. 5: 172–180.

Ingham, E. and J. Fisher. 2005. The role of macrophages in osteolysis of total joint replacement. Biomat. 26: 1271–1286.

Jarcho, M. 1981. Calcium phosphate ceramics as hard tissue prosthetics. Clin. Orthop. Relat. Res. 157: 259–78.

Junker, R., A. Dimakis, M. Thoneick and J.A. Jansen. 2009. Effects of implant surface coatings and composition on bone integration: a systematic review. Clin. Oral Impl. Res. 20: 185–206.

Kashiwagi, K., T. Tsuji and K. Shiba. 2009. Directional BMP-2 functionalization of titanium surfaces. Biomat. 30: 1166–1175.

Khalil-Allafi, J., B. Amin-Ahmadi and M. Zare. 2010. Biocompatibility and corrosion behavior of the shape memory NiTi alloy in the physiological environments simulated with body fluids for medical applications. Mat. Sci. Eng. C30: 1112–1117.

Kodama, A., S. Bauer, A. Komatsu, H. Asoh, S. Ono and P. Schmuki. 2009. Bioactivation of titanium surfaces using coatings of TiO2 nanotubes rapidly pre-loaded with synthetic hydroxyapatite. Acta Biomater. 5: 2322–2330.

Kokubo, T. and H. Takadama. 2006. How useful is SBF in predicting *in vivo* bone bioactivity? Biomat. 27: 2907–2915.

La Berge, M. 1998. Wear. pp. 364–405. In: J. Black and G. Hastings [eds.]. Handbook of Biomaterials Properties. Chapman & Hall, London.

Lausmaa, J. 2001. Mechanical, thermal, chemical and electrochemical surface treatment of titanium. pp. 171–230. In: P. Tengvall, M. Textor and P. Thomsen [eds.]. Titanium in Medicine. Springer–Verlag, Berlin, Heidelberg, New York.

Lee, S.H. and H. Shin. 2007. Matrices and scaffolds for delivery of bioactive molecules in bone and tissue engineering. Adv. Drug Del. Rev. 59: 339–359.

Le Guehennec, L., A. Soueidan, P. Layrolle and Y. Amouriq. 2007. Surface treatments of titanium dental implants for rapid osseointegration. Dent. Mater. 23: 844–854.

Liu, H. and T. Jay Webster. 2007. Nanomedicine for implants: A review of studies and necessary experimental tools. Biomat. 28: 354–369.

Loesberg, W.A., J. te Riet, F.C. van Delft, P. Schon, C.G. Figdor, S. Speller et al. 2007. The threshold at which substrate nanogroove dimensions may influence fibroblast alignment and adhesion. Biomat. 28: 3944–3951.

Long, W.T.M.D. 2005. The clinical performance of metal on metal as an articulation surface in total hip replacement. Iowa Orthop. J. 25: 10–16.

Lu, X. and Y. Leng. 2005. Electrochemical micromachining of titanium surfaces for biomedical applications. J. Mat. Proc. Tech. 169: 173–178.

Lu, X., H.P. Zhang, Y. Leng, L. Fang, S. Qu, B. Feng et al. 2010. The effects of hydroxyl groups on Ca adsorption on rutile surfaces: a first-principles study. J. Mater. Sci. Mater. Med. 21: 1–10.

Madore, C., O. Piotrowski and D. Landolt. 1999. Through-mask electrochemical micromachining of titanium. J. Electrochem. Soc. 146: 2526–2532.

Marti, A. 2000. Cobalt-based alloys used in bone surgery. Int. J. Care Injured 31: 18–21.

Mathew, M.T., C. Nagelli, R. Pourzal, A. Fischer, M.P. Laurent, J.J. Jacobs et al. 2014. Tribolayer formation in a metal-on-metal (MoM) hip joint: An electrochemical investigation. J. Mech. Behav. Biomed. Mater. 29: 199–212.

Mayman, D.J., A. González Della Valle, E. Lambert, J. Anderson, T. Wright, B. Nestor et al. 2007. Late fiber metal shedding of the first and second-generation Harris Galante acetabular component. A Report of 5 Cases. J. Arthroplasty 22: 624–629.

Mehrabi, K., M. Bruncko and A.C. Kneissl. 2012. Microstructure, mechanical and functional properties of NiTi-based shape memory ribbons. J. Alloy Compd. 526: 45–52.

Moran, E., I. Byren and B.L. Atkins. 2010. The diagnosis and management of prosthetic joint infections. J. Antimicrobial Chemotherapy 65: 45–54.

Morra, M., C. Cassinelli, G. Cascardo, P. Cahalan, L. Cahalan, M. Fini et al. 2003. Surface engineering of titanium by collagen immobilization. Surface characterization and *in vitro* and *in vivo* studies. Biomat. 24: 4639–4654.

Morra, M. 2006. Biochemical modification of titanium surfaces: peptides and ECM proteins. Eur. Cells Mater. 12: 1–15.

Nanci, A., J.D. Wuest, L. Peru, V. Sharma, S. Zalzal and M.D. McKee. 1998. Chemical modification of titanium surfaces for covalent attachment of biological molecules. J. Biomed. Mater. Res. 40: 324–335.

Nevins, M., M. Carmelo, M.L. Nevins, P. Schupbach and D.M. Kim. 2012. Connective tissue attachment to laser-microgrooved abutments: A human histologic case report. Int. J. Periodontics Rest. Dent. 32: 385–392.

Niinomi, M. 2002. Recent metallic materials for biomedical applications. Metall. Mater. Trans. A 33A: 477–486.

Ortega-Saenz, J.A., M. Alvarez-Vera and M.A.L. Hernandez-Rodriguez. 2013. Biotribological study of multilayer coated metal-on-metal hip prostheses in a hip joint simulator. Wear 301: 234–242.

Pan, G., C. Balagna, L. Martino and S. Spriano. 2014a. Microstructure and transformation temperatures in rapid solidified Ni–Ti alloys. Part I: The effect of cooling rate. J. Alloy Compd. 589: 628–632.

Pan, G., C. Balagna, L. Martino, J. Pan and S. Spriano. 2014b. Microstructure and transformation temperatures in rapid solidified Ni–Ti alloys. Part II: The effect of copper addition. J. Alloy Compd. 589: 633–642.

Pettersson, M., S. Tkachenko, S. Schmidt, T. Berlind, S. Jacobson, L. Hultman et al. 2013. Mechanical and tribological behavior of silicon nitride and silicon carbon nitride coatings for total joint replacements. J. Mech. Behav. Biom. Mater 25: 41–47.

Pham, V.-H., S.-H. Lee, Y. Li, H.-E. Kim, K.-H. Shin and Y.-H. Koh. 2013. Utility of tantalum (Ta) coating to improve surface hardness *in vitro* bioactivity and biocompatibility of Co–Cr. Thin Solid Films 536: 269–274.

Papagelopoulos, P.J., A.A. Partsinevelos, G.S. Themistocleous, A.F. Mavrogenis, D.S. Korres and P.N. Soucacos. 2006. Complications after tibial plateau fracture surgery. Injury 37: 475–484.

Puleo, D.A., R.A. Kissling and M.S. Sheu. 2002. A technique to immobilize bioactive proteins, including bone morphogenetic protein-4(BMP-4), on titanium alloy. Biomat. 23: 2079–2087.

Quirynen, M., H.C. van der Mei, C.M.L. Bollen, A. Schotte, M. Marechal, G.I. Doornbusch et al. 1993. An *in vivo* study of the influence of the surface roughness of implants on the microbiology of supra- and subgingival plaque. J. Dent. Res. 72: 1304–1309.

Raffaini, G. and F. Ganazzoli. 2012. Molecular modelling of protein adsorption on the surface of Ti dioxide polymorphs. Phil. Trans. R. Soc. A 370: 1444–1462.

Rahmati, B., A.A.D. Sarhan, E. Zalnez, Z. Kamiab, A. Dabbagh, D. Choudhury et al. 2016. Development of tantalumoxide(Ta-O)thin film coating on biomedical Ti-6Al-4V alloy to enhance mechanical properties and biocompatibility. Cer. Int. 42: 466–480.

Rizzi, G., A. Scrivani, M. Fini and R. Giardino. 2004. Biomedical coatings to improve the tissue-biomaterial interface. Int. J. Artif. Organs 27: 649–57.

Robin, A. and J.L. Rosa. 2000. Corrosion behavior of niobium, tantalum and their alloys in hot hydrochloric and phosphoric acid solutions. Int. J. Ref. Met. Hard Mat. 18: 13–21.

Romanò, C.L., S. Scarponi, E. Gallazzi, D. Romanò and L. Drago. 2015. Antibacterial coating of implants in orthopedics and trauma: A classification in an evolving panorama. J. Orthopaedic Surg. Res. 10: 157–166.

Russell, S.M. 2000. Nitinol melting and fabrication. pp. 1–10. In: S.M. Russell and A.R. Pelton [eds.]. Proceedings of the International Conference SMST-2000. Pacific Grove California.

Saldivar Garcia, A.J., A. Mani Medrano and A. Salinas Rodriguez. 1999. Formation of HCP Martensite during the isothermal aging of an FCC Co27Cr5Mo0.05C orthopaedic implant alloy. Metall. Mater. Trans. 30A: 1177–1184.

Schmalzried, T.P. and J.J. Callaghan. 1999. Current concept review—wear in total hip and knee replacements. J. Bone Joint Surg. Am. 81: 114–136.

Schupbach, P. and R. Glauser. 2007. The defense architecture of the human periimplant mucosa: A histological study. J. Prosthetic Dent. 97: 15–25.

Schwarz, F., M. Wieland, Z. Schwartz, G. Zhao, F. Rupp, J. Geis-Gerstorfer et al. 2009. Potential of chemically modified hydrophilic surface characteristics to support tissue integration of titanium dental implants. J. Biomed. Mater. Res. B Appl. Biomater. 88: 544–57.

Shetty, V.D. and R.N. Villar. 2006. Development and problems of metal on metal hip arthroplasty. P. I. Mech. Eng. H 220: 371–377.

Spriano, S., E. Vernè, M.G. Faga, S. Bugliosi and G. Maina. 2005. Wear 259: 919–925.

Spriano, S. and S. Bugliosi. 2006. Medical prosthetic devices presenting enhanced biocompatibility and wear resistance, based on cobalt alloys and process for their preparation. WO 2006/038202A2.

Spriano, S., S. Ferraris, C.L. Bianchi, C. Cassinelli, P. Torricelli, M. Fini et al. 2010. Bioactive titanium surfaces. pp. 269–294. In: P.N. Sanchez [ed.]. Titanium Alloys: Preparation, Properties and Applications. Nova Science Publishers, Inc.

Spriano, S. and S. Ferraris. 2014. How can topographical surface features affect the interaction of implants with soft tissue? 3rd Int. Conf. On Health Science and Biomedical System HSBS14, Florence, Italy.

Spriano, S., S. Ferraris, G. Pan, C. Cassinelli and E. Vernè. 2015. Multifunctional titanium: surface modification process and biological response. J. Mechanics in Medicine and Biology 15: 1540001.

Stephens, P. 2014. Antibiotic resistance now global threat, WHO Warns 2014, http://www.bbc.com/news/health-27204988.

Stiehler, M., M. Lind, T. Mygind, A. Baatrup, A. Dolatshashi-Pirouz, H. Li et al. 2008. Morphology, proliferation and osteogenic differentiation of mesenchymal stem cells cultured on titanium, tantalum and chromium surfaces. J. Biomed. Mater. Res. A 86A: 448–458.

Trampuz, A. and W. Zimmerli. 2005. Prosthetic joint infections: update in diagnosis and treatment. Swiss Med. Wkly 135: 243–251.

Varoni, E.M., M. Iriti and L. Rimondini. 2012. Plant products for innovative biomaterials in dentistry. Coatings 2: 179–194.

Vogely, H. Ch., C.J.M. Oosterbos, E.W.A. Puts, M.W. Nijhof, P.G.J. Nikkels, A. Fleer et al. 2000. Effects of hydroxyapatite coating on Ti-6 Al-4V implant-site infection in a rabbit tibial model. J. Orthop. Res. 18: 485–493.

Walboomers, X.F., L.A. Ginsel and J.A. Jansen. 2000. Early spreading events of fibroblast on microgrooved substrates. J. Biomed. Mat. Res. 51: 529–534.

Wang, A., A. Essner, V.K. Polineni, C. Stark and J.H. Dumbleton. 1998. Lubrication and wear of ultra-high molecular weight polyethylene in total joint replacement. Tribol. Int. 31: 17–33.

Ward, M.B., A.P. Brown, A. Cox, A. Curry and J. Denton. 2010. Microscopical analysis of synovial fluid wear debris from failing CoCr hip prostheses. J. Phys. Conf. Ser. 241: 012022.

Watnick, P. and R. Kolter. 2000. Biofilm, city of microbes. J. Bacteriology 182: 2675–2679.

Wennerberg, A. and T. Albrektsson. 2009. Effects of titanium surface topography on bone integration: a systematic review. Clin. Oral. Impl. Res. 20: 172–184.

Wimmer, M.A., J. Loos, R. Nassutt, M. Heitkemper and A. Fischer. 2001. The acting wear mechanisms on metal-on-metal hip joint bearings: *in vitro* results. Wear 250: 129–139.

www.smith-nephew.com.

www.zimmer.com.

Xiao, S.J., G. Kenausis and M. Textor. 2001. Biochemical modification of titanium surfaces. pp. 171–230. *In*: P. Tengvall, M. Textor and P. Thomsen [eds.]. Titanium in Medicine. Springer–Verlag, Berlin, Heidelberg, New York.

Yamada, S., D. Heymann, J.-M. Bouler and G. Daculsi. 1997. Osteoclastic resorption of calcium phosphate ceramics with different hydroxyapatite/beta tri-tricalcium phosphate ratios. Biomat. 18: 1037–1041.

Yan, Y., A. Neville and D. Dowson. 2007. Tribo-corrosion properties of cobalt-based medical implant alloys in simulated biological environment. Wear 263: 1105–1111.

Yoshinari, M., K. Matsuzaka, T. Inoue, Y. Oda and M. Shimono. 2003. Effects of multigrooved surfaces on fibroblast behaviour. Biomedical Materials Research 65A: 359–368.

Zhao, G., Z. Schwartz, M. Wieland, F. Rupp, J. Geis-Gerstorfer, D.L. Cochran and B.D. Boyan. 2005. High surface energy enhances cell response to titanium substrate microstructure. J. Biomed. Mater. Res. A. 74: 49–58.

Zhao, G., A.L. Raines, M. Wieland, Z. Schwartz and B.D. Boyan. 2007. Requirement for both micron- and submicron scale structure for synergistic responses of osteoblasts to substrate surface energy and topography. Biomat. 28: 2821–2829.

Zinelis, S., N. Silikas, A. Thomas, K. Syres and G. Eliades. 2012. Surface characterization of SLActive dental implants. Eur. J. Esthet. Dent. Spring. 7: 72–92.

Zinger, O., G. Zhao, Z. Schwartz, J. Simpson, M. Wieland, D. Landolt and B. Boyan. 2005. Differential regulation of osteoblasts by substrate microstructural features. Biomat. 26: 1837–1847.

Fe-based Superconductors
Crystallochemistry, Band Structure and Phase Diagrams

Alberto Martinelli

INTRODUCTION

The discovery of Fe-based superconductors (Fe-SC) with relatively high superconducting transition temperature (T_c) represents a fundamental result in the field of the solid state physics and a milestone in the history of superconductivity. In fact, these materials exhibit the highest T_c after Cu-SC. The former superconducting compounds belonging to this class of materials, LaFePO and LaNiPO (Kamihara et al. 2006; Watanabe et al. 2007), where synthesized since 2006, but did not attracted much interest, due to their relatively low superconducting transition temperatures $(T_c \leq 4$ K). The breakthrough occurred in February 2008, when a $T_c \sim 26$ K was reported for LaFeAs($O_{1-x}F_x$) (Kamihara et al. 2008); this discovery definitely triggered the extensive interdisciplinary researches on Fe-SC, relaunching the research in superconductivity. A few weeks later, a $T_c \sim 55$ K was measured for SmFeAs($O_{1-x}F_x$) compositions (Zhi-An et al. 2008), a value that still represent the highest T_c attained in the Fe-SCs. A new family of compounds was discovered after a few months, when a $T_c \sim 38$ was achieved by hole-doping (K-substitution) BaFe$_2$As$_2$ (Rotter et al.).

SPIN-CNR, Corso Perrone 24, 16152 Genova, Italy.
E-mail: alberto.martinelli@spin.cnr.it

After several years, Fe-SC are still the subject matter of many researches and new developments are ongoing.

The discovery of Fe-SC represents also a turning point, since it overtook the general belief that magnetism is always detrimental for superconductivity (Paglione and Greene 2010) compounds containing cationic species characterized by a nominally large magnetic moment (such as Fe^{2+}). Hence a deeper knowledge of these materials is expected to broaden our knowledge and understanding of unconventional superconductivity, providing new possibilities for discovering high temperature superconductors and possibly raising T_c.

A delicate and tangled interplay between magnetic, superconductive and crystallo-chemical properties characterizes the fascinating physics of this class of compounds. Hence the understanding of their normal state properties (i.e., their properties above T_c) is a fundamental step in the development of a superconductivity theory. The observed values of T_c, exceeding the upper theoretical limit predicted by the BCS theory, provide evidence that these materials can be classified as unconventional superconductors, that is, phonons do not play a major role here. With a few exceptions, superconductivity emerges upon the destabilization of the magnetic ground state, usually achieved producing electron- or hole-doping of the parent compound by chemical substitution; in some cases superconductivity can also emerge by applying pressure. In particular superconductivity can arise upon the complete suppression of magnetic ordering in some systems, whereas in other cases a coexistence between these two ground states is observed within an under-doped compositional field. Hence the correct description of the phase diagram region where a crossover between magnetism and superconductivity takes place stands out as a fundamental issue. In this context, the microscopic origin for the pairing interaction has not yet been ascertained: the prevailing scenario suggests that the electron pairing interactions responsible for superconductivity are produced by the same in-plane (π,π) spin fluctuations driving magnetic ordering, that is, magnetism and superconductivity compete for the same conduction electrons.

So far a great deal of experimental data are available and several Fe-SC compounds are known that can be classified according to their structural properties and grouped according to their stoichiometries. Hereafter the so-called 11-, 122- and 1111-type compounds are described, since these compositions have been studied in more depth; other relevant compounds belong to the 111-, 112- and 245-type families. In particular, this brief review outlines the progress of research concerning the crystallo-chemical properties, the band structure and the phase diagram of Fe-SC materials. A particular attention is devoted to the interplay between magnetism and superconductivity, representing one of the most prominent issues in the physics of Fe-SC. Needless to say, it is challenging to comprehensively cover all these topics; detailed review articles covering these and other different aspects of the physics of these materials are available in literature (for example see Johnston 2010; Johrendt 2011; Bascones et al. 2015; Paglione and Greene 2010; Inosov 2015; Dai et al. 2012; Wilson 2010; Dai 2015; Fernandes and Schmalian 2012; Lumsden and Christianson 2010; Fernandes et al. 2014; Carretta et al. 2013). The reader is referred to these works for a complete description of all the features characterizing Fe-SCs.

Crystallochemistry

From the crystallo-chemical point of view, the Fe-SC materials (and Cu-SC compounds as well) violate several of the Matthias rules developed for finding new superconducting compounds (Mazin 2010); in particular Fe-SC compounds are layered and not cubic (violation of rule 1), they can contain oxygen (violation of rule 3) and finally they are based on Fe, an element showing magnetism (violation of rule 4). In Fe-SC compounds iron nominally occurs as Fe^{2+} cationic species, that is, with a $[Ar]3d^6$ ground electronic configuration, whereas in the parent Cu-SC compounds copper occurs as $3d^9$ Fe^{2+} cationic species; then in the first case Fe^{2+} has one electron more than half filling, whereas in the second case Cu^{2+} has one electron less than complete filling.

Edge-sharing tetrahedral layers constitute the fundamental building block of Fe-SC materials, with Fe atoms arranged according to a planar square network and capped by four chalcogen (typically Se and Te) or pnictogen (typically As or P) atoms at the corners of the tetrahedron. These layers display an atomic arrangement similar to that observed in α-PbO (litharge), being compressed along the c-axis. The α-PbO structure-type is constituted of edge-sharing tetrahedral layers stacked along the c-axis; these layers are fluorite-type, since they can be formally derived from the fluorite type structure by a selective deletion of half of the tetrahedra occurring in the CaF_2 unit cell (Fig. 1).

Three main families stand out in the Fe-SC class of materials, formally distinguished by their respective stoichiometric compositions and structural features:

1) 11-type: essentially constituted of β-FeSe$_{1-x}$ and the solid solution Fe$_{1+y}$(Te$_{1-x}$Se$_x$). β-FeSe$_{1-x}$ is the simplest Fe-based superconductor, being isotypic with α-PbO (space group $P4/nmm$ -129; Fig. 2); on cooling it undergoes a $P4/nmm \rightarrow Cmme$ structural transition located between 70 K and 100 (Margadonna et al. 2008; Pomjakushina et al. 2009; McQueen et al. 2009a; Khasanov et al. 2010). Remarkably no evidence for short- or long-range magnetic ordering has been ever detected in

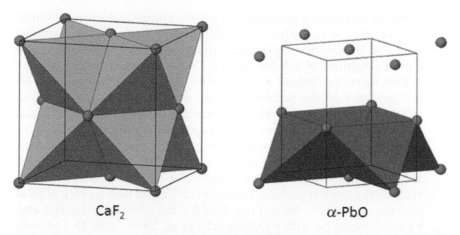

CaF$_2$ α-PbO

Fig. 1: Representation of the fluorite CaF$_2$ and α-PbO structure-types evidencing their strict relationship.

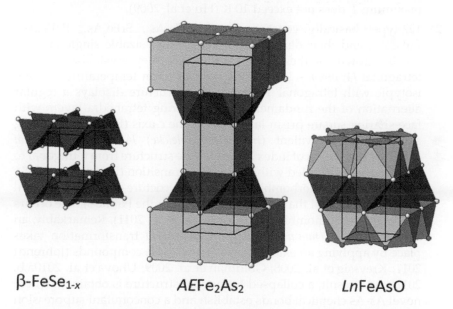

β-FeSe$_{1-x}$ AEFe$_2$As$_2$ LnFeAsO

Fig. 2: The crystal structures for the 11-, 122- and 1111-type families of Fe-SC; note that only polyhedra coordinated by cationic species are represented.

this compound (Margadonna et al. 2008; Pomjakushina et al. 2009; McQueen et al. 2009a,b). At ambient pressure the T_c of β-FeSe$_{1-x}$ is ~ 8.5 K, but can be increased up to ~ 37 K under an applied pressure of 8.9 GPa (Medvedev et al. 2009). β-Fe$_{1+y}$Te, a compound with a strongly defective Cu$_2$Sb structure-type, is usually included in this family, given the increased T_c measured for selected terms of the solid solution $Fe_{1+y}(Te_{1-x}Se_x)$. Actually β-$Fe_{1+y}Te$ exhibits an antiferromagnetic ground state and no trace for superconductivity has been ever detected. The crystal structures of β-$FeSe_{1-x}$ and β-$Fe_{1+y}Te$ could look quite similar at first sight, but indeed a remarkable difference is associated with the geometry of the tetrahedral cage centred by the chalcogen; in fact in β-$FeSe_{1-x}$ tetrahedral layers are compressed along the c-axis, conversely in β-$Fe_{1+y}Te$ they are elongated along the same axis. This subtle, but notable structural difference is retained in the solid solution $Fe_{1+y}(Te_{1-x}Se_x)$, as evidenced by several local structure investigations, indicating significant different heights for Se and Te atoms (Louca et al. 2010; Joseph et al. 2010; Tegel et al. 2010) and hence a lack of local structure relaxation. Indeed the T_c of β-$FeSe_{1-x}$ can be enhanced also by substituting Se with S (Mizuguchi et al. 2009) up to 15.5 (onset), but this system attracted much less attention. S-substitution can also induce superconductivity in β-$Fe_{1+y}Te$, although in this case the maximum T_c does not exceed 10 K (Hu et al. 2009).

2) 122-type: basically constituted of $CaFe_2As_2$, $SrFe_2As_2$, $BaFe_2As_2$, $EuFe_2As_2$ and their derivatives. High-quality sizable single crystals can be grown for these type-compounds; they crystallize into the tetragonal $I4/mmm$ - 139 space group at room temperature and are isotypic with tetragonal $ThCr_2Si_2$. This structure displays a regular alternation of the fundamental edge sharing tetrahedra layers with face-sharing square prism layers along the c-axis (Fig. 2). On cooling a translation-equivalent (*translationengleiche*) $I4/mmm \rightarrow Fmmm$ structural transition of index 2 changes the structure from tetragonal to orthorhombic, coupled with a magnetic transition (Rotter et al. 2008), leading to an orthorhombic β-$ThCr_2Si_2$ structure-type. Very careful studies ascertained that magnetic ordering at the Fe sub-lattice occurs below the orthorhombic distortion (Kim et al. 2011). Remarkably, an isomorphous $I4/mmm \rightarrow I4/mmm$ structural transformation takes place by applying an external pressure in these compounds (Johrendt 2011; Kreyssig et al. 2008; Goldman et al. 2009; Uhoya et al. 2010a,b, 2011). As a result, a collapsed tetragonal structure is obtained, where novel As-As chemical bonds establish and a concomitant suppression of magnetic ordering occurs; at low temperatures this collapsed phase becomes superconducting. Remarkably, several heavy fermion compounds exhibiting superconductivity, cooperative magnetism and normal metallic behaviour crystallize with the same $ThCr_2Si_2$ structure-type.

3) 1111-type: the prototypical composition is *Ln*FeAsO (*Ln*: lanthanide) even though also *AE*FeAsF (*AE*: alkaline earth) compositions have been investigated. The structure displays a regular alternation of

the fundamental edge sharing tetrahedra layers (centred by Fe^{2+}) with face-sharing anti-square prism layers (centred by Ln^{3+}) stacked along the c-axis (Fig. 2). Actually, the representation of Fig. 2 is quite unusual and follows from the convention to represent coordination polyhedra around the cationic species only, that is, Ln^{3+} and Fe^{2+} (note that the same kind of coordination polyhedra forms around the anionic species, i.e., As^{3-} and O^{2-}). Conversely, in almost all structural representations, the tetrahedral character tends to be sticked out: hence the structure of 1111-type compounds is sketched as a stacking of two different edge-sharing tetrahedral layers, the former centred by the cationic Fe^{2+} species, the latter by the anionic O^{2-} species. At room temperature they are isotypic with ZrCuSiAs; this structure-type has a relationship with the α-PbO one. In fact when empty interstitial sites in the a-PbO structure are fully occupied, the PbClF structure-type forms (inside which 111-type compounds crystallize), characterized by the occurrence of empty tetrahedral sites; when the tetrahedral sites are fully occupied the ZrCuSiAs type structure is obtained. These compounds undergo a tetragonal to orthorhombic 1st order $P4/nmm \rightarrow Cmme$ translation-equivalent structural transition of index 2 at $T_s \sim 140$–180 K. A few tens of degrees below, the structural transformation is followed by a 2nd order magnetic transition, involving spin ordering at the Fe sub-lattice. At low temperature also the magnetic Ln^{3+} species order antiferromatically (Kimber et al. 2008; Zhao et al. 2008a,c; Ryan et al. 2009). Some of these pure compounds become superconducting under high pressure, such as LaFeAsO and SmFeAsO (Okada et al. 2008; Takahashi et al. 2009), but not CeFeAsO (Zocco et al. 2011). Conversely, superconductivity cannot be induced by applying chemical pressure, for example, by partial substitution of La with Y, characterized by a smaller ionic radius (Tropeano et al. 2009; Martinelli et al. 2009). In LnFeAs(O,F) compounds with a magnetic Ln species (such as Ce, Pr, Nd and Sm), the T_c value is significantly higher than in the corresponding LaFeAs(O,F) system, even though magnetic Ln – Fe interactions seem not to be crucial for T_c enhancement (Maeter et al. 2009).

Other notable families belonging to Fe-SC class of materials includes the 111-type compounds, mainly represented by LiFeAs and NaFeAs, the 245-type compounds, characterized by a general formula $A_2Fe_4Se_5$ (A = K, Rb, Cs, Tl, K) (Bao 2015), and the 112-type (La, Ca)FeAs$_2$ compound. As for the 11-, 122- and 1111-type compounds, all these materials are structurally based on Fe centred edge-sharing tetrahedral layers.

Iron in these compounds occurs as Fe^{2+} cationic species, as revealed by bond valence sum calculations (Martinelli et al. 2008).

Remarkably, superconductivity in Fe-SC compounds is quite resistant to chemical substitution and a relatively large amount of disorder can be accommodated at the Fe sub-lattice, even though a detrimental effect on superconductivity is always present. For comparison, in Cu-based superconductors a very low amount of substitutional dopants at the Cu sub-lattice induces the complete suppression of superconductivity. The robustness of superconductivity to chemical substitution in Fe-SC is related to the pairing mechanism, as elucidated by theoretical analyses (*vide infra*).

The crystal structures of all these phases (Fig. 2) are characterized by the occurrence of quite regular coordination polyhedra (tetrahedral, square prism and double anti-square prism), suggesting the electronic states reported in Fig. 3 for $CaFe_2As_2$ (selected as representative) in the light of the valence shell electron pair repulsion (VSEPR) theory. It is evident a net charge transfers among the different planes and layers; similar argumentations holds for both β-$FeSe_{1-x}$ and 1111-type compounds (Martinelli 2015).

The origin of the structural transition in β-$FeSe_{1-x}$, 122- and 1111-type materials is still debated. Lattice degrees of freedom do not drive symmetry breaking: the symmetry mode analysis clearly shows that the displacive B_{2g} mode should drive the $P4/nmm \rightarrow Cmme$ structural transition, but this soft mode is not involved in any atomic displacement at the occupied sites in the crystal structures of β-$FeSe_{1-x}$ and 1111-type compounds (Martinelli 2013). On the other hand several theoretical investigations suggest that symmetry breaking could be originated by electronic (spin, charge, orbital) degree of freedom (Fang et al. 2008; Xu et al. 2008). More specifically, at present two main mechanisms have been proposed to explain the structural transformations (and the coupled magnetic ordering) in Fe-SC compounds: (1) orbital degrees of freedom drive the structural transition

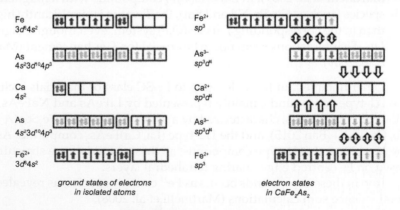

Fig. 3: Electron states in $CaFe_2As_2$.

and induces magnetic anisotropy, thus triggering the magnetic transition (Krüger et al. 2009; Lv et al. 2009; Chen et al. 2010); (2) spin fluctuations drive the structural transition and induce orbital ordering (Fernandes et al. 2012).

The tetragonal to orthorhombic structural transformation determines a lattice rotation by 45° along the c axis; in addition the orthorhombic a and c axes are a factor $\sqrt{2}$ larger than the tetragonal ones. This has been a source of confusion in the description of the magnetic orderings, because both the tetragonal and orthorhombic notations are adopted. Hence in the first case $Q_{magnetic} = (\frac{1}{2},\frac{1}{2})$, whereas in the second case $Q_{magnetic} = (1,0)$.

With the exception of β-FeSe$_{1-x}$ characterized by a superconducting ground state, at room pressure superconductivity can be induced in 122- and 1111-type pure compounds by three different kinds of doping attained by chemical substitution:

1) electron doping: an increase in the negative charge at the FeAs layers is obtained for example by Co- or F-substitution in the Ba(Fe$_{1-x}$Co$_x$)$_2$As$_2$, RE(Fe$_{1-x}$Co$_x$)AsO and REFeAs(O$_{1-x}$F$_x$) systems.

2) hole doping: for example the positive charge in the FeAs layer is increased in the (Ba$_{1-x}$K$_x$)Fe$_2$As$_2$, (La$_{1-x}$Sr$_x$)FeAsO and REFeAs(O$_{1-x\ x}$) systems by K-, Sr-substitution or vacancies at the O sub-lattice.

3) isovalent doping: by replacing Fe with Ru in the 122 systems such as Ba(Fe$_{1-x}$Ru$_x$)$_2$As$_2$, As with P in BaFe$_2$(As$_{1-x}$P$_x$)$_2$ and LaFe(As$_{1-x}$P$_x$) O systems; remarkably, by replacing Se with Te or S in β-FeSe$_{1-x}$ the superconductive transition temperature can be increased.

In any case, the destabilization of the magnetic ground state appears as a key factor for inducing superconductivity in these materials. On this basis, the prevailing scenario suggests that the electron-pairing interactions are produced by the same magnetic interactions driving magnetic ordering, that is, magnetism and superconductivity compete in these materials for the same conduction electrons.

Remarkably, superconducting properties in Fe-SC are strongly dependent on their structural properties; in fact highest T_c are measured for a chalcogen or pnictogen anion height of 1.38 Å above the Fe plane (Mizuguchi et al. 2010) or a tetrahedral bond angle of 109°47′ (Horigane et al. 2009), which is the ideal tetrahedral bond angle value. In this context, density functional calculations show that the Fermi surface nesting for Fe$_{1+y}$(Te$_{1-x}$Se$_x$) compositions strongly varies with the chalcogen height (i.e., the distance between the Se/Te atom and the Fe plane), regardless of the chemical identity of the chalcogen atom (Kumar et al. 2012). The relevant role played by the anion height is particularly evident in doped 1111-type compounds; in fact the substitution of La by Ce, Pr, Nd and Sm induces chemical pressure and progressively decreases the ionic radius at the *Ln* site. As a result, the anion height increases to ~ 1.38 Å and T_c achieves its maximum value (for optimally doped $NdFeAs(O_{1-x}\square_x)$, $GdFeAs(O_{1-x}\square_x)$, $SmFeAs(O_{1-x}F_x)$ T_c is peaked at about 55 K); for heavier rare earth, such as Tb and Dy, the anion height further increases, moving away from the optimal value, and T_c decreases (Mizuguchi et al. 2010). Chemical pressure can also be induced by partial substitution of La with Y, even though in this case structural disorder plays a detrimental role (Tropeano et al. 2009). Indeed theoretical investigations recognized that the position of the pnictogen in 1111-type compounds is the key factor determining both T_c and the form of the superconducting gap (Kuroki et al. 2009).

Band Structure

In their normal state Fe-SC are semimetals (Ma and Lu 2008), that is, the centres of momenta of the filled valence band and empty conduction bands are displaced horizontally from each other, but slightly overlapping in energy; as a result some electrons from the valence band spill over to the conduction band (Fig. 4). This is a remarkable difference compared to Cu-based superconductors, which in the normal state are insulators.

First band structure calculations predicted the possibility for a collinear antiferromagnetic ordering due to a spin density wave originated by the nesting of the hole and electron Fermi surfaces (Dong et al. 2008). This hypothesis was soon corroborated by neutron diffraction analysis, evidencing the occurrence of an antiferromagnetic spin ordering at low temperature in LaFeAsO, associated to a structural transformation and conforming to the theoretical predictions (de la Cruz et al. 2008). As a rule the antiferromagnetic structure in 122- and 1111-type compounds is characterized by an in–plane $Q_{magnetic}$ = (1,0) magnetic propagation wave-vector (orthorhombic unit cell with $c > a > b$), with an ordered Fe moment value anomalously small (~ $0.3 - 1$ μ_B). Spin ordering develops in a collinear stripe-like structure, with ordered Fe-moments oriented along the a-axis in the orthorhombic a-b plane; antiferromagnetic spin coupling occurs along the longer a-axis, whereas along the shorter b-axis, ferromagnetic coupling is present (Fig. 5; Zhao et al. 2008a, b, c; Chen et al. 2008; Huang et al. 2008a,b). This antiferromagnetic order is progressively suppressed by doping.

The Fermi surface topology and band structure are rather similar for the 11-, 122-, and 1111-type compounds. Precisely, the Fe-SC are 3-dimensional materials in principle, but their actual quasi 2-dimensional behaviour is related to the weak electron hopping along the c-axis. In all these materials the electronic structure near the Fermi level is largely

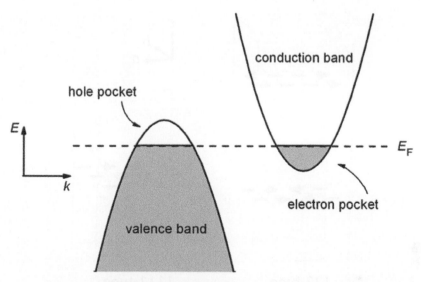

Fig. 4: Schematic representation of the electronic structure of a semimetal with electron and hole pockets separated in momentum space; filled electron states are displayed as shaded areas, evidencing the spillover of electrons from the valence to the conduction band.

dominated by Fe $3d$ electrons, the contribution of pnictogen or chalcogen atoms being very small, and five bands cross the Fermi level. Then all five Fe $3d$ orbitals contribute charge carriers, as a consequence of the close packing experienced by Fe atoms in the tetrahedral layer. The Fermi surface of layered Fe-SC consists of quasi-two-dimensional electron and hole Fermi surfaces. Fermi surface calculations reveal that in the momentum space there are three concentric hole pockets around the Brillouin zone center Γ at (0,0) of the tetragonal unit cell and two electron pockets about the M corner point at (π,π) (Fig. 6; Ma and Lu 2008; Singh and Du 2008; Kuroki et al. 2008; Mazin et al. 2008; Graser et al. 2009).

Remarkably in CeFePO electronic correlations are dominated by Ce $4f$ electrons rather than Fe $3d$ electrons; as a result this compound is a paramagnetic heavy fermion (Brüning et al. 2008), not a superconductor as LaFePO. Conversely in CeFeAsO the Fe $3d$ orbits produce a major contribution to the Fermi level (Liu et al. 2011) and superconductivity can be induced by proper doping.

Theoretical calculations reveal that the Fermi surface structure of β-Fe$_{1+y}$Te is very similar to those of other Fe-SC compounds (Subedi et al. 2008); despite this, β-Fe$_{1+y}$Te is characterized by an antiferromagnetic ground state whose diagonal double stripe spin ordering is strongly different from that occurring in other Fe-SC compounds, characterized by an in-plane magnetic wave-vector $\mathbf{Q}_{\text{magnetic}} = (\frac{1}{2},0)$ (Martinelli et al. 2010).

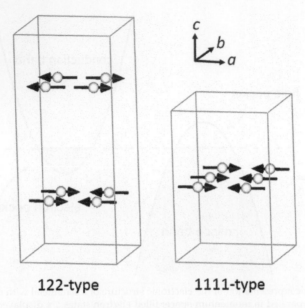

Fig. 5: Spin ordering in 122- and 1111-type compounds: Fe-moments are oriented along the *a*-axis in the orthorhombic *a-b* plane; spin coupling is antiferromagnetic along the longer *a*-axis and ferromagnetic along the shorter *b*-axis.

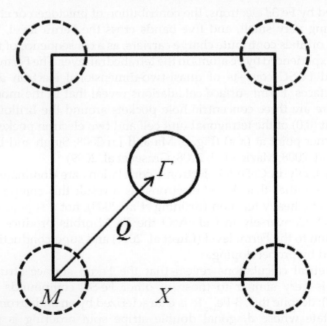

Fig. 6: Simplified representation of the Fermi surface topology, consisting of electron and hole pockets of about the same size at the *M* and *Γ* points respectively; nesting between hole and electron pockets is provided by the wave vector $Q_{nesting} = (\pi,\pi)$.

According to angle-resolved photoemission spectroscopy (ARPES) measurements and band structure calculations the Fermi surface topology of Fe-SC is characterized by an in-plane nesting wave-vector $Q_{nesting} = (\pi,\pi)$ $\equiv$ ($\frac{1}{2}$,$\frac{1}{2}$), that is a mapping of the quasi-two-dimensional Fermi surface sections onto each other is obtained if some portions of the Fermi surface is displaced in the k-space by the nesting vector. In principle this feature can favour a spin or a charge density wave state. However, in a spin density wave state the total spin density is not uniform, but its magnitude and/ or direction periodically varies in space; nevertheless the spin ordering is antiferromagnetic. The detection of spin orderings with in-plane $Q_{magnetic} =$ ($\frac{1}{2}$,$\frac{1}{2}$) in 122- and 1111-type compounds strongly supports the spin density wave scenario (de la Cruz et al. 2008).

Two different superconducting pairing mechanisms seem to occur in Fe-SC, depending on the dominating fluctuations: a $s_{\pm}$ pairing state, when antiferromagnetic spin fluctuations dominate, and a s_{++} pairing state, when orbital fluctuations dominate. Theoretical studies suggest that the electron pairing could be mediated by inter-pocket scattering from Γ to M (Kuroki et al. 2008), providing the occurrence of a $s_{\pm}$ wave superconducting state where antiferromagnetic spin fluctuations dominate (Mazin et al. 2008; Chubukov et al. 2008); $\pm$ indicates that the superconducting gap changes its sign between the different sheets of the Fermi surface. Carrier doping varies the chemical potential and the Fermi surface, thus inducing a Lifshitz transition (change of the Fermi surface topology without a structural transition). On account of electron-doping, hole-pockets around the Γ point shrink and the electron pockets about the M point expand as the doping increases; as the hole-pocket disappears, lowering below the Fermi level, the Lifshitz transition occurs and an electron pocket emerges at the Γ point. Then with over-doping spin fluctuations become too weak to mediate pairing. The opposite is observed with hole-doping. This mechanism is in agreement with what is observed, for example, in the $Ba(Fe_{1-x}Co_x)_2As_2$ system, where heavy doping destroys the superconductive state.

On the other hand, later investigations revealed that the superconducting transition temperature increases when nesting is worsened or even suppressed by doping in materials such as $LaFeAs(O_{1-x}H_x)$, where H acts as an electron donor, and $K_xFe_2Se_2$ (Iimura et al. 2012; Qian et al. 2011); in these cases inter-pocket scattering from Γ to M is suppressed. Such behaviour can be possibly ascribed to a different s_{++} wave superconducting state induced in this case by orbital fluctuations (Kontani and Onari 2010); moreover, the s_{++} wave state can become dominant over the $s_{\pm}$ wave state in the presence of impurities.

ARPES measurements and band structure calculations indicate that the electronic structure has a multi-orbital character (Graser et al. 2009; Zhang et al. 2011). When carried out as a function of temperature, ARPES analyses evidence an energy splitting of two orthogonal bands with dominant d_{xz} and d_{yz} character associated with the structural transition (Yi et al. 2011). In particular these measurements realized below the structural transition temperature reveal that the electronic structure along Γ-X and Γ-Y directions are different, breaking the C_4 rotational symmetry and thus inducing the structural and magnetic transition.

The nature of the ordered magnetic state in these compounds is still debated, and different interpretations were proposed. A first scenario supports an itinerant character of the antiferromagnetic state. This hypothesis is corroborated by the experimental finding that the nesting wave-vector for the electron and hole Fermi surface pockets $\mathbf{Q}_{nesting} = (\pi,\pi)$ is consistent with the in-plane tetragonal magnetic wave-vector $\mathbf{Q}_{magnetic} = (\pi,\pi)$ observed in 122- and 1111-type compounds. This leads to a spin density wave instability (Dong et al. 2008; Singh and Du 2008; Mazin et al. 2008). Alternative theories based on local moments have been proposed, in which magnetic frustration is thought to be induced by near-neighbour and next-to-near-neighbour interactions among local Fe moments (Yildirim 2008; Ma et al. 2008; Si and Abrahams 2008), whereas ferro-orbital ordering has been suggested to drive both structural and magnetic transitions (Lv et al. 2009; Lee et al. 2009). By inelastic neutron scattering analyses it has been established that magnetism in Fe-SC results from a complicated mixture of local and itinerant characters (Zhao et al. 2009). On the other hand, ARPES analyses evidence that the Fermi surface at the Brillouin-zone center is dominated by a single d_{xz} or d_{yz} orbital at low temperatures, thus supporting the orbital ordering scenario (Shimojima et al. 2010; Zhang et al. 2012; Chen et al. 2011). Noteworthy, density functional theory calculations suggest that the orthorhombic polymorph is more stable than the tetragonal one only by taking into account magnetic order.

Remarkably, the physics of Fe-SC includes Dirac cones, a band structure that describes how electrons behave as massless relativistic Dirac fermions at particular positions of the momentum space. Dirac cone like electronic band structure, whose occurrence is related to the metallic nature of the magnetic ground state, have been experimentally detected for both 122- and 1111-type materials (Richard et al. 2010; Pallecchi et al. 2011).

Phase Diagrams

The starting point for the understanding of most materials properties is the knowledge of the phase equilibria involved in the investigated systems. Phase diagrams provide information concerning the stability and the possible coexistence of distinct phases in thermodynamic equilibrium as a function of specified state variables. For a recent review on phase diagrams of Fe-SC see (Martinelli et al. 2015). Despite the outstanding theoretical and experimental investigations, it is not yet clear if a universal phase diagram can be established for these materials. In Fig. 7 a schematic phase diagram is drawn, highlighting the features characterizing these systems.

From the crystallographic point of view, most investigations report a tetragonal-to-orthorhombic structural transformation that is progressively suppressed with carrier doping, up to its complete suppression (Avci et al. 2012; Bernhard et al. 2012; Nomura et al. 2008; Zhao et al. 2008c; Margadonna et al. 2009). Conversely accurate structural investigations point to a strong reduction of the orthorhombic distortion, with the retention of the structural transition even for optimally doped compositions (Martinelli et al. 2011). It has been often reported that superconductivity can be obtained by suppressing symmetry breaking, but this argumentation is not strictly correct. In fact, in almost all systems a fully superconductive state can be obtained for under-doped compositions retaining the orthorhombic structure. In this context, the case of β-FeSe$_{1-x}$ is representative, since this phase undergoes a tetragonal-to-orthorhombic structural transformation on cooling and it is characterized by a superconductive ground state (Margadonna et al. 2008; Pomjakushina et al. 2009). Nonetheless in the 122- and 1111-type systems the highest T_c is generally measured upon hindering the orthorhombic symmetry.

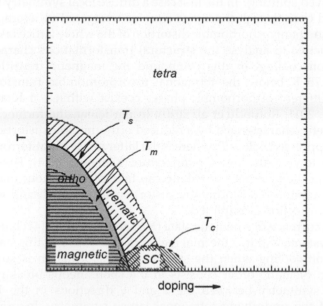

Fig. 7: Schematic phase diagram of Fe-SC, showing the most credited phase relationships between the tetragonal, orthorhombic, nematic, magnetic and superconducting (SC) phases. The structural magnetic and superconductive transition temperatures are indicated by T_s, T_m and T_c, respectively.

At this point it must be observed that it is not unusual to bump into Fe-SC phase diagrams where boundaries defining the orthorhombic phase seem very unlikely (for example boundaries with abrupt change of slope). Then, in some cases, phase diagrams apparently do not comply with the phase rule and other thermodynamic principles that must control the relationships among the different phases. Moreover, in this context, it is noteworthy that the tetragonal to orthorhombic transformation is usually evaluated by the selective peak splitting of the tetragonal {hhl} diffraction lines; such a splitting progressively decreases with doping and hence its detection can become challenging for relatively highly-doped compositions or when the diffractometer has a limited instrumental resolution. Several works report diffraction data for samples where the suppression of the structural transformation with doping is claimed, but actually the diffraction lines collected at low temperature display a marked broadening (coupled with a decreased intensity) absent in the data collected at higher temperature. The origin of this feature is hardly ever analysed, since the custom is "no splitting, no symmetry breaking"; obviously this shallow conclusion can bias towards erroneous results. So a question arises: what is the physical origin of broadening? More precisely: is the broadening produced by lattice microstrain or by a reduced orthorhombic distortion determining a strong peak convolution and hence an unresolved splitting? In the first case a diffuse local symmetry breaking is involved even though the lattice remains tetragonal on average; in the second case a faint orthorhombic distortion of the whole lattice takes place. It is instructive to analyse the structural transformations characterizing $BaFe_2As_2$ on cooling. In this compound the magnetic transition takes place ~ 0.75 K below the tetragonal to orthorhombic transformation, but two different orthorhombic phases coexist within ~ 1 K around T_s (Kim et al. 2011; Rotundu et al. 2010). In particular, the former phase is paramagnetic, characterized by a reduced orthorhombic distortion and is rapidly suppressed below T_s; whereas the latter phase is antiferromagnetic and stable down to the lowest temperature (Kim et al. 2011). These results suggest that the symmetry breaking can take place without magnetism, leading to a reduced orthorhombic distortion; as spin ordering sets in the structural distortion is amplified.

The occurrence of a nematic state (Fernandes et al. 2014) characterizes all these systems within a thermal region placed above T_s; this state is also observed on cooling when the structural transformation is suppressed by doping. Nematic order consists in the spontaneous breaking of the electronic symmetry between the x and y directions in the Fe-plane, but not of the underlying tetragonal lattice symmetry. In practice the tetragonal rotational symmetry is broken on account of a $C_4 \rightarrow C_2$ point-group reduction, but the translational symmetry is retained because the lattice maintains a tetragonal-like metric. Then, when the nematic state is established, the physical properties (transport, magnetic and optical properties) display a lower symmetry than the average crystallographic ones and as a consequence the Neumann's principle is violated. Under these circumstances, it must be pointed out that the real lattice symmetry is determined by the symmetry operations that really occur within it, not by its metric. Electronic and magnetic nematic degrees of freedom are believed to be associated with the origin of structural transitions and their interplay with magnetism as well as with the unconventional superconductivity in Fe-based materials. The nematic state most likely has an electronic origin and is driven by the same fluctuations that induce superconductivity and magnetic ordering (Fernandes et al. 2014). Noteworthy, lattice microstrains have been observed in the tetragonal phase in a thermal region standing just above T_s, pre-empting the structural transition, where also the nematic phase is located; the occurrence of microstrains indicates a diffuse local breaking of the tetragonal symmetry, at least at the microscopic scale (Martinelli et al. 2012).

Remarkably, in most of these systems a magnetic ground state is present and superconductivity can develop only upon its weakening or suppression; likely the proximity between the superconductive and magnetic states appears to be a fundamental prerequisite for high-temperature superconductivity.

Most of the phase diagrams are characterized by a microscopic coexistence between the antiferromagnetic and superconductive states within more or less restricted compositional ranges: the coexistence field is rather wide in the 122-type compounds, but quite narrow or even null in the 1111-type systems (Sanna et al. 2009; Sanna et al. 2010), even though, in some cases, it can be induced by applying pressure. This microscopic phase coexistence is probably an intrinsic property of the 122- and 1111-type systems and suggests the existence of the $s_\pm$ superconducting order. The magnetic and superconductive phase fields are never separated by a paramagnetic field, suggesting that both ground states compete for the same conduction electrons. Instead, the proximity to a magnetic quantum critical point (a phase transition taking place at 0 K that can be driven by tuning parameters such as doping, pressure or magnetic field) suggests that the electron pairing could be produced by the same magnetic interactions driving the magnetic ordering. The quantum phase transition at the boundary between the antiferromagnetic and superconductive states represents one of the most fascinating features characterizing the Fe-SC phase diagrams (Giovannetti et al. 2011). At present it has not yet established neither its origin nor whether a quantum critical phase lies beneath the superconducting dome or the criticality is avoided by the transition to the superconducting state (Abrahams and Si 2011). What makes the quantum criticality important in Fe-SC is the possible strong overlap between superconductivity and the quantum critical phase, not present in the Cu-based superconductors.

Magnetic ordering is always observed upon the breaking of the tetragonal symmetry, but never in advance of the establishment of the orthorhombic structure. Conversely, in several cases, a fully superconductive state ground state where magnetism is completely suppressed has been detected in samples characterized by an orthorhombic symmetry. The coexistence of magnetism and superconductivity raises some questions. In fact theoretical investigations reveal that nodes should appear in the magnetic state for s_{++} superconductivity, but not for $s_{\pm}$ one (Parker et al. 2009); in addition phase coexistence is prevented for s_{++} pairing, but the $s_{\pm}$ state can coexist with magnetism (Fernandes et al. 2010). Indeed the experimental observation of microscopic coexistence of magnetism and superconductivity in several Fe-Sc systems is consistent with the $s_{\pm}$, but not the s_{++} state; in these cases it can be concluded that antiferromagnetic spin fluctuations dominate. Evidence of a magnetic pairing mechanism are also provided by inelastic neutron scattering analysis (Inosov et al. 2010). On the other hand a s_{++} pairing state cannot be ruled out, since theoretical calculations demonstrated that in some specific cases inter-pocket scattering is suppressed and orbital fluctuations become dominant. Thus in these systems magnetism and superconductivity are not expected to coexist.

Concluding Remarks

The development of new materials is strictly related to a clear understanding of the properties of the already known materials and of the fundamental physics dominating these same properties. Despite the intense efforts carried out in the last years, the highest $T_c \sim 55$ K for Fe-SC is still that one obtained after only a few weeks since their discovery, at least for what concerns bulk materials, and the liquid nitrogen limit has not yet been crossed. Remarkably, an outstandingly high T_c value (~ 100 K) is reported by recent experiments carried out on single-layer FeSe films grown on doped $SrTiO_3$ (Ge et al. 2015). Nonetheless these materials exhibit the highest T_c among all known superconducting materials, aside Cu-SC.

The Cu- and Fe-SC materials share common ingredients for high T_c; moreover, they are characterized by very rich phase diagrams, displaying spin and charge ordering, insulating, metallic, nematic and superconductive phases, arising from a delicate and tangled interplay among competing physical interactions, where crystallo-chemical properties play a major role as well. In general the phase diagrams of the Fe-SC and other unconventional superconductor systems, such as cuprates and heavy-fermion superconductors, are closely similar (Paglione and Greene 2010). A proximity between the superconductive and magnetic states is present and superconductivity develops after the weakening or the complete suppression of magnetic ground state. This proximity suggests that magnetic fluctuations are responsible for paring: as a consequence magnetism and superconductivity seemingly compete for the same conduction electrons. Moreover spin density wave and nematicity emerged in the last few years as a fundamental component of the physics of Cu-SC compounds in the pseudo-gap state (Vojta 2009; Chang et al. 2012). In particular, recent investigations on the $YBa_2Cu_3O_{7-x}$ system ascertained that even in this system a competition between charge and spin density wave states with superconductivity takes place, with a crossover from the density wave state to the superconductive one strongly resembling the crossover observed in Fe-SC (Chang et al. 2012).

Remarkably several $LnFePO$ and $LnFeAsO$ compounds were already known several years before the discovery of their superconducting properties (Zimmer et al. 1995; Quebe et al. 2000), a somehow repetition of what happened with $(La,Ba)_2CuO_4$ and MgB_2, prepared in chemical laboratories years before the discovery of their superconductive state. This episode clearly evidences the need for a more strict cooperation among solid state chemists and physicists, in order to achieve significant progresses. Despite the deep advances in the understanding the superconducting state of the last years, superconductive materials are effectively expected to be discovered by chance, rather than anticipated by challenging theoretical deductions.

Generally speaking, the research on Fe-SC materials improved the knowledge and gave new insights into the physics of the superconductive and electron correlated materials. For example the proximity to a magnetic state, a proper Fermi surface topology and a layered structure look like essential components for the achievement of superconductivity (Mazin 2010). Regardless the powerful techniques applied for analysing their properties, some of the physics of Fe-SC is still controversial. Therefore the complexity of the Fe-SC materials indeed requires further systematic work, despite the comprehensive investigations regarding their crystal-chemical properties, band structures, spin and orbital physics, phase diagrams, pairing symmetry and other physical properties.

Keywords: Superconductivity, Fe-based superconductors, Crystallochemistry, Phase diagrams, Band structure, Magnetic structure, Nematicity

References

Abrahams, E. and Q. Si. 2011. Quantum criticality in the iron pnictides and chalcogenides. J. Phys.: Condens. Matter 23: 223201.

Avci, S., O. Chmaissem, D.Y. Chung, S. Rosenkranz, E.A. Goremychkin, J.P. Castellan et al. 2012. Phase diagram of $Ba_{1-x}K_xFe_2As_2$. Phys. Rev. B 85: 184507.

Bao, W. 2015. Structure, magnetic order and excitations in the 245 family of Fe-based superconductors. J. Phys.: Condens. Matter 27: 023201.

Bascones, E., B. Valenzuela and M.J. Calderón. 2015. Magnetic Interactions in Iron Superconductors: A review, C. R. Physique (in press).

Bernhard, C., C.N. Wang, L. Nuccio, L. Schulz, O. Zaharko, J. Larsen et al. 2012. Muon spin rotation study of magnetism and superconductivity in $Ba(Fe_{1-x}Co_x)_2As_2$ single crystals. Phys. Rev. B 86: 184509.

Brüning, E.M., C. Krellner, M. Baenitz, A. Jesche, F. Steglich and C. Geibel. 2008. CeFePO: A heavy fermion metal with ferromagnetic correlations. Phys. Rev. Lett. 101: 117206.

Carretta, P., R. De Renzi, G. Prando and S. Sanna. 2013. A view from inside iron-based superconductors. Phys. Scr. 88: 068504.

Chang, J., E. Blackburn, A.T. Holmes, N.B. Christensen, J. Larsen, J. Mesot et al. 2012. Direct observation of competition between superconductivity and charge density wave order in $YBa_2Cu_3O_{6.67}$. Nature Physics 8: 871–876.

Chen, Y., J.W. Lynn, J. Li, G. Li, G.F. Chen, J.L. Luo et al. 2008. Magnetic order of the iron spins in NdFeAsO. Phys. Rev. B 78: 064515.

Chen, C.-C., J. Maciejko, A.P. Sorini, B. Moritz, R.R.P. Singh and T.P. Devereaux. 2010. Orbital order and spontaneous orthorhombicity in iron pnictides. Phys. Rev. B 82: 100504(R).

Chen, F., Y. Zhang, J. Wei, B. Zhou, L. Yang, F. Wu et al. 2011. Electronic structure reconstruction of CaFe2As2 in the spin density wave state. J. Phys. Chem. Solids 72: 469–473.

Chubukov, A.V., D.V. Efremov and I. Eremin. 2008. Magnetism, superconductivity, and pairing symmetry in iron-based superconductors. Phys. Rev. B 78: 134512.

Dai, P., J. Hu and E. Dagotto. 2012. Magnetism and its microscopic origin in iron-based high-temperature superconductors. Nature Physics 8: 709–718.

Dai, P. 2015. Antiferromagnetic order and spin dynamics in iron-based superconductors. Rev. Mod. Phys. 87: 855–897.

Dong, J., H.J. Zhang, G. Xu, Z. Li, G. Li, W.Z. Hu et al. 2008. Competing orders and spin-density-wave instability in $La(O_{1-x}F_x)FeAs$. Europhysics Letters 83: 27006.

De la Cruz, C., Q. Huang, J.W. Lynn, Jiying Li, W. Ratcliff II, J.L. Zarestky et al. 2008. Magnetic order close to superconductivity in the iron-based layered $La(O_{1-x}F_x)FeAs$ systems. Nature 453: 899–902.

Fang, C., H. Yao, W.-F. Tsai, J. Hu and S.A. Kivelson. 2008. Theory of electron nematic order in LaFeAsO. Phys. Rev. B 77: 224509.

Fernandes, R.M., D.K. Pratt, W. Tian, J. Zarestky, A. Kreyssig, S. Nandi et al. 2010. Unconventional pairing in the iron arsenide superconductors. Phys. Rev. B 81: 140501(R).

Fernandes, R.M., A.V. Chubukov, J. Knolle, I. Eremin and J. Schmalian. 2012. Preemptive nematic order, pseudogap, and orbital order in the iron pnictides. Phys. Rev. B 85: 024534; Erratum Phys. Rev. B 85: 109901.

Fernandes, R.M. and J. Schmalian. 2012. Manifestations of nematic degrees of freedom in the magnetic, elastic, and superconducting properties of the iron pnictides. Supercond. Sci. Technol. 25: 084005.

Fernandes, R.M., A.V. Chubukov and J. Schmalian. 2014. What drives nematic order in iron-based superconductors? Nature Physics 10: 97–104.

Ge, J.-F., Z.-L. Liu, C. Liu, C.-L. Gao, D. Qian, Q.-K. Xue et al. 2015. Superconductivity above 100 K in single-layer FeSe films on doped $SrTiO_3$. Nature Materials 14: 285–289.

Giovannetti, G., C. Ortix, M. Marsman, M. Capone, J. van den Brink and J. Lorenzana. 2011. Proximity of iron pnictide superconductors to a quantum tricritical point. Nature Communications 2: 398.

Goldman, A.I., A. Kreyssig, K. Prokeš, D.K. Pratt, D.N. Argyriou, J.W. Lynn et al. 2009. Lattice collapse and quenching of magnetism in $CaFe_2As_2$ under pressure: A single-crystal neutron and x-ray diffraction investigation. Phys. Rev. B 79: 024513.

Graser, S., T.A. Maier, P.J. Hirschfeld and D.J. Scalapino. 2009. Near-degeneracy of several pairing channels in multiorbital models for the Fe pnictides. New Journal of Physics 11: 025016.

Horigane, K., H. Hiraka and K. Ohoyama. 2009. Relationship between structure and superconductivity in $FeSe_{1-x}Te_x$. J. Phys. Soc. Jpn. 78: 074718.

Hu, R., E.S. Bozin, J.B. Warren and C. Petrovic. 2009. Superconductivity, magnetism, and stoichiometry of single crystals of $Fe_{1+y}(Te_{1-x}S_x)_z$. Phys. Rev. B 80: 214514.

Huang, Q., J. Zhao, J.W. Lynn, G.F. Chen, J.L. Luo, N.L. Wang et al. 2008a. Doping evolution of antiferromagnetic order and structural distortion in $LaFeAsO_{1-x}F_x$. Phys. Rev. B 78: 054529.

Huang, Q., Y. Qiu, Wei Bao, M.A. Green, J.W. Lynn, Y.C. Gasparovic et al. 2008b. Neutron-diffraction measurements of magnetic order and a structural transition in the parent $BaFe_2As_2$ compound of FeAs-based high-temperature superconductors. Phys. Rev. Lett. 101: 257003.

Iimura, S., S. Matsuishi, H. Sato, T. Hanna, Y. Muraba, S.W. Kim et al. 2012. Two-dome structure in electron-doped iron arsenide superconductors. Nature Communications 3: 943.

Inosov, D.S., J.T. Park, P. Bourges, D.L. Sun, Y. Sidis, A. Schneidewind et al. 2010. Normal-state spin dynamics and temperature-dependent spin-resonance energy in optimally doped $BaFe_{1.85}Co_{0.15}As_2$. Nature Physics 6: 178–181.

Inosov, D.S. 2015. Spin fluctuations in iron pnictides and chalcogenides: From antiferromagnetism to superconductivity. C. R. Physique 17: 60–89.

Johnston, D.C. 2010. The puzzle of high temperature superconductivity in layered iron pnictides and chalcogenides. Advances in Physics 59: 803–1061.

Johrendt, D. 2011. Structure–property relationships of iron arsenide superconductors. J. Mater. Chem. 21: 13726.

Joseph, B., A. Iadecola, A. Puri, L. Simonelli, Y. Mizuguchi, Y. Takano et al. 2010. Evidence of local structural inhomogeneity in $FeSe_{1-x}Te_x$ from extended x-ray absorption fine structure. Phys. Rev. B 82: 020502(R).

Kamihara, Y., H. Hiramatsu, M. Hirano, R. Kawamura, H. Yanagi, T. Kamiya et al. 2006. Iron-based layered superconductor: LaOFeP. J. Am. Chem. Soc. 128: 10012–10013.

Kamihara, Y., T. Watanabe, M. Hirano and H. Hosono. 2008. Iron-based layered superconductor $La[O_{1-x}F_x]FeAs$ (x = 0.05–0.12) with Tc = 26 K. J. Am. Chem. Soc. 130: 3296–3297.

Khasanov, R., M. Bendele, K. Conder, H. Keller, E. Pomjakushina, V. Pomjakushin. 2010. Iron isotope effect on the superconducting transition temperature and the crystal structure of $FeSe_{1-x}$. New J. Phys. 12: 073024.

Kim, M.G., R.M. Fernandes, A. Kreyssig, J.W. Kim, A. Thaler, S.L. Bud'ko et al. 2011. Character of the structural and magnetic phase transitions in the parent and electron-doped $BaFe_2As_2$ compounds. Phys. Rev. B 83: 134522.

Kimber, S.A.J., D.N. Argyriou, F. Yokaichiya, K. Habicht, S. Gerischer, T. Hansen et al. 2008. Magnetic ordering and negative thermal expansion in PrFeAsO. Phys. Rev. B 78: 140503(R).

Kontani, H. and S. Onari. 2010. Orbital-fluctuation-mediated superconductivity in iron pnictides: Analysis of the five-orbital hubbard-holstein model. Phys. Rev. Lett. 104: 157001.

Kreyssig, A., M.A. Green, Y. Lee, G.D. Samolyuk, P. Zajdel, J.W. Lynn et al. 2008. Pressure-induced volume-collapsed tetragonal phase of $CaFe_2As_2$ as seen via neutron scattering. Phys. Rev. B 78: 184517.

Krüger, F., S. Kumar, J. Zaanen and J. van den Brink. 2009. Spin-orbital frustrations and anomalous metallic state in iron-pnictide superconductors. Phys. Rev. B 79: 054504.

Kumar, J., S. Auluck, P.K. Ahluwalia and V.P.S. Awana. 2012. Chalcogen height dependence of magnetism and Fermiology in $FeTe_xSe_{1-x}$. Supercond. Sci. Technol. 25: 095002.

Kuroki, K., S. Onari, R. Arita, H. Usui, Y. Tanaka, H. Kontani et al. 2008. Unconventional pairing originating from the disconnected Fermi surfaces of superconducting $LaFeAsO_{1-x}F_x$. Phys. Rev. Lett. 101: 087004.

Kuroki, K., H. Usui, S. Onari, R. Arita and H. Aoki. 2009. Pnictogen height as a possible switch between high-Tc nodeless and low-Tc nodal pairings in the iron-based superconductors. Phys. Rev. B 79: 224511.

Lee, C.-C., W.-G. Yin and W. Ku. 2009. Ferro-orbital order and strong magnetic anisotropy in the parent compounds of iron-pnictide superconductors. Phys. Rev. Lett. 103: 267001.

Liu, J., B. Luo, Z.Y. Sun, H. Fu and K. Yao. 2011. *Ab initio* investigation of the noncollinear magnetic structure of CeFeAsO. Phys. Rev. B 84: 115123.

Louca, D., K. Horigane, A. Llobet, R. Arita, S. Ji, N. Katayama et al. 2010. Local atomic structure of superconducting $FeSe_{1-x}Te_x$. Phys. Rev. B 81: 134524.

Lumsden, M.D. and A.D. Christianson. 2010. Magnetism in Fe-based superconductors. J. Phys.: Condens. Matter 22: 203203.

Lv, W., J. Wu and P. Phillips. 2009. Orbital ordering induces structural phase transition and the resistivity anomaly in iron pnictides. Phys. Rev. B 80: 224506.

Ma, F. and Z.-Y. Lu. 2008. Iron-based layered compound LaFeAsO is an antiferromagnetic semimetal. Phys. Rev. B 78: 033111.

Ma, F.J., Z.Y. Lu and T. Xiang. 2008. Arsenic-bridged antiferromagnetic superexchange interactions in LaFeAsO. Phys. Rev. B 78: 224517.

Maeter, H., H. Luetkens, Yu. G. Pashkevich, A. Kwadrin, R. Khasanov, A. Amato et al. 2009. Interplay of Rare Earth and Iron magnetism in ReOFeAs with Re = La, Ce, Pr, and Sm: A muon spin relaxation study and symmetry analysis. Phys. Rev. B 80: 094524.

Margadonna, S., Y. Takabayashi, M.T. McDonald, K. Kasperkiewicz, Y. Mizuguchi, Y. Takano et al. 2008. Crystal structure of the new $FeSe_{1-x}$ superconductor. Chem. Commun. 5607–5609.

Margadonna, S., Y. Takabayashi, M.T. McDonald, M. Brunelli, G. Wu, R.H. Liu et al. 2009. Crystal structure and phase transitions across the metal-superconductor boundary in the $SmFeAsO_{1-x}F_x$ ($0 \leq x \leq 0.20$) family. Phys. Rev. B 79: 014503.

Martinelli, A., M. Ferretti, P. Manfrinetti, A. Palenzona, M. Tropeano, M.R. Cimberle et al. 2008. Synthesis, crystal structure, microstructure, transport and magnetic properties of SmFeAsO and $SmFeAs(O_{0.93}F_{0.07})$. Supercond. Sci. Technol. 21: 095017.

Martinelli, A., A. Palenzona, M. Tropeano, C. Ferdeghini, M.R. Cimberle and C. Ritter. 2009. Neutron powder diffraction investigation of the structural and magnetic properties of $(La_{1-y}Y_y)FeAsO$. Phys. Rev. B 80: 214106.

Martinelli, A., A. Palenzona, M. Tropeano, C. Ferdeghini, M. Putti, M.R. Cimberle et al. 2010. From antiferromagnetism to superconductivity in $Fe_{1+y}(Te_{1-x}Se_x)$ ($0 \leq x \leq 0.20$): a neutron powder diffraction analysis. Physical Review B 81: 094115.

Martinelli, A., A. Palenzona, M. Tropeano, M. Putti, C. Ferdeghini, G. Profeta et al. 2011. Retention of the tetragonal to orthorhombic structural transition in F-substituted SmFeAsO: a new phase diagram for $SmFeAs(O_{1-x}F_x)$. Phys. Rev. Lett. 106: 227001.

Martinelli, A., A. Palenzona, M. Putti and C. Ferdeghini. 2012. Microstructural evolution throughout the structural transition in 1111 oxy-pnictides. Phys. Rev. B 85: 224534.

Martinelli, A. 2013. Symmetry-mode and spontaneous strain analysis of the structural transition in $Fe_{1+y}Te$ and REFeAsO compounds. J. Physics: Condens. Matter 25: 125703.

Martinelli, A. 2015. Crystallochemistry of Fe-Based superconductors: Interplay between chemical, structural and physical properties in the Fe(Se,Te) and 1111-Type systems. J. Supercond. Nov. Magn. 28: 1103–1106.

Martinelli, A., F. Bernardini and S. Massidda. 2015. The phase diagrams of iron-based superconductors: theory and experiments. Comptes Rendus Physique (in press).

Mazin, I.I. 2010. Superconductivity gets an iron boost. Nature 464: 183–186.

Mazin, I.I., D.J. Singh, M.D. Johannes and M.H. Du. 2008. Unconventional superconductivity with a sign reversal in the order parameter of $LaFeAsO_{1-x}F_x$. Phys. Rev. Lett. 101: 057003.

McQueen, T.M., Q. Huang, V. Ksenofontov, C. Felser, Q. Xu, H. Zandbergen et al. 2009a. Extreme sensitivity of superconductivity to stoichiometry in $Fe_{1+d}Se$, Phys. Rev. B 79: 014522.

McQueen, T.M., A.J. Williams, P.W. Stephens, J. Tao, Y. Zhu, V. Ksenofontov et al. 2009b. Tetragonal-to-Orthorhombic Structural Phase Transition at 90 K in the Superconductor $Fe_{1.01}Se$. Phys. Rev. Lett. 103: 057002.

Medvedev, S., T.M. McQueen, I.A. Troyan, T. Palasyuk, M.I. Eremets, R.J. Cava et al. 2009. Electronic and magnetic phase diagram of β-$Fe_{1.01}Se$ with superconductivity at 36.7 K under pressure. Nature Materials 8: 630–633.

Mizuguchi, Y., F. Tomioka, S. Tsuda, T. Yamaguchi and Y. Takano. 2009. Substitution effect on FeSe superconductor. J. Phys. Soc. Jpn. 78: 074712.

Mizuguchi, Y., Y. Hara, K. Deguchi, S. Tsuda, T. Yamaguchi, K. Takeda et al. 2010. Anion height dependence of Tc for the Fe-based superconductor. Supercond. Sci. Technol. 23: 054013.

Nomura, T., S.W. Kim, Y. Kamihara, M. Hirano, P.V. Sushko, K. Kato et al. 2008. Crystallographic phase transition and high-Tc superconductivity in LaFeAsO:F. Supercond. Sci. Technol. 21: 125028.

Okada, H., K. Igawa, H. Takahashi, Y. Kamihara, M. Hirano, H. Hosono et al. 2008. Superconductivity under High Pressure in LaFeAsO. J. Phys. Soc. Jpn. 77: 113712.

Paglione, J. and R.L. Greene. 2010. High-temperature superconductivity in iron-based materials. Nature Physics 6: 6452011.

Qian, T., X.-P. Wang, W.-C. Jin, P. Zhang, P. Richard, G. Xu et al. 2011. Absence of a holelike Fermi surface for the iron-based $K_{0.8}Fe_{1.7}Se_2$ superconductor revealed by angle-resolved photoemission spectroscopy. Phys. Rev. Lett. 106: 187001.

Pallecchi, I.F. Bernardini, M. Tropeano, A. Palenzona, A. Martinelli, C. Ferdeghini et al. 2011. Magnetotransport in La(Fe,Ru)AsO as a probe of band structure and mobility. Phys. Rev. B 84: 134524.

Parker, D., M.G. Vavilov, A.V. Chubukov and I.I. Mazin. 2009. Coexistence of superconductivity and a spin-density wave in pnictide superconductors: Gap symmetry and nodal lines. Phys. Rev. B 80: 100508(R).

Pesin, D. and A.H. MacDonald. 2012. Spintronics and pseudospintronics in graphene and topological insulators. Nature Materials 11: 409–416.

Pomjakushina, E., K. Conder, V. Pomjakushin, M. Bendele and R. Khasanov. 2009. Synthesis, crystal structure, and chemical stability of the superconductor $FeSe_{1-x}$. Phys. Rev. B 80: 024517.

Quebe, P., L.J. Terbüchte and W. Jeitschko. 2000. Quaternary rare earth transition metal arsenide oxides RTAsO (T = Fe, Ru, Co) with ZrCuSiAs type structure. J. All. Comp. 302: 70–74.

Richard, P., K. Nakayama, T. Sato, M. Neupane, Y.-M. Xu, J.H. Bowen et al. 2010. Observation of Dirac cone electronic dispersion in BaFe2As2. Phys. Rev. Lett. 104: 137001.

Rotter, M., M. Tegel and D. Johrendt. 2008. Superconductivity at 38 K in the Iron Arsenide $(Ba_{1-x}K_x)Fe_2As_2$. Phys. Rev. Lett. 101: 107006.

Rotundu, C.R., B. Freelon, T.R. Forrest, S.D. Wilson, P.N. Valdivia, G. Pinuellas et al. 2010. Heat capacity study of $BaFe_2As_2$: Effects of annealing. Phys. Rev. B 82: 144525.

Ryan, D.H., J.M. Cadogan, C. Ritter, F. Canepa, A. Palenzona and M. Putti. 2009. Coexistence of long-ranged magnetic order and superconductivity in the pnictide superconductor $SmFeAsO_{1-x}F_x$ (x = 0, 0.15), Phys. Rev. B 80: 220503(R).

Sanna, S., R. De Renzi, G. Lamura, C. Ferdeghini, A. Palenzona, M. Putti et al. 2009. Magnetic-superconducting phase boundary of $SmFeAsO_{1-x}F_x$ studied via muon spin rotation: Unified behavior in a pnictide family. Phys. Rev. B 80: 052503.

Sanna, S., R. De Renzi, T. Shiroka, G. Lamura, G. Prando, P. Carretta et al. 2010. Nanoscopic coexistence of magnetic and superconducting states within the FeAs layers in CeFeAs $O_{1-x}F_x$. Physical Review B 82: 060508.

Shimojima, T., K. Ishizaka, Y. Ishida, N. Katayama, K. Ohgushi, T. Kiss et al. 2010. Orbital-dependent modifications of electronic structure across the magnetostructural transition in $BaFe_2As_2$. Phys. Rev. Lett. 104: 057002.

Si, Q.M. and E. Abrahams. 2008. Strong correlations and magnetic frustration in the high T_c iron pnictides. Phys. Rev. Lett. 101: 076401.

Singh, D.J. and M.H. Du. 2008. Density functional study of $LaFeAsO_{1-x}F_x$: A low carrier density superconductor near itinerant magnetism. Phys. Rev. Lett. 100: 237003.

Subedi, A., L. Zhang, D.J. Singh and M.H. Du. 2008. Density functional study of FeS, FeSe, and FeTe: Electronic structure, magnetism, phonons, and superconductivity. Phys. Rev. B 78: 134514.

Takahashi, H., H. Okada, K. Igawa, Y. Kamihara, M. Hirano, H. Hosono et al. 2009. High-pressure studies on superconductivity in LaFeAsO1–xFx and $SmFeAsO_{1-x}F_x$. J. Supercond. Nov. Magn. 22: 595–598.

Tegel, M., C. Löhnert and D. Johrendt. 2010. The crystal structure of $FeSe_{0.44}Te_{0.56}$. Solid State Commun. 150: 383–458.

Tropeano, M., C. Fanciulli, F. Canepa, M.R. Cimberle, C. Ferdeghini, G. Lamura et al. 2009. Effect of chemical pressure on spin density wave and superconductivity in undoped and 15% F-doped $La_{1-y}Y_yFeAsO$ compounds. Phys. Rev. B 79: 174523.

Uhoya, W., A. Stemshorn, G. Tsoi, Y.K. Vohra, A.S. Sefat, B.C. Sales et al. 2010a. Collapsed tetragonal phase and superconductivity of $BaFe_2As_2$ under high pressure. Phys. Rev. B 82: 144118.

Uhoya, W., G. Tsoi, Y.K. Vohra, M.A. McGuire, A.S. Sefat, B.C. Sales et al. 2010b. Anomalous compressibility effects and superconductivity of $EuFe_2As_2$ under high pressures. J. Phys.: Condens. Matter 22: 292202.

Uhoya, W.O., J.M. Montgomery, G.M. Tsoi, Y.K. Vohra, M.A. McGuire, A.S. Sefat et al. 2011. Phase transition and superconductivity of $SrFe_2As_2$ under high pressure. J. Phys.: Condens. Matter 23: 122201.

Vojta, M. 2009. Lattice symmetry breaking in cuprate superconductors: stripes, nematics, and superconductivity. Advances in Physics 58: 699–820.

Watanabe, T., H. Yanagi, T. Kamiya, Y. Kamihara, H. Hiramatsu, M. Hirano et al. 2007. Nickel-based oxyphosphide superconductor with a layered crystal structure. LaNiOP. Inorg. Chem. 46: 7719–7721.

Wilson, J.A. 2010. A perspective on the Fe-based superconductors. J. Phys.: Condens. Matter 22: 203201.

Xu, C., M. Muller and S. Sachdev. 2008. Ising and spin orders in the iron-based superconductors. Phys. Rev. B 78: 020501(R).

Yi, M., D. Lu, J.-H. Chu, J.G. Analytis, A.P. Sorini, A.F. Kemper et al. 2011. Symmetry-breaking orbital anisotropy observed for detwinned $Ba(Fe_{1-x}Co_x)_2As_2$ above the spin density wave transition. PNAS 108: 6878–6883.

Yildirim, T. 2008. Origin of the 150 K anomaly in LaFeAsO: competing antiferromagnetic interactions, frustration, and a structural phase transition. Phys. Rev. Lett. 101: 057010.

Zhang, Y., F. Chen, C. He, B. Zhou, B.P. Xie, C. Fang et al. 2011. Orbital characters of bands in the iron-based superconductor $BaFe_{1.85}Co_{0.15}As_2$. Phys. Rev. B 83: 054510.

Zhang, Y., C. He, Z.R. Ye, J. Jiang, F. Chen, M. Xu et al. 2012. Symmetry breaking via orbital-dependent reconstruction of electronic structure in detwinned NaFeAs. Phys. Rev. B 85: 085121.

Zhao, J., Q. Huang, C. de la Cruz, J.W. Lynn, M.D. Lumsden, Z.A. Ren et al. 2008a. Lattice and magnetic structures of PrFeAsO, $PrFeAsO_{0.85}F_{0.15}$, and $PrFeAsO_{0.85}$. Phys. Rev. B 78: 132504.

Zhao, J., W. Ratcliff, II, J.W. Lynn, G.F. Chen, J.L. Luo, N.L. Wang et al. 2008b. Spin and lattice structures of single-crystalline $SrFe_2As_2$. Phys. Rev. B 78: 140504(R).

Zhao, J., Q. Huang, C. de la Cruz, S. Li, J.W. Lynn, Y. Che et al. 2008c. Structural and magnetic phase diagram of CeFeAsO$_{1-x}$F$_x$ and its relation to high-temperature superconductivity. Nature Materials 7: 953–959.

Zhao, J., D.T. Adroja, D.-X. Yao, R. Bewley, S. Li, X.F. Wang et al. 2009. Spin waves and magnetic exchange interactions in CaFe$_2$As$_2$. Nature Physics 5: 555–560.

Zhi-An, R., L. Wei, Y. Jie, Y. Wei, S. Xiao-Li, Zheng-Cai et al. 2008. Superconductivity at 55 K in Iron-Based F-Doped Layered Quaternary Compound Sm[O$_{1-x}$F$_x$]FeAs. Chin Phys. Lett. 25: 2215–2216.

Zimmer, B.I., W. Jeitschko, J.H. Albering, R. Glaum and M. Reehuis. 1995. The rate earth transition metal phosphide oxides LnFePO, LnRuPO and LnCoPO with ZrCuSiAs type structure. J. Alloy. Compd. 229: 238–242.

Zocco, D.A., R.E. Baumbach, J.J. Hamlin, M. Janoschek, I.K. Lum, M.A. McGuire et al. 2011. Search for pressure-induced superconductivity in CeFeAsO and CeFePO iron pnictides. Phys. Rev. B 83: 094528.

2.4

Electronic and Magnetic Properties of Highly Disordered Fe-based Frank Kasper-Phases in View of First Principles Calculation and Experimental Study

J. Cieslak* and *J. Tobola*

What are the Frank-Kasper phases and Why is the σ-phase so Curious?

Frank-Kasper phases (FK) have been originally called 'Complex alloy structures regarded as sphere packing's' by their authors (Frank and Kasper 1958; Frank and Kasper 1959). These structures are mainly formed by transition metals, but there are also known examples of these phases which contain other elements (e.g., Al or Au), surprisingly even polymers can crystallize in σ-structure (Lee et al. 2010). The most characteristic feature of these phases is the topological arrangement of atoms, ordered in such a way that each of them is located in the center of a polyhedron

AGH University of Science and Technology, Faculty of Physics and Applied Computer Science, al. Mickiewicza 30, 30-059 Krakow, Poland.
* Corresponding author: Jakub.Cieslak@fis.agh.edu.pl

with triangular walls. There are four possible polyhedra having 20, 24, 26 or 28 faces, which correspond to the following coordination numbers: CN = 12, 14, 15 and 16. Up to now, there have been at least 27 such phases discovered in various systems, whereas some authors (Sikrić et al. 2010) suggested 84 topologically possible structures undergoing definition of the FK phase. So, at least 57 new phases are still waiting for discovery.

The most popular of all FK phases is definitely the σ-phase, which was discovered in 1927 in a FeCrNi alloy (Bain and Griffits 1927); 24 years later its crystallographic structure was determined (Bergman and Shoemaker 1951; Begman and Shoemaker 1954). It is now the most frequently identified and the most intensively studied of all FK phases. Like other FK phases it appears in a limited temperature range (e.g., 500–830°C in the FeCr system or 1500–1600°C in the FeMo one). The σ-phase remains stable below the temperature range of its transformation and it dissolves at higher temperatures. Its mechanical properties (high brittleness, high hardness) are typical for all other FK phases. Currently, there are known c.a. 50 binary (and several hundreds of ternary or other multi-component) systems wherein the σ-phase was identified (Hall and Algie 1966). Structural parameters of this phase, similar for all systems, are presented in Table 1 on the example of the FeCr sigma phase.

Technological applications of σ-phases (as well as other FK phases) is rather limited, although some attempts have been undertaken to employ its unusual hardness (Shibuya et al. 2014; Kuhn et al. 2014). In most cases, however, this phase has to be avoided in steels operating at high temperatures, since its presence as inclusions in the matrix drastically

Table 1: Atomic crystallographic positions, numbers of NN atoms and distances between NN atoms for the five lattice sites of the FeCr σ-phase. The average values with standard deviations (in parentheses) are indicated, when more than one distance is possible. The corresponding weighted mean values are given in the last row.

	Crystallographic positions	A	B	C	D	E	Total
		NN numbers					
A	2i (0.000, 0.000, 0.000)	-	4	-	4	4	12
B	4f (0.400, 0.400, 0.000)	2	1	2	4	6	15
C	8i (0.740, 0.066, 0.000)	-	1	5	4	4	14
D	8i (0.464, 0.131, 0.000)	1	2	4	1	4	12
E	8j (0.183, 0.183, 0.252)	1	3	4	4	2	14
		NN distances (Å)					average
A		—	2.605	—	2.366	2.547	2.506
B		2.605	2.519	2.420	2.695	2.864(5)	2.702
C		—	2.420	2.75(2)	2.487(4)	2.766(6)	2.655
D		2.366	2.695	2.487(4)	2.422	2.55(2)	2.572
E		2.547	2.864	2.766(6)	2.549(19)	2.28(2)	2.640

deteriorates the mechanical properties of the material (Hsieh and Weite 2012; Elmer et al. 2007). Unfortunately, the σ-phases can occur in many types of steels. In such situations, efforts focus mostly on delaying their appearance, on slowing their transformation rate (usually by adding proper dopants to the alloy) (Blachowski et al. 2000) or by their dissolution at higher temperatures.

From a scientific point of view, FK phases constitute a very interesting object of investigations, as well as a really challenging subject for experiment and theory. These structures are always two- or more-component alloys and occur in rather limited range of concentrations. Usually they crystallize in cubic, tetragonal, hexagonal or rhombohedral structures with the unit cell containing from 7 (Z phase, e.g., Zr_4Al_3) to 93 atoms (ν phase, e.g., $Mn_{81.5}Si_{18.5}$), which are spread over several (or even dozen) non-equivalent crystallographic positions (forming sublattices). Although FK phases are topologically ordered, they are characterized by extremely high degree of chemical disorder, which is usually different for different sublattices. In practice, it means selective occupations of individual sublattices by various elements, constituting the alloy. A compilation of the literature data can be found in (Sinha 1972).

In this rich family of FK phases and in view of the aforementioned criteria, the σ-phase can be regarded as a typical representative of all FK phases. It crystallizes in a tetragonal unit cell (D^{14}_{4h}, $P4_2/mnm$) containing 30 atoms distributed over five sublattices (generally referred to as A, B, C, D and E) selectively occupied by alloying elements. Structural studies of the σ-phase by X-ray diffraction (XRD) allow the identification of the crystal structure via determination of lattice parameters and atomic positions. Due to the fact that the elements constituting σ-phases have often close atomic numbers, the distinction between sublattices is practically impossible in such cases and neutron diffraction (ND) becomes a very useful tool (Cieslak et al. 2008a). An additional serious difficulty for structural studies is the absence of an efficient method to get single crystals of the σ-phases, so that the measurements are limited to the powder diffraction studies (XRD or ND). One can only rely on fortunate finding monocrystalline grains in the polycrystalline material of the σ-phase, but this method allows researchers to get single crystals smaller than ~100 microns in diameter.

Experimental magnetic studies have shown that all iron-containing binary σ-phases (FeCr, FeV, FeMo and FeRe) exhibit weak magnetic properties and complex behaviors (Parsons 1960; Read and Thomas 1966; Read et al. 1968; Dubiel et al. 2010a; Cieslak 2014a; Cieslak 2015a). In most cases, they manifest themselves only at low temperatures (less than 80 K), with the exception of the FeV system in which magnetic ordering can be observed even at room temperature (RT) for the Fe-richest samples.

Recently, a spin glass state was detected at low temperatures in these phases, being additionally of a re-entrant type in the case of FeCr and FeV systems (Barco et al. 2012).

Other experimental methods, such as Mössbauer spectroscopy (MS) or nuclear magnetic resonance (NMR), may probe local and selective properties of the σ-phase. Using the MS technique, measurements are possible in limited cases, where one of the alloying elements is Mössbauer isotope (mainly ^{57}Fe). Similarly, the NMR technique can be used for selected isotopes only. On the whole, both MS and NMR measurements confirmed the random distribution of elements on the sublattices in FeX σ-phases (Dubiel et al. 2010a; Cieslak et al. 2012b; Cieslak et al. 2013a). These methods also made it possible to determine magnetic ordering temperatures and allowed researchers to establish various hyperfine parameters characterizing iron on different sites in FK phases. It should be noted, however, that due to a chemical disorder and complex crystal structures appearing in most cases, reliable interpretation of experimental findings was possible only using the results of the electronic structure calculations (see below).

What are we able to Calculate and What is Worth Computing for FK Phases?

By undertaking complementary experimental and theoretical analysis of a crystalline system, one has to propose the model of electronic structure and next, to decide to what extent experimental data can be used as input for calculation, and to what extent they must be calculated independently to allow comparison with measured data.

The answer to this question is directly related to the purpose of performed calculations. One may intend to test different computational methods (or different approaches) in a well-defined system (with known physical quantities) by comparing theoretical and experimental values. One may also want to learn more about the system using available computational methods, when experimental description of the system is insufficient. Finally, one may attempt to accomplish both of these goals as far as possible. At first glance, the situation seems to be the most attractive when conclusions are fully drawn on the basis of calculations. In such cases we refer to so-called first principles calculations (or *ab initio* computations). However, even in the aforementioned cases, our calculations always contain a certain amount of empirical data (e.g., the elementary charge and mass of the electron, physical constants, etc.). In materials science, one usually performs calculations for a given, empirically determined, crystal structure. Note that such calculations can be carried out for other hypothetical structures and energetically the most favorable one can be compared to the real structure. Similarly, values of lattice parameters—usually available experimentally—can be used in electronic structure calculations, especially with complex multicomponent crystal structures. Theoretical determination of lattice parameters and crystallographic positions of individual atoms is also possible numerically, by minimizing the formation energy of the system as a function of these parameters. Analysis of known systems confirms that the *ab initio* techniques (generally based on DFT methodology) in principle give results consistent with experimental data, and they were found to be very useful particularly in all those cases, where the analyzed system is hardly measurable for various reasons. However, in all cases when crystal data are available for complex multi-atom systems, they are usually used in computations to save time and energy.

The calculations of the electronic structure are subject to various approximations. On one hand these approximations are due to the calculation method itself. On the other one, approximations concern the models chosen to describe the system.

Description of the chemically disordered system must generally be based on the average values of physical quantities. This is due to a multitude of possible relative arrangements of atoms constituting the system, having a slightly different electronic structure in each case. Taking into account the influence of not only the nearest, but also more distant atomic neighbors, we face the necessity of constructing multiplied unit cells (so-called supercells) to perform calculations on a vast number of different atomic arrangements (just to cover the most representative cases). However, the application of more and more complex supercells in general becomes impossible due to time-consuming computations, limits of available memory, stability of algorithms, etc.

As an example, let us consider electronic structure calculations of simple binary disordered alloy $A_{1-x}B_x$. In order to obtain the average values of the physical quantities two approaches can be proposed. The first one is based on the knowledge of the crystal structure and the probabilities of sublattices occupations by individual atoms. In addition, it should be assumed that the properties of a given atom mainly depend on the number of the respective neighbors (*a* or *b* in our case) in its nearest vicinity, and to a lesser extent, on their relative arrangements. Since the probability distribution P of finding the specific number of atoms of given type, (a_i, b_i), in the first, second and further coordination shells (denoted by index i) $P_{NNi} = P(a_iA + b_iB)$, can be obtained analytically (e.g., assuming a binomial distribution), one can calculate average properties of atoms characterized by probabilities of known configurations of neighbors. In practice, the calculations of the electronic structure are performed in a simple unit cell (neglecting the atoms' assignment to sublattices) for a finite number of ordered structures, where individual crystal positions are occupied randomly (but with probabilities specific to analyzed structure). This means the omission of the symmetry of crystallographic sites (due to presence of two atoms A and B), while still keeping translational symmetry. With this approach, we are able to analyze different properties of atoms as a function of the number of their neighbours of a given type, belonging to specified coordination shells. In addition, knowing the probability density of the occurrence of such arrangements, one has the possibility of averaging the results obtained for different configurations.

The second approach is based on the so-called coherent potential approximation (CPA), which is generally incorporated into one of the DFT methods, devoted to ordered compounds. In our calculations the CPA approach has been combined with the Korringa-Kohn-Rostoker (KKR) technique (Kaprzyk and Mijnarends 1986; Butler et al. 1990; Bansil et al. 1999; Stopa et al. 2004). The CPA model allows researchers to treat random atoms distribution on selective sites (one site in discussed $A_{1-x}B_x$ binary alloy) via construction of the effective medium, defined by the averaging of Green's functions ascribed to A and B atoms, namely G_A and G_B, over the concentrations, i.e., $G_{cp} = (1-x)G_A + xG_B$ and this condition is solved in a self-consistent way. As the CPA model does not rely on simple averaging of physical properties of atoms constituting the disordered alloy, it is not always easy to correlate the properties of the alloy that resulted from KKR-CPA calculations with the respective properties of the parent compounds. It is worth noting that within the CPA model, we are able to maintain the symmetry of the unit cell in the whole range of alloy composition, for which ground state properties can be computed (e.g., density of states, magnetic moments, etc.). On the other hand, a few important features measured by the local experimental techniques (MS or NMR) cannot be directly investigated by the KKR-CPA method, since all properties are averaged and there is no direct insight into the effect of local atomic arrangements. So, in the case of theoretical investigations of FK phases, the aforementioned approaches based on supercells computations as well as the CPA model are highly complementary, since the average values obtained using both types of electronic structure calculations are consistent.

Summarizing, the first approach requires more input data and consequently more time-consuming computations due to larger and variable unit cells, but it allows for much more comprehensive description of electronic, magnetic and hyperfine properties of the material. On the other hand, the second approach can give an insight into the properties, which can be measured as averaged (e.g., magnetization, lattice parameters, etc.).

For example, consider a binary alloy (e.g., *bcc* structure) $A_{30}B_{70}$ crystallizing in the hypothetical structure in which each atom has eight nearest neighbors and next six neighbors in a subsequent coordination shell. Let us limit the discussion only to the first two coordination shells (NN and NNN), and suppose further that only one sublattice is present. In this case, each site can be occupied by both types of atoms, that is, A with a probability of 3/10 and B with a probability of 7/10. In the CPA approach, calculations should be performed for one case only, but taking into account all possible configurations within the CPA self-consistent model. In the supercells approach one should consider various numbers of atoms in both coordination shells. In general, for two types of atoms and, in total, $6 + 8 = 14$ possible positions to be occupied, there are 2^{14} different combinations each of which is equally probable. However, starting from the assumption that only the number of neighbors of a given type (and not their arrangement) in each coordination shell affects the electronic structure of the atom, one can find only 63 different configurations of NN-NNN neighbors. Please note that the probabilities of such configurations are different to each other and can be easily determined from binominal distribution. By taking for further analysis only those configurations with probability greater than 0.01 (24 of 63 in the case of $A_{30}B_{70}$ system), one accounts for 94% of all possibilities. In practice, the number of analyzed unit cells is even smaller, since during calculations performed for a given unit cell, one can take into account simultaneously several specific probabilities, corresponding to different positions of atoms with various NN values.

Microscopic Properties—Mössbauer Measurements and Hyperfine Parameters

As already mentioned, the experimental methods employed for the investigation of materials can be roughly classified as global and local methods. The former include, for example, diffraction methods (XRD, ND), in which the useful signal is derived from the averaged structure (in some areas of the sample). Then, each part of this structure (so called coherence domain) is the source of signal that carries information about the average parameters characterizing this part of the sample. Conversely, in other methods, called the local ones (e.g., MS or NMR), signals are generated by single-atom probes, electronic states of which depend mainly on their local atomic neighborhoods. In this situation, each probe is the source of different and specific partial signal. The integration of these partial signals is carried out only at the stage of collecting the information from the whole analyzed area of the sample.

In principle, experimental studies of FK phases may be carried out using both local and global methods. In MS technique, the presence of Mössbauer isotopes (usually ^{57}Fe) in the analyzed phase is essential. In practice, one can investigate most of the representatives of the FK phases, for example γ (Laves phases), μ, μ', χ, R, D, K or W. In the case of NMR technique, the range of studied phases is even higher. Unfortunately, the correct interpretation of the experimental results is not easy, because the atom-probe may be located in a number of non-equivalent crystal positions, and additionally with various configurations of the nearest neighbors. Actually, such coincidence of different types of disorder makes Mössbauer spectrum hard to analyze, being composed of a large number of superimposed (overlapped) subspectra, whose unambiguous separation is not possible without additional information.

For example, the σ-phase has five non-equivalent sublattices with iron present on all these sites. Consequently, the resulting Mössbauer spectrum should consist of five subspectra, each of them must have some more or less complicated character due to the chemical disorder and large number of different configurations of the nearest neighbors. In the paramagnetic state (e.g., at room temperature), each subspectrum should have a form of quadrupole doublet generally described by four parameters: the amplitude, line width G, isomer shift IS and quadrupole splitting QS. Finally, taking into account the background of measurement, Mössbauer spectrum of the σ-phase should be described using *a priori* 21 independent parameters.

Unfortunately, the differences between the IS (and QS) values describing individual subspectra are often small, compared to the spectral line widths. Consequently, the measured spectrum does not show visible structure, which is the basis of the unique determination of particular hyperfine parameters. Hopefully, it is possible to make some reduction of the number of these parameters, when assuming that the G values for all subspectra are identical, thereby reducing five independent parameters to one. Moreover, site occupancies can be determined from diffraction experiments, so one can assume that the relative amplitudes of individual subspectra are proportional to the amount of iron in the corresponding sublattices. This additionally reduces the number of independent parameters characterizing the spectrum by next four, but still 13 other adjustable parameters cannot be uniquely determined based on the measured spectrum only.

At this stage of the analysis it is reasonable to acquire further parameters through electronic structure calculations. Since the CPA approach provides the average values only, one must therefore perform calculations for a number of ordered approximants (supercells), and then average the results based on the known probability distributions as described above. Electron densities at the nuclei (ρ_e) and atomic potentials (*V*) calculated for each atom in the unit cell, available from KKR and KKR-CPA methods are particularly interesting for such analysis. The differences between the ρ_e values of Fe atoms on different sublattices are proportional to the differences between the corresponding values of *IS* (note that MS enables to measure only the relative values of *IS*). Based on the knowledge of electric fields (derived from atomic potentials *V*) one can construct a lattice component of the electric field gradient tensor (EFG) for each atom in the crystal structure, and on that basis, the *QS* value corresponding to each sublattice can be determined. Finally, using results of the electronic structure calculations the number of adjustable parameters describing Mössbauer spectrum can be reduced to five, only. Consequently, one can unambiguously determine the remaining five parameters as a result of the fitting procedure (see Fig. 1). The aforementioned procedure combining experimental data derived from additional techniques (XRD and ND) and the first principles electronic structure calculations (KKR and KKR-CPA), have been successfully applied to analyze Mössbauer spectra of different binary and ternary Fe-containing σ-phases, both in paramagnetic (Cieslak et al. 2008c; Cieslak et al. 2010b; Cieslak et al. 2012b; Cieslak et al. 2013a) and magnetic states (Cieslak et al. 2010a; Cieslak et al. 2012a; Cieslak et al. 2015a).

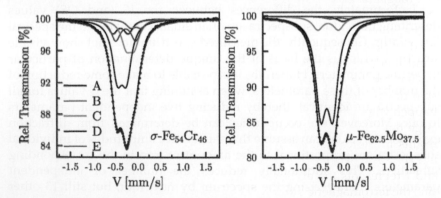

Fig. 1: Mössbauer spectra of the FeCr σ-phase and FeMo μ-phase measured at room temperature and fitted using the results of the electronic structure calculations.

Electronic Structure of FeX σ-phases

The electronic structure of σ-FeX (X = Cr, V, Mo, Re) alloys (Cieslak et al. 2008c; Cieslak et al. 2010b; Cieslak et al. 2012b; Cieslak et al. 2013a) is essentially formed by *d*-states of both transition metal elements. As presented in Fig. 2, the density of states (DOS) in the vicinity of the Fermi level is dominated by *d*-Fe states. Whatever the considered FeX σ-phase, the largest contributions to DOS at E_F, $n(E_F)$, come from Fe(*8i*) (sites C) and Fe(*4f*) atoms (sites B), however in the latter, the occupancy of iron is much lower than the one of X element. Roughly speaking, there are two important factors that presumably give rise to the largest DOS near the Fermi level on Fe atoms (Table 2), namely the coordination number as well as the interatomic distances (Table 3). Indeed, CN is the highest for sites C (CN = 15) and B (CN = 14), where the largest $n(E_F)$ was calculated from non-spin polarized KKR and/or KKR-CPA methods. The opposite is true for the A (CN = 12) and D (CN = 12) sites, which generally exhibit the lowest values of $n(E_F)$. In this regard, the E site appears to be the intermediate case, since in spite of the high coordination number (CN = 14), identical to the B sites, the Fe-DOS values are either comparable to those detected on A and D sites (FeV and FeCr phases) or they are similar to those computed on B or C sites (FeMo and FeRe phases). It is also interesting to note that the overall DOS shape of FeV and FeCr are quite similar and recalls three peaks structure with the shallow minimum present below the Fermi energy (better seen in FeCr system). In both cases the valence states are of bonding character, since *d*-Fe and *d*-X states are well overlapped (very close shape), which is well seen especially in lower part of valence states. In contrast, the strong *d*-X peak appearing above the Fermi level has non-bonding character, since it is associated with very low *d*-Fe DOS in this energy range.

Magnetic moments on Fe atoms in FeX-phases are presumably of itinerant nature, since their presence can be quite well explained in view of the Stoner criterion (Table 2) (Cieslak et al. 2010a), "which is commonly used to predict ferromagnetic state in solid systems within the itinerant band models. This criterion roughly means that the total energy of the system can be decreased due to splitting of subbands for spin-up and spin-down electrons via exchange interactions. On the whole, it may happen when density of states (DOS) on individual atoms are sufficiently large that their spin-polarisation yielding local magnetic moment onset, results in the overall energy gain. Actually, when the Stoner condition $n_d(E_F) \cdot I > 1$, where $n_d(E_F)$ denotes *d*-like density of states at the Fermi level and I represents the exchange parameter, is satisfied on selected atomic site, the ferromagnetic order may arise in the system." Inspecting the *d*-Fe

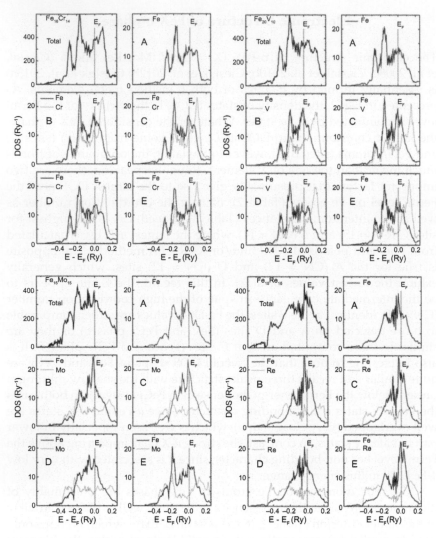

Fig. 2: Non-spin polarized KKR-CPA DOS in FeX σ-phases. Total as well as Fe- and X-decomposed DOS are plotted. DOS contributions on five sublattices are labelled by A, B, C, D and E. The Fermi energy (E_F) marked by vertical line is shifted to zero.

contributions to DOS and accounting for the Stoner exchange parameter, the Fe magnetic moments should systematically be born on C and B sites, which is supported by DOS at E_F on respective sublattices reaching the Stoner limit. On E sites, the situation is more complex, since spontaneous magnetic moments on Fe atoms are only expected in FeMo (Cieslak et al. 2015a), and FeRe phases (Cieslak et al. 2014a), where the interatomic distances are large, whereas it is not the case in FeCr and FeV phases, where

Table 2: Site-decomposed non-polarised DOS at the Fermi level, $n(E_F)$ expressed in states/Ry (per spin) in the σ-FeX (X = V, Cr, Mo and Re) phases. The corresponding Stoner products, $S = n_d(E_F) \cdot I$, for Fe are also given using the computed Stoner parameter $I_{Fe} = 0.036$ Ry. All $n_d(E_F)$ values must be doubled (two spin directions) to compare with the Stoner limit $n_d(E_F) \cdot I = 1$.

σ-phase		$Fe_{20}V_{10}$		$Fe_{16}Cr_{14}$		$Fe_{15}Mo_{15}$		$Fe_{15}Re_{15}$	
Sites	*Atoms*	$n_d(E_F)$	*S*	$n_d(E_F)$	*S*	$n_d(E_F)$	*S*	$n_d(E_F)$	*S*
A (2a, CN = 12)	Fe	13.8	0.99	11.6	0.84	12.7	0.92	14.6	1.05
B (4f, CN = 14)	Fe	16.9	1.22	13.5	0.97	14.2	1.02	13.7	0.98
	X	9.0	0.59	11.6	0.74	6.6	0.85	5.8	0.32
C (8i, CN = 15)	Fe	18.8	1.35	18.2	1.31	20.1	1.65	19.4	1.40
	X	5.5	0.36	10.8	0.79	5.0	0.30	7.1	0.40
D (8i, CN = 12)	Fe	12.9	0.93	13.4	0.96	14.9	1.18	22.2	1.60
	X	-	-	13.5	0.88	4.9	0.29	6.5	0.36
E (8j, CN = 14)	Fe	11.4	0.82	11.9	0.87	13.9	1.07	20.3	1.44
	X	4.3	0.28	9.7	0.63	5.9	0.35	5.8	0.32

Table 3: Average interatomic distances (in Angstroms) determined for five crystallographic sites in FeX-σ-phases.

Site/Phase	$Fe_{20}V_{10}$	$Fe_{16}Cr_{14}$	$Fe_{15}Mo_{15}$	$Fe_{15}Re_{15}$
A (2a, CN = 12)	2.529	2.501	2.602	2.589
B (4f, CN = 14)	2.730	2.700	2.824	2.807
C (8i, CN = 15)	2.686	2.656	2.788	2.763
D (8i, CN = 12)	2.554	2.525	2.639	2.625
E (8j, CN = 14)	2.666	2.636	2.762	2.743

smaller magnetic moments (induced by strong magnetic surrounding) are present. On A site, the Stoner limit is fulfilled only in FeRe phase, while on D site it happens both in FeMo and FeRe phases. Noteworthy, the Stoner limit can also be verified for the transition metal X atom in FeX σ-phases using the calculated parameters for the respective atoms: $I_V = 0.033$ Ry, $I_{Cr} = 0.034$ Ry, $I_{Mo} = 0.030$ Ry and $I_{Re} = 0.028$ Ry as well as the $n_d(E_F)$ values determined for X atoms (Table 3). In view of our calculations the Stoner limit is never reached for X atoms in these σ-phases, but the highest Stoner product was found in FeV phase for D site. So, all X atoms—if any—keep rather induced magnetic moments.

Table 4 presents magnetic moments computed in ferromagnetic state for the aforementioned FeX σ-phases (Cieslak et al. 2010a; Cieslak et al. 2012a). Actually, the self-consistent KKR-CPA calculations have taken into account all specific electronic structure features of individual systems and

Table 4: The local magnetic moments (in μ_B) of Fe and X atoms calculated on five sublattices in FeX σ-phases, assuming ferromagnetic state.

Site/Atoms	$Fe_{20}V_{10}$		$Fe_{16}Cr_{14}$		$Fe_{15}Mo_{15}$		$Fe_{15}Re_{15}$	
	Fe	V	Fe	Cr	Fe	Mo	Fe	Re
A (2a, CN = 12)	0.64	-	0.89	-	1.05	-	0.74	-
B (4f, CN = 14)	2.21	−0.28	1.78	0.28	2.43	−0.05	2.85	−0.04
C (8i, CN = 15)	2.25	−0.24	1.56	0.18	2.73	−0.68	2.81	−0.02
D (8i, CN = 12)	0.93	−0.68	0.99	0.05	1.37	−0.07	1.55	−0.06
E (8j, CN = 14)	1.53	−0.55	1.06	0.11	1.87	−0.17	2.48	−0.04

some differences with respect to the simple Stoner analysis based on non-spin polarized DOS and exchange parameters I, are plausible. However, the final values of the Fe magnetic moments in FeX phases, well reflect the conclusions and trends arisen from the Stoner analysis. On the whole, one sees that the computed magnetic moments of Fe atoms well correspond to the CN as well as the interatomic distances criteria, that is, the calculated Fe magnetic moment is the largest when the CN and the interatomic distance are also the largest. However, other factors related to the specific electronic interactions are also important, since for FeRe system the average Fe moments are found larger (besides A site) than the values found in FeMo system, where the interatomic distances are larger. Such behavior can be partly connected with less intense X-DOS in the vicinity of E_F as well as overall broadening of valence states when comparing transition metal X element of the FeX σ-phases from 3d, 4d and 5d series. However, one should bear in mind that in FeRe series of compounds, also the relativistic effects are expected to play an important role in magnetic properties, so the simple comparison with FeX phase having lighter X elements can be partly misleading.

Magnetic Properties

FK phases most often remain paramagnetic at room temperature (RT), while at lower temperatures, magnetic ordering may appear. The type of magnetic structure markedly depends on two factors: the type of FK phase and the elements constituting the system. Concerning the latter, all Fe-containing σ-phases were found to have magnetic ordering (Figs. 3 and 4) with the ground state characterized by a spin-glass behavior.

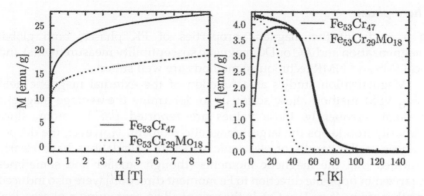

Fig. 3: Magnetization of the σ-FeCr and σ-FeCrMo samples, as a function of the external magnetic field, measured at T = 4K (left) and as a function of temperature at the constant magnetic field 100 Oe (right). On the latter, two different curves for ZFC and FC are clearly visible for σ-FeCrMo sample.

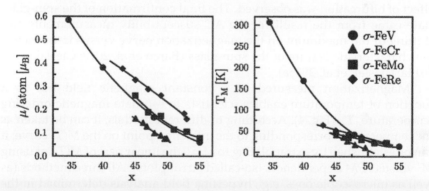

Fig. 4: Average magnetic moments and magnetic ordering temperatures in the full range of the existence of four iron containing σ-phases as a function of the dopant concentration.

Conversely, attempts to find magnetic properties in Fe-free σ-phases have failed so far (to our best knowledge). A good example of various magnetic properties depending on the type of crystal structure (but not on constituents) can be found in the FeMo system, wherein there are at least three FK phases, appearing at different ranges of concentrations and temperatures (Fernandez 1982). Apart from the above-mentioned σ-phase with spin-glass structure, there are also a μ-phase with antiferromagnetic structure and an R-phase, which was found to be non-magnetic at least down to 2K (Cieslak, unpublished data).

The σ-phase

In a study of the magnetic properties of FK phases, both global (magnetization and AC or DC magnetic susceptibility measurements) and local (MS and NMR techniques) methods are well suited.

Magnetization studies as a function of the external magnetic field using VSM method allow scientists to determine the average magnetic moment. Frequently, those values are reported per Fe atom, since primarily iron keeps the largest magnetic moment. However, the deeper analysis of the results of electronic structure calculations have clearly indicated that the magnetic moments (though smaller and sometimes polarized in opposite direction to Fe moment direction) were also induced on other atoms (e.g., V or Cr). In the case of all Fe-containing σ-phases, low temperature M(H) curves are not saturated even at a value of H = 15T (see Fig. 3), suggesting that their magnetic structure is of the spin-glass type. It has been postulated as a result of a series of measurements on samples cooled with (FC) or without (ZFC) external magnetic field, for which the effect of bifurcation was observed. The final confirmation of the spin-glass state came from the results of the AC susceptibility measurements and shifting of the maximum on the magnetization curve versus frequency— the characteristic feature of the spin-glass (Barco et al. 2012; Cieslak et al. 2014a; Cieslak et al. 2015a).

Magnetization measured in a constant magnetic field but as a function of temperature enables scientists to estimate magnetic ordering temperature, T_M (Fig. 4). According to different models, it can be taken as the temperature corresponding to the inflection point on the $M(T)$ curve, it can be established by extrapolating to $M = 0$ the linear part of $M(T)$, or using M^2 versus H/M dependences (so-called Arrot plots). All three methods (as well as microscopic ones, e.g., hyperfine field analysis determined in the MS measurements carried out in function of temperature), give similar results within few Kelvins.

Measurements of the magnetic properties of the Fe-containing σ-phases, FeX (X = Cr, V, Mo, and Re) (Cieslak et al. 2005; Cieslak et al. 2008b; Cieslak et al. 2009; Cieslak et al. 2014a; Cieslak et al. 2015a) clearly indicate, that they vanish by decreasing the Fe concentration in the system: one can observe the decrease of the average magnetic moment as well as of the magnetic ordering temperature. On the other hand, comparison of these systems shows that they are characterized by quite similar values of T_M, usually not exceeding 80 K (Fig. 4 and Table 5). Also, the average values of magnetic moments (determined from magnetization data) are rather small, being in the range of 0.1–0.3 μ_B/atom. However, the σ-$Fe_{1-x}V_x$ system occurring in much wider range of Fe concentration (32 < x < 56), may be an exception going beyond the weak magnetism for small V contents. As seen in

Table 5: Selected physical quantities characterizing σ-$Fe_{53}X_{47}$ (X = V, Cr, Mo and Re) phases.

	V	Cr	Mo	Re
Fe-occupancy [%]				
A	93.2	87.8	100	91.8
B	18.2	27.2	12.9	13.3
C	35.2	40.5	25.5	26.7
D	96.3	89.2	100	86.2
E	34.9	33.5	41.9	56.2
IS [mm/s]				
A	-	-	-	-
B	0.341	0.351	0.303	0.26
C	0.201	0.216	0.205	0.246
D	0.012	0.023	0.023	0.067
E	0.115	0.113	0.113	0.141
QS [mm/s]				
A	0.351	0.342	0.28	0.598
B	0.292	0.242	0.404	0.616
C	0.282	0.181	0.329	0.587
D	0.209	0.21	0.208	0.352
E	0.454	0.454	0.464	0.64
Lattice parameters [Å]				
a	8.93693	8.79077	9.17772	9.10255
c	4.55742	4.61971	4.79404	4.74447
Average magnetic moment μ [μ_B]				
	0.19	0.12	0.20	0.29
T_M [K]				
	52	25	37	45

Fig. 4, for the Fe-richest samples, T_M is higher than RT and the average magnetic moments may even reach the value of 0.6 μ_B/atom.

Examining the impact of third element on physical properties of σ-phases, primarily the influence of impurities on the kinetics of the α-σ phase transformation has already been intensively studied (Sauthoff 1981; Yano 2000 and references therein). It is known that its formation can be accelerated or delayed depending on added elements. It has been also established that beyond certain ranges of concentration of the constituents, the σ-phase does not appear at all. However, physical properties of such multicomponent systems were poorly studied up to now.

Interesting magnetic properties were observed in the ternary systems, σ-FeCrX where X is a third transition metal element. As evidenced from the magnetization measurements (Cieslak, unpublished data), ternary systems are also characterized by spin-glass state, but small concentrations of the third component (c.a. few at.%), cause initially an increase and then a decrease of the average magnetic moment and ordering temperature. In most cases, the range of existence of σ-phase in the given system is very limited, but in some cases there is the possibility to fully replace one element by another. This may happen when the σ-phase is present in both parent systems XY and XZ, and then often in a $XY_{1-x}Z_x$ for any values of x. For example, the σ-phase is present both in the FeCr and CoCr systems, therefore one can analyze their properties when replacing Fe atoms by Co ones in the entire concentration range. As it turned out, such substitution leads to the gradual disappearance of the magnetic properties of σ-phase in the FeCrCo system. Similar conclusions can be drawn for FeVNi system, where Fe atoms are replaced with Ni - the system becomes non-magnetic while replacing half of the Fe content. Likewise, FeCrMo system can be analyzed (σ-phase appears in binary FeCr and FeMo systems), but here the Fe content remains constant, while chromium is replaced by molybdenum: an enhancement of both the average magnetic moment and the T_M value is observed in the system with the increase of Mo content (Fig. 3) (Cieslak et al. 2015a). This is probably due to the expansion of the crystal lattice caused by substitution.

The above-mentioned effects have been observed by the experimental method which allows analysis of the average magnetization of the system. However, especially in the case of FK phases, where electronic and magnetic behaviours of atoms strongly depend on their crystallographic positions, namely atomic surroundings, well adapted local methods such as NMR and MS suffer from insufficient resolution to distinguish several sublattices and great number of atomic configurations. Such circumstances hinder unambiguous interpretation of the experimental results.

From that reason spin-polarized electronic structure calculations by KKR method were performed for a representative number of ordered supercells. The results were analyzed based on the probability distributions of neighbors of a specific type in the close vicinity of the considered atom. Such computations yielded the values of magnetic moments and hyperfine fields for each atom in the unit cell, separately. Taking results obtained for atoms on five individual sublattices, the magnetic moments μ and hyperfine fields H were arranged versus the number of Fe atoms in the first coordination shell, NN_{Fe}. A linear dependence of μ and H with NN_{Fe} was derived (Fig. 5). Consequently, accounting for the probabilities of sublattices occupancies by individual elements (e.g., Fe, Cr and Mo)

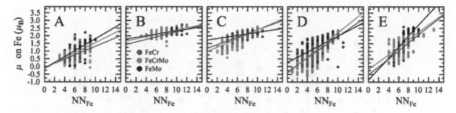

Fig. 5: Magnetic moments on Fe atoms belonging to different sublattices of the σ-phase as a function of the number of Fe atoms in the first coordination shell, NN_{Fe}, calculated for three different systems.

one can average the theoretical results and compare them with the experimental values.

Examples of the KKR results are presented in Fig. 5. As one can see, the computed magnetic moments and their changes with NN_{Fe} are characteristic for each sublattice. It is caused by different average distances from the NN atoms for each site, as well as the various CN and arrangements of atoms belonging to the preferentially occupied sublattices. Consequently, it results in a different spatial distribution of constituting elements in the nearest neighborhood. On the whole, one observes that the calculated magnetic moments are quite scattered, reaching the values more than $2\mu_B$. Therefore, the resulting average value of μ is significantly greater than this obtained experimentally. One of the reasons for observed disagreement might be the ferromagnetic (FM) model assumed in our computations. Indeed the above-mentioned systems presumably exhibit much complex magnetic structures (Cieslak et al. 2010a; Cieslak et al. 2012a; Cieslak et al. 2014a; Cieslak et al. 2015a) and different models of magnetic ordering should also be taken into account to improve agreement.

Group symmetry analysis (Cieslak et al. 2010a) applied to all sublattices of the σ-phase allows for the existence of antiferromagnetic ordering (AFM) for certain sites (B and C) only, which obviously should reduce the average value of the total magnetic moment. Since purely antiferromagnetic arrangement in the case of chemically disordered alloy cannot be handled and the antiparallel ordering (APM) is present rather than the AFM one, the KKR and KKR-CPA calculations were performed with the initial structure of magnetic moment directions as suggested by the symmetry analysis. The results indicated the co-existence of FM and APM orderings depending on sites and clearly showed the reduction of the average total magnetic moments. On the whole, theoretical results were found in good agreement with experimental data for the σ-FeCr system, but in the case of σ-FeV system the calculated magnetisation was slightly lower than the measured points. The latter can be tentatively explained by co-existence of the areas with FM and APM arrangements (Fig. 6).

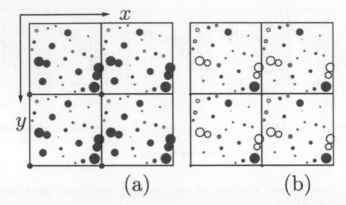

Fig. 6: Magnetic moments μ of Fe and Cr atoms in the unit cell, as seen along z axis, for (a) FM structure and (b) APM one. Full and empty circles represent magnetic moments parallel and antiparallel to the z direction, respectively. The size of red (Fe) and green (Cr) circles is proportional to the values of μ.

One should, however, pay attention to several important aspects resulting from our theoretical investigations:

(i) The symmetry analysis admits the presence of the AFM arrangement in σ-phases, but does not indicate necessity of its occurrence.

(ii) The calculations carried out for the APM ordering actually did not change meaningfully the values of magnetic moments of individual atoms (especially on Fe) in comparison to the results obtained for the FM arrangement. Observed reduction of the average total magnetic moment is caused mainly by their antiparallel alignments.

(iii) The total energy calculated for the APM structure is only slightly smaller than the corresponding value determined for the FM structure, being then energetically comparable.

(iv) The applied approaches do not take into account noncollinearity of magnetic moments, which may occur in real systems. Theoretical results for more complex magnetic arrangements would probably affect the average total magnetic moments.

(v) Random filling of sublattices (different occupancies for different sites) applied to electronic structure calculations and using probabilities determined experimentally, result in constitution of regions (clusters) with higher and lower Fe concentrations (see Fig. 6). It shows that Fe atoms belonging to the clusters with higher Fe content exhibit significantly larger magnetic moments, since they depend on the NN_{Fe} regardless of the FM/APM structure.

(vi) The calculations are restricted to magnetic moments orderings within the unit cell (determined values depend primarily on the immediate neighborhood of the atom), whereas measurements can rather 'see' the long-distance magnetic properties larger than several cells, which turned out to be the spin-glass in the FeX σ-phases.

Conclusion. Since the computed values of magnetic moments only weakly depend on the atomic configuration, a reduction in the average magnetization can be achieved not only by changing orientation of some of these moments (e.g., the APM model), but also by changing orientation of the entire clusters (domains) within the model of spin-glass. Probably in this case we are dealing with so-called cluster spin-glass determined by the type of chemical disorder present in the structure (Mydosh 2015). Therefore, magnetic moments determined experimentally (averaged per atom) are underestimated, and the theoretical values achieved from KKR and KKR-CPA are expected to be closer to the local atomic moments. However, this conclusion requires further theoretical and experimental study.

The μ-phase

The μ-phase (Pearson symbol: hR13, space group: $C_{3d}^5 - R\bar{3}m$, archetype: Fe_7W_6) has a rhombohedral structure and hosts 13 atoms distributed over 5 lattice sites with CN ranging from 12 to 16 (Forsyth and d'Altre da Veiga 1962). Magnetic properties of the μ-phase were discovered so far in three systems: FeNb ($T_N \approx 280\,K$), FeTa ($T_N \approx 320\,K$) (Ahmed et al. 1983) and FeMo ($T_N \approx 120\,K$) (Cieslak et al. 2015c). In these systems, magnetic moments of atoms belonging to sublattice *6h* are coupled antiferromagnetically along the rhombohedral axis (111) (Ahmed et al. 1983).

Electronic structure calculations carried out in the μ-FeMo phase showed that atoms on *6h* positions actually order antiferromagnetically. This sublattice is almost fully occupied by Fe atoms, although Mo-atoms may be found on this site (with small probability), too. Other positions, that is, *1a* and three nonequivalent *2c* sites are also preferentially occupied, that is, *1a* and one of *2c* sublattices by Fe and other two *2c* sites by Mo atoms (Cieslak et al. 2014b). This particular arrangement of neighbors reduces the magnetic moment on these Fe(*6h*) atoms, having Mo atoms in their closest surrounding. Furthermore, it was found that the magnetic moments of Fe on the other sublattices are coupled ferromagnetically. These values of the magnetic moment are small and strongly dependent on neighbor arrangements for *1a* (vaying from 0 to 2 μ_B), whereas quite large (2–3 μ_B) and weakly dependent on NN_{Fe} for the three *2c* sublattices (Cieslak et al. 2015c).

Formation Energy and Site Occupancy

As it was already mentioned, FK phases are chemically disordered systems, which means that the translational symmetry of the lattice is maintained, but the sublattices are randomly occupied by two (or more) elements, usually with different probabilities. Occupancy of the crystal sublattice can be determined based on diffraction techniques (mainly XRD and ND). XRD, although prevalent, however has certain limitations, since it gives information from the area relatively close to the sample surface. Furthermore, it weakly distinguishes elements with close atomic number. ND technique delivers more information about crystallographic sites (e.g., filled by atoms with close Z) from the whole volume of the sample.

On the basis of diffraction measurements, preferential occupation of the crystal lattice sites by alloying elements can be determined. In principle, the general rule regarding all FK-phases is that larger atoms prefer to occupy positions where there is more space (the distances from the nearest neighbors are larger) (Laves 1942; Duwez and Bean 1951; Watson and Bennett 1984; Seiser et al. 2011). This leads to a situation in which the individual sites are occupied by elements with different probabilities (it rarely happens that a single site is fully occupied by only one type of atom). Detailed diffraction study of the σ-phase revealed that within a particular sublattice there is no vacancy and the distribution of elements is random. The sublattice occupancy may be affected by various factors, depending on the symmetry, NN distance, and used elements. For the mentioned FeCr σ-phases, the sublattice occupancy proved to be practically independent neither on the history of the precursor (*bcc* alpha) of the σ-phase, nor the transformation temperature, nor heating time. The samples were pre-rolled or transformed as cast, annealed shortly (two days) or several months, at higher and lower temperatures, but in all cases the σ-phase was characterized by practically the same sublattice occupancies (Cieslak et al. 2008a). On the other hand, a careful examination of the μ-phase in the FeMo system showed that preferential occupations, and thus unit cell parameters, depend on the temperature at which the transformation process was carried out (Cieslak et al. 2014b).

Total energy, E_{tot}, is calculated using KKR or KKR-CPA methods for the set of atoms (e.g., for unit cell). It can be understood as the work to be done to build the system from single elements, atom by atom, at 0 K. On that basis one can define a formation energy, E_{form}, as the difference between the total energy (E_{tot}) of the investigated phase and a sum total energies of its constituents (as bulk):

$$E_{form}(A_m B_n) = E_{tot}(A_m B_n) - m\, E_{tot}(A) - n\, E_{tot}(B) \tag{1}$$

In principle all E_{tot} values in the Eq. 1 should be determined for elements A and B, assuming the same crystal structure as that of the investigated $A_m B_n$ phase. Unfortunately, in the case of FK phases their one-component analogs do not crystallize in reality. Of course, such systems can be analyzed purely theoretically and one can perform calculations even for hypothetical phases. However, treating these values as a reference of the formation energy may create conceptual problems. An alternative way is to determine the total energy of single elements for real structures occurring at 0 K. Such defined formation energy can be understood as the work to be done to change the pure elemental crystalline structures (but not single atoms) to form the considered alloy system (also at 0 K). As already mentioned, in fact, such systems are formed always at a higher temperature, either by phase transition (e.g., α-σ in the FeCr system) or by sintering pure metals (e.g., Fe and Mo in the σ-FeMo system).

The theoretical determination of the preferences of sublattice occupation based on the electronic structure calculations is not an easy task. One should take into account that such calculations are carried out for the system at 0 K, while FK phases are formed at elevated temperatures of at least several hundred Celsius degrees. Therefore, not only the formation energy should be analyzed, but rather Gibbs free energy, $G = E - TS$, containing entropy scaled by a temperature of the system. In principle, total entropy consists of four contributions, namely configuration, S_{conf}, magnetic, S_{mag}, phonon, S_{ph}, and electronic S_{el}, which should be analyzed to investigate relative crystal stability.

Configuration entropy is related to the degree of disorder of the system and for random systems it can be expressed by the known formula (2) where index i denotes the constituent element, index j denotes sublattice, and x stands for site occupancy:

$$S_{conf} = -k_B \sum_i \sum_j x_{ij} \ln(x_{ij}) \tag{2}$$

Since it is determined only by the site-dependent concentrations of constituent atoms, it can be calculated for any configuration of sublattice occupancies. Magnetic entropy can be described using Eq. 3, where μ denotes magnetic moment:

$$S_{mag} = k_B \sum_i \sum_j x_{ij} \ln(|\mu_{ij}| + 1) \tag{3}$$

In order to calculate this contribution, magnetic moments of atoms in various atomic surroundings should be known. In practice, these values are established only for a finite number of specific atomic arrangements. On that basis, one can determine other magnetic moment values, for example, by interpolation, using different models of magnetic structure. The simplest one quite well describes binary systems with crystal structures such as *fcc* or *bcc*, including small concentration of additional elements (called dopants). It implies that each substituted atom in the first (second, third ...) coordination shell changes property (e.g., magnetic moment or hyperfine field) of the atom on a given site exactly in the same way. In other words, the property X on this site depends only on the number (but not on their arrangement) of dopants in subsequent coordination shells: $X = X_0 - n_1 \cdot X_1 - n_2 \cdot X_2 - ...$ where n_i is the number of atoms and X_i stays for the correction from single atom of a given type in the i-th coordination shell. In the case of multi-component FK phases atoms (besides the nearest ones) usually do not form well-separated and equally-distant coordination shells so the model should have a form $X = X_0 - \Sigma_i n(r_i) \cdot X(r_i)$, where $n(r_i)$ and $X(r_i)$, are now continuous functions of distance (summation is truncated to the arbitrary chosen number of neighbors). Unfortunately, such approach is not well adapted to the investigated FK-phases, probably due to the high concentration of alloyed elements and also because of a more complex relationship between the number of atoms n_i and the change of the parameter X.

Another model assumes that the magnetic entropy in the unit cell is determined by occupancy of the sites, x_1, x_2, ... x_k so it should be a continuous function of the number of atoms of different types on all sites. Although the analytical form of such function is not known, one can approximate it using the polynomial of n-th degree, $W(x_1, x_2, ... x_k)$, the coefficients of which can be determined by fitting to the S_{mag} values, calculated for selected unit cells with known atomic arrangements. Unfortunately, since one should take into account the occupancy of each site separately, the number of coefficients to be determined rapidly increases with the degree of the polynomial, n, and in practice such analysis seems to be reliable for $n < 3$. Applying this model, for example, to the ternary σ-FeCrCo system one can find the differences between the values determined using KKR calculation and the values of such determined polynomial to be ~0.05 k_B/atom, while the values for the different arrangements of the atoms differ each other ~0.3 k_B/atom, on average. As can be seen, determination of the magnetic entropy in this case might be subjected to a considerable error. In addition, its value can be affected by the change of the magnetic moments of atoms with an increase of the temperature caused, for example, by lattice expansion.

The analysis of phonon entropy would require more experimental and theoretical data on lattice dynamics in the σ-FeX phases. However, S_{phon} is expected to be less influential (not significantly affecting the total value of entropy), since S_{phon} determined, e.g., for the σ-FeCr is ~0.1 k_B/atom against the much larger value of S_{conf} ~0.5 k_B/atom. Moreover, the S_{phon} values determined from synchrotron radiation and *ab initio* modeling (Dubiel et al. 2010b) in the σ-FeCr (and the corresponding *bcc*-alpha) phases are quite similar, so their difference presumably neither affects the alpha-σ transformation process nor physical properties of the σ-FeCr phases.

Electronic entropy at low temperature is a linear function of T, but this relationship saturates for temperatures well above the Debye temperature to become constant and independent on an analyzed system (stoichiometry, atomic arrangements, etc.). Therefore, comparing different phases one can neglect S_{el} difference in the analysis of relative stability in high temperature range, where FK-phases are constituted. So, considering the site preferences during formation of different phases it is sufficient to rely on their total energies as well as magnetic and configuration entropies only.

Similar to the magnetic entropy, the total energy is also a function of site occupancies. In a simplified approach, it can be determined for the unit cells characterized by the absence of disorder on sublattices. Such assumption significantly simplifies the calculations that should be carried out limiting them to all the possible configurations of fully occupied sites. For example, for binary σ-phase (5 sublattices and 2 possible elements) there are $2^5 = 32$ different cases and $3^5 = 243$ configurations in the ternary system. On the basis of such calculated values of total energy one should construct the diagrams of formation energy versus the concentrations of each element in the unit cell. The envelope determining the bottom of the E_{form} points on these diagrams (so-called convex-hulls) indicates energetically the most favorable configurations in the range of considered concentrations (Kabliman et al. 2009; Pavlu et al. 2010; Mathieu et al. 2013).

Real crystalline systems may be more precisely described when the disorder on sublattices is taken into account. This can be done, for example, by calculating the total energy for one selected atomic arrangement and setting it as a reference system. Then, as a comparison, similar calculations should be performed for other atomic arrangements, differing from the reference one in a pre-specified way (e.g., by variation of the occupancy on one site only). Finally, such computations have to be performed for the number of so modified arrangements, sufficient to determine the shape of E_{tot} function versus the occupancy variation on all sublattices. In the next step one can approximate its shape using an analytical function, for example, second order polynomial in the five dimensional space (due to A-E sites), which proved to be very effective. The values of total energy per one atom in FK phases are of order of tens keV/atom, while the differences in energies for different arrangements of the atoms in the unit cell are about one eV/atom. However, adjustment of this relationship by the mentioned polynomial allowed scientists to minimize differences between the polynomial and the calculated values to few meV/atom. Actually, such calculations carried out for a number of selected arrangements are time-consuming, but it results in the formation energy in the form which can be easily used for further manipulations (e.g., for computations of its lowest values).

Knowledge of analytical form of S_{conf} (strictly) and S_{mag} (approximately) as well as E_{form} (also with a good approximation, see Fig. 7) as a function of the site occupancies allows scientists to express thermodynamic Gibbs function of the system analytically—vs. site occupancies and temperature. For example, for the FeCr σ-phase having five independent lattice sites this relationship is a 6-dimensional function, and an 11-dimensional one for three component σ FeCrCo system. Further analysis of this relationship, and in particular the determination of its lowest value enables scientists to determine sublattice occupancy by constituents vs. temperature.

Such calculations were performed, for example, for two component σ-FeCr and σ-FeV systems (Fig. 8) (Cieslak et al. 2012c; Cieslak et al. 2013b) as well as μ-FeMo (Fig. 9) (Cieslak et al. 2015d) and good agreement with the experimental data obtained from ND and XRD measurements was found (see Fig. 10).

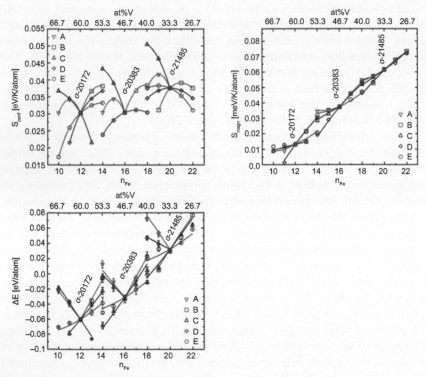

Fig. 7: Configuration and magnetic entropies as well as formation energies around three chosen reference arrangements in the σ-FeV system. Labels denote Fe atom populations on subsequent sublattices, whereas V atoms constitute the rest.

Similar calculations have been done also for the ternary FK systems (Cieslak et al. 2015b). They were analyzed in the same way, but the complexity and amount of performed calculations was much greater. In the case of the three component system there exist much more degrees of freedom to arrange atoms on sublattices (which should be reflected in the calculation). Actually, graphical presentation of the 11-dimensional function that should be analyzed fails in the most cases. But, the fact that the occupancies of the sublattices calculated in this way agree quite well

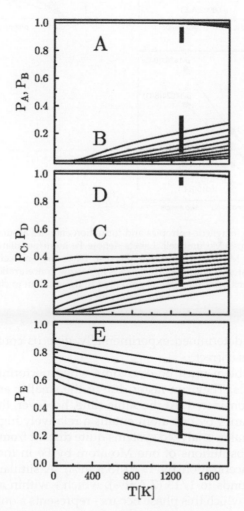

Fig. 8: Fe sublattice occupancies in the σ-$Fe_{100-x}V_x$ system ($32 < x < 56$), determined by electronic structure calculations (solid lines) and measured using ND technique (vertical bars).

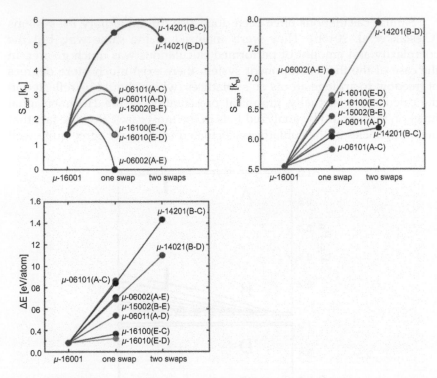

Fig. 9: Configuration, magnetic entropies and formation energies around μ-16001 reference arrangement in the μ-FeMo unit cell. Labels denote Fe atom populations on subsequent sublattices, whereas Mo atoms constitute the rest. Since the unit cell of the μ-phase is relatively small (13 atoms) and this phase exists in narrow concentration range, it is more realistic to swap atoms between sites (as presented in figure) than to change number of Fe atoms in the unit cell.

with the values determined experimentally, give us confidence that the adopted model is correct.

Another problem must be overcome when performing calculations in small unit cells. They run faster due to the smaller amount of atoms that should be considered. At the same time, however, the change of the sublattice occupancy by one atom entails a relatively high concentration change, which makes analyzed systems quite distant from those in reality. For example, substitutions of one Mo atom by Fe in the FeMo σ-phase changes the concentration by 1/30 (3.3 at%). A similar change in the μ-phase corresponds to 1/13 (7.7 at%), which - within a small range of concentration in which this phase occurs - represents a quite large skip. In this case, however, one can simply swap the atoms between sublattices, which does not change the total concentration of any of them, and allows scientists to achieve the desired effect. Calculations made on the basis of

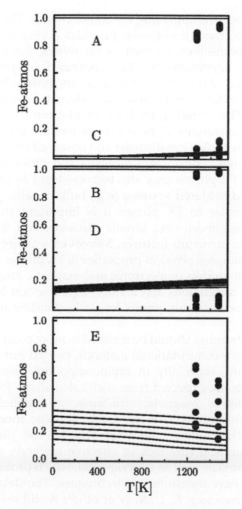

Fig. 10: Fe sublattice occupancies in the μ-FeMo system, determined by electronic structure calculations (solid lines) and measured using ND technique (circles).

this method allowed scientists to also determine the site occupancy in satisfactory agreement with measured data (Cieslak et al. 2015d).

Summary and Perspectives

Frank-Kasper (FK) phases are very interesting, important and complex systems, both from fundamental and application points of view, since their physical properties and crystal stability are highly determined by substitutional disorder, appearing on all sublattices. The analysis of this class of materials, though not easy, provides a very useful knowledge on FK phases themselves, as well as on others, for which they may be regarded as approximants. The combined *ab initio* calculations with experimental data on crystal structure have allowed scientists to interpret hyperfine interactions in selected FK-phases (e.g., FeX sigma-phases). The detailed analysis of electronic structure features for different arrangements of magnetic moments have shed a light on interplay between sublattices disorder and observed spin-glass behaviors. Because of the high coordination numbers and different configurations of neighbors the FK phases may also be considered as indicative models of topologically disordered systems (e.g., bulk metallic glasses) or even quasicrystals. Similar to FK phases it is important to emphasize the role of atomic neighborhoods, atomic distances and their relationship to local electronic structure features. Moreover, a variety of parameters responsible for complex physical properties of FK phases allows scientists to verify different models of electronic and magnetic interactions. On the whole, captured correlations and relationships relevant to the analysis of FK phases are good starting points for understanding of more complex systems.

Their further studies should be directed to more comprehensive use of the complementary computational methods, carried out using supercells and CPA medium, especially in multicomponent disordered systems. Interesting results are expected from such calculations for paramagnetic-like or non-collinear magnetic structures, in parallel supported by experimental determination of the local magnetic moments and their arrangements. The serious challenge will be the phonon spectrum analysis, so theoretical (the lack of effective methods of calculation in disordered systems) as well as experimental, which probably requires the development of new measurement techniques. The detailed analysis of less common phases as χ, R, D, K, W or others is still waiting for interest and research.

Acknowledgments

This work was supported by the Ministry of Science and Higher Education of Polish Government as well as the National Science Centre (Grant DEC-2012/05/B/ST3/03241).

Keywords: Frank-Kasper phases, Disordered systems, Electronic properties, Electronic structure calculations, AC magnetic susceptibility, Mössbauer spectroscopy, X-ray diffraction, Neutron diffraction

References

Ahmed, M.S., G.C. Hallam and D.A. Read. 1983. Antiferromagnetism of the FeNb and FeTa μ-phases. 37: 101–109.

Bain, E.C. and W.E. Griffits. 1927. An introduction to the iron-chromium nickel alloys. Trans. AIMME 75: 166–213.

Bansil, A., S. Kaprzyk, P.E. Mijnarends and J. Tobola. 1999. Electronic structure and magnetism of $Fe_{3-x}V_xX$ (X = Si, Ga, and Al) alloys by the KKR-CPA method. Phys. Rev. B. 60: 13396–13412.

Barco, R., P. Pureur, G.L.F. Fraga and S.M. Dubiel. 2012. Extended scaling in the magnetic critical phenomenology of the σ-phase $Fe_{0.53}Cr_{0.47}$ and $Fe_{0.52}V_{0.48}$ alloys. J. Phys.- Condens. Matter 24: 046002.

Bergman, B.G. and D.P. Shoemaker. 1951. The space group of the σ-FeCr crystal structure. J. Chem. Phys. 19: 515.

Bergman, G. and D.P. Shoemaker. 1954. The σ-phase revisited: an alternative pathway to a trial structure. Acta Cryst. 7: 857–865.

Blachowski, A., J. Cieslak, S.M. Dubiel and B. Sepiol. 2000. Influence of titanium on the kinetics of the σ-phase formation in a coarse-grained Fe-Cr alloys. Intermetallics 8: 963–966.

Butler, W.H., P. Dederichs, A. Gonis and R. Weaver [eds.]. Applications of Multiple Scattering Theory to Materials Science (MRS Symposia Proceedings vol. 253). 1992. MRS, Pittsburgh, PA.

Cieslak, J., B.F.O. Costa, S.M. Dubiel, M. Reissner and W. Steiner. 2005. Magnetic properties of a nanocrystalline σ-Fe-Cr alloy. J. Phys.-Condens. Matter 17: 2985.

Cieslak, J., M. Reissner, S.M. Dubiel, J. Wernisch and W. Steiner. 2008. Influence of composition and annealing conditions on the site-occupation in the σ-phase of Fe–Cr and Fe–V systems. J. Alloy Compd. 460: 20–25.

Cieslak, J., M. Reissner, W. Steiner and S.M. Dubiel. 2008. On the magnetism of the σ-phase Fe-Cr alloys. Phys. Stat. Sol. (a) 205: 1794–1799.

Cieslak, J., J. Tobola, S.M. Dubiel, S. Kaprzyk, W. Steiner and M. Reissner. 2008. Electronic structure of a σ-Fe-Cr compound. J. Phys.: Condens. Matter 20: 235234.

Cieslak, J., B.F.O. Costa, S.M. Dubiel, M. Reissner and W. Steiner. 2009. Magnetic ordering above room temperature in the σ-phase of $Fe_{66}V_{34}$. J. Magn. Magn. Mater 321: 2160–2165.

Cieslak, J., J. Tobola, S.M. Dubiel and W. Sikora. 2010. Magnetic properties of σ-Fe-Cr alloys as calculated with the charge- and spin-self-consistent KKR(CPA) method. Phys. Rev. B 82: 224407.

Cieslak, J., J. Tobola and S.M. Dubiel. 2010. Electronic structure of the σ-phase of paramagnetic Fe-V alloys. Phys. Rev. B. 81: 174203.

Cieslak, J., J. Tobola and S.M. Dubiel. 2012. Theoretical study of magnetic properties and hyperfine interactions in σ-Fe-V alloys. Intermetallics 22: 7–12.

Cieslak, J., S.M. Dubiel, J. Przewoznik and J. Tobola. 2012. Structural and hyperfine characterization of σ-phase Fe-Mo alloys. Intermetallics 31: 132–136.

Cieslak, J., J. Tobola and S.M. Dubiel. 2012. Study of phase stability in the σ-Fe-Cr system. Intermetallics 24: 84–88.

Cieslak, J., S.M. Dubiel, J. Tobola and J. Zukrowski. 2013. Experimental and theoretical study of the σ-phase Fe-Re alloys. Mater. Chem. Phys. 139: 590–595.

Cieslak, J., J. Tobola and S.M. Dubiel. 2013. Formation energy in σ-phase Fe-V alloys. J. Phys. Chem. Solids 74: 1303–1307.

Cieslak, J., S.M. Dubiel, M. Reissner and J. Tobola. 2014. Discovery and characterization of magnetism in σ-phase intermetallic Fe-Re compounds. J. Appl. Phys. 116: 183902.

Cieslak, J., J. Przewoznik and S.M. Dubiel. 2014. Structural and electronic properties of the μ-phase Fe-Mo compounds. J. Alloy Compd. 612: 465–470.

Cieslak, J., M. Reissner and S.M. Dubiel. 2016. Magnetism of σ-phase Fe-Mo alloys: its characterization by magnetometry and Mössbauer spectrometry. J. Magn. Magn. Mater. 401: 751–754.

Cieslak, J. and S.M. Dubiel. 2016. Site occupancies in sigma-phase Fe–Cr–X (X = Co, Ni) alloys: Calculations versus experiment. Comp. Mat. Sci. 122: 229–239.

Cieslak, J. Magnetic properties of μ-FeMo phase. To be published.

Cieslak, J. Formation Energy in μ-FeMo system. To be published.

Dubiel, S.M., J.R. Tozoni, J. Cieslak, D.C. Braz, E.L. Gea Vidoto and T.J. Bonagamba. 2010. Sublattice magnetism in σ-phase $Fe_{100-x}V_x$ (x = 34.4, 39.9, and 47.9) studied via zero-field ^{51}V NMR. Phys. Rev. B. 81: 184407.

Dubiel, S.M., J. Cieslak, W. Sturhahn, M. Sternik, P. Piekarz, S. Stankov and K. Parlinski. 2010. Vibrational properties of alpha- and σ-phase Fe-Cr alloy. Phys. Rev. Lett. 104: 155503.

Duwez, P. and S.R. Bean. 1951. Symposium on the nature, occurrence, and effects of the sigma-phase. Proc. ASTM 110: 48–60.

Elmer, J.W., T.A. Palmer and E.D. Specht. 2007. Direct observations of σ-phase formation in duplex stainless steels using *in-situ* synchrotron X-ray diffraction. Metall. Mater. Trans. 38A: 464–475.

Fernandez Guillermet, A. 1982. The Fe-Mo (Iron-Molybdenum) system. Bulletin of Alloy Phase Diagrams 3(3): 359–367.

Forsyth, J.B. and L.M. d'Altre da Veiga. 1962. The Structure of the μ-Phase Co_7Mo_6. Acta Cryst. 15: 543–546.

Frank, F.C. and J.S. Kasper. 1958. Complex alloy structures regarded as sphere packings. I. Definitions and basic principles. Acta Cryst. 11: 184–190.

Frank, F.C. and J.S. Kasper. 1959. Complex alloy structures regarded as sphere packings. II. Analysis and classification of representative structures. Acta Cryst. 12: 483–499.

Hall, E.O. and S.H. Algie. 1966. The σ-phase. Metall. Rev. 11: 61–88.

Hsieh, C.-C. and W. Weite. 2012. Overview of intermetallic sigma (σ) phase precipitation in stainless steels. ISRN Metallurgy 2012: 1–16.

Kabliman, E.A., A.A. Mirzoev and A.L. Udovskii. 2009. First-principles simulation of an ordered σ-phase of the Fe-Cr system in the ferromagnetic state. The Physics of Metals and Metallography 108(5): 435–440.

Kaprzyk, S. and P.E. Mijnarends. 1986. A simple linear analytic method for Brillouin zone integration of spectral functions in the complex energy plane. J. Phys. C: Solid State Phys. 19: 1283–1292.

Kuhn, B., M. Talik, L. Niewolak, J. Zurek, H. Hattendorf, P.J. Ennis et al. 2014. Development of high chromium ferritic steels strengthened by intermetallic phases. Mater. Sci. Eng. A 594: 372–380.

Laves, F. and H.J. Wallbaum. 1942. Über den Einfluß geometrischer Faktoren auf die stöchiometrische Formel metallischer Verbindungen, gezeigt an der Kristallstruktur des KNa_2. Anorg. Allg. Chem. 250: 110–120.

Lee, S., M.J. Bluemle and F.S. Bates. 2010. Discovery of a Frank-Kasper σ-phase in sphere-forming block copolymer melts. Science 330: 349–353.

Mathieu, R., N. Dupin, J.-C. Crivello, K. Yaqoob, A. Breidi, J.-M. Fiorani et al. 2013. CALPHAD description of the Mo–Re system focused on the σ-phase modelling. Calphad 43: 18–31.

Mydosh, J.A. 2015. Spin glasses: redux: an updated experimental/materials survey. Rep. Prog. Phys. 78: 052501.

Parsons, D. 1960. Variation of magnetic moment with composition in the iron-vanadium σ-phase. Nature 185: 839–840.

Pavlů, J., J. Vřešťál and M. Šob. 2010. *Ab initio* study of formation energy and magnetism of σ-phase in Cr–Fe and Cr–Co systems. Intermetallics 18: 212–220.

Read, D.A. and E.H. Thomas. 1966. The magnetic properties of σ-phase alloys. IEEE Trans. Magn. MAG-2: 415–419.

Read, D.A., E.H. Thomas and J.B. Forsythe. 1968. Evidence of itinerant electron ferromagnetism in σ-phase alloys. J. Phys. Chem. Solids 29: 1569–1572.

Sauthoff, G. and W. Speller. 1981. Kinetics of sigma phase precipitation in Fe-Cr-Si alloys. Z. Metallkd 72: 457–461.

Seiser, B., R. Drautz and D.G. Pettifor. 2011. TCP phase predictions in Ni-based superalloys: structure maps revisited. Acta Mater 59: 749–763.

Shibuya, M., Y. Toda, K. Sawada, H. Kushima and K. Kimura. 2014. Effect of precipitation behavior on creep strength of 15% Cr ferritic steels at high temperature between 923 and 1023 K. Mater. Sci. Eng. A. 592: 1–5.

Sikirić, M.D., O. Delgado-Friedrichsb and M. Dezac. 2010. Space fullerenes: A computer search for new Frank–Kasper structures. Acta Cryst. A 66: 602–615.

Sinha, A.K. 1972. Topologically close-packed structures of transition metal alloys. Prog. Mater. Sci. 15: 79–185.

Stopa, T., S. Kaprzyk and J. Tobola. 2004. Linear aspects of the Korringa–Kohn–Rostoker formalism. J. Phys.: Condens. Matter. 16: 4921.

Watson, R.E. and L.H. Bennett. 1984. Transition-metal alloy formation. The occurrence of topologically close packed phases: I. Acta Metall. 32: 477–489.

Yano, K. and K. Abiko. 2000. Role of carbon and nitrogen on the transformation of the sigma phase in highly purified Fe-50 mass % Cr alloys. Mater. Trans. JIM 41: 122–129.

2.5

Atomistic Modeling to Design Favoured Compositions for the Metallic Glass Formation

J.B. Liu, J.H. Li and B.X. Liu*

INTRODUCTION

Traditionally, new materials used to be detected by time-consuming experiments. With increasing of computing power, simulations of materials behaviors have become possible and can speed up experimentation. More ambitiously, computational materials science is trying to understand and predict the formation-structure-property correlations. In this chapter, we show our effort to develop an atomistic theory, giving guidance to design favoured compositions for synthesizing metallic glasses.

In 1959, Duwez et al. obtained the Au-Si amorphous alloy, that is, metallic glass (MG) by liquid melt quenching (LMQ) (Duwez et al. 1960; Klement et al. 1960). Since then, numerous metallic glasses have been obtained in binary, ternary and multi-component alloy systems (Anantharaman 1984; Güntherodt and Beck 1981; Luborsky 1983). In the early stages, metallic glasses obtained by LMQ were mostly in the form of thin films or foils. In late 1980s, researchers found some multi-component

Key Laboratory of Advanced Materials (MOE), School of Materials Science and Engineering, Tsinghua University, Beijing 100084, China.
* Corresponding author: dmslbx@mail.tsinghua.edu.cn

alloy systems (typically 3–5 elements), in which LMQ could readily produce metallic glasses with their sizes on the order of centimeter, and are known as bulk metallic glasses (BMG) (Inoue 1998a; Kawazoe et al. 1997; Liaw and Buchanan 2006; Miller and Liaw 2007; Wang et al. 2004b). From then on, investigations of metallic glasses have attracted increasing attention because of their fundamental scientific importance and potential engineering application.

In early 1980s, some other techniques, such as ion beam mixing (Liu et al. 1983a; Liu et al. 1983b), and solid-state amorphization (Clemens et al. 1984; Schwarz and Johnson 1983), were introduced. In these techniques, the metallic glasses were formed under a far-from-equilibrium condition driven by external force (Martin and Bellon 1997). In fact, these techniques are capable of synthesizing metallic glasses in equilibrium immiscible binary alloy systems with negative heat of formation (ΔH_f), as well in many equilibrium immiscible binary alloy systems with positive heat of formation (ΔH_f) (Liu et al. 2001; Suryanarayana 1999).

For the metallic glasses, the fundamental scientific issues are related to their processing, structure and property. Concerning the processing, i.e., to produce or obtain the desired metallic glasses, one has to know in which alloy system and in what composition range, formation of metallic glasses is favored. Over the past decades, researchers have defined a parameter, namely glass-forming ability (GFA), which relates to ease or difficulty of metallic glass formation. In this chapter, we would show the development an atomistic theory capable of predicting an alloy composition range/ region energetically favoured for metallic glass formation in binary/ ternary alloy systems and predicting amorphization driving force for the possible glassy alloy.

The experimental observations of the formation of metallic glasses in the binary alloy systems are firstly summarized. Some empirical criteria/ rules to predict metallic glass formation are then discussed. Thirdly, description of developing an atomistic theory is presented. Briefly, the present authors proposed to take the interatomic potential of an alloy system as starting base to develop the theory. The constructed potentials are applied to run Molecular Dynamics (MD) or Monte Carlo (MC) simulations, in which solid solution models are used in simulations to compare relative stability of solid solution vs. its competing disordered counterpart, as a function of composition. The simulation results led to a unanimous verdict, that is, based on an n-body potential of an alloy system, simulations reveal the physical origin of metallic glass formation, and quantitatively determine an alloy composition range/region of the binary/ternary alloy system, within which metallic glass formation is energetically favoured, thus predicting the glass formation range/region (GFR) for the binary/ternary alloy systems.

Experimental Observations of the Binary Metallic Glass Formation

In LMQ, an alloy melt is rapidly quenched to freeze in its initial liquid state, thus forming a metallic glass through a liquid-to-solid phase transformation. The LMQ technique has extensively been developed and produced a large number of metallic glasses.

Since 1980s, ion beam mixing (IBM) of multiple metal layers (Liu et al. 1983a; Liu et al. 1983b), and solid-state amorphization (SSA) (Clemens et al. 1984; Schwarz and Johnson 1983) have been employed to study the metallic glass formation. Differing from LMQ, metallic glass formation by these techniques is through a solid-solid phase transformation. Consequently, these techniques have produced a variety of metallic glasses in both equilibrium miscible ($\Delta H_f < 0$) and immiscible ($\Delta H_f > 0$) binary alloy systems (Liu et al. 2001; Suryanarayana 1999).

Binary Metallic Glass Formation by Liquid Melt Quenching

The earliest method of LMQ is a gun technique proposed by Duwez et al. In this method, a molten alloy is atomized by expulsion from a shock tube and propelled against the surface of a substrate with high thermal conductivity (Duwez et al. 1960; Klement et al. 1960). Later, some other methods have been developed, such as twin-roll quenching, melting spinning, etc. These techniques are capable of producing continuous amorphous filaments, ribbons or sheets (Anantharaman 1984). Another LMQ technique is the copper mold casting and it has so far produced a number of metallic glasses, including the bulk metallic glasses (BMG), whose size could be up to millimeter–centimeter (Johnson 1999; Luborsky 1983).

In LMQ, a cooling speed higher than 10^6 K/s is typically needed. In recent years, a cooling speed up to 10^7–10^8 K/s has been achieved (Suryanarayana 1999). The list of metallic glasses obtained by LMQ has ever been lengthening and up-to-now metallic glasses have been obtained in over 100 binary alloy systems. Based on LMQ data, the binary alloy systems have been classified into glass-forming and non-glass-forming systems. Table 1 lists some of the binary alloy systems and respective composition ranges, within which metallic glasses have been obtained.

Binary Metallic Formation by Ion Beam Mixing

Ion beam mixing (IBM) of multiple metal layers was introduced in 1980 for studying the formation of metal-silicides (Tsaur et al. 1980), and a little later, was employed to study the metastable alloy phase formations

Table 1: Glass-forming range (GFR, at% B) of the A-B binary metal systems obtained by liquid melt quenching.

System	GFR	Reference	System	GFR	Reference
Al-Ca	9–11, 55–85	(Guo et al. 2004; Inoue et al. 1994)	Cu-Ti	25–65	(Brunelli et al. 2001; Koster et al. 1995)
Al-Ce	7–12	(Akihisa et al. 1988)	Cu-Zr	10–80	(Brunelli et al. 2001; Eifert et al. 1984; Inoue and Zhang 2004; Wang et al. 2004a; Xu et al. 2004; Yang et al. 2009)
Al-Ho	9–12	(Inoue 1998b)	Fe-Hf	8–10	(Ryan et al. 1987)
Al-Mn	15–25	(Fukamichi and Goto 1991)	Fe-Pr	45–90	(Croat 1981)
Al-Pr	9–11	(Inoue 1998b; Massalski et al. 1986)	Fe-Zr	9–13, 57–82	(Altounian et al. 1985; Buschow 1984)
Al-Tb	9–14	(Bacewicz and Antonowicz 2006; Inoue 1998b; Massalski et al. 1986)	Ge-Zr	81–88	(Inoue et al. 1984; Inoue et al. 1982)
Al-Yb	9–13	(Inoue 1998b; Massalski et al. 1986)	Hf-Ni	30–70	(Buschow and Beekmans 1979; Zhang et al. 1994)
Al-Ti	34–54	(Akiyama et al. 1993)	Mg-Y	15–16	(Horikiri et al. 1994)
Al-Zr	45–53, 63–74	(Gudzenko and Polesya 1975)	Nb-Ni	40–62.5	(Leonhardt et al. 1999; Xia et al. 2006b; Zhu et al. 2006)
Be-Ti	59–63	(Tanner and Ray 1979)	Ni-Ta	30–50	(Uhlig et al. 1993; Zhang et al. 2008)
Be-Zr	50–70	(Hasegawa and Tanner 1978; Tanner and Ray 1979)	Ni-Zr	20–80	(Inoue et al. 1990)
Co-Gd	62–71	(Zhang et al. 2009)	Pd-Zr	65–70	(Eifert et al. 1984; Jastrow et al. 2004; Saida et al. 2008)
Co-Tb	45–65	(Hassanain et al. 1995)	Pt-Zr	70–80	(Eifert et al. 1984; Lee et al. 2006; Saida et al. 2008)
Co-Zr	42–80	(Buschow 1984)	Rh-Zr	66–80	(Eifert et al. 1984)
Cu-Hf	30–70	(Inoue and Zhang 2004; Jia and Xu 2009; Zhang et al. 1994)	Ti-Zr	15–22	(Whang 1984)

(Liu 1986; Liu and Jin 1997). In studying a binary A-B alloy system, the A-B multilayered films with the desired over composition A_xB_{1-x} (x = 0 ~ 1) are prepared by alternately depositing A and B metals onto an inert substrate (e.g., NaCl, SiO_2, etc.). The deposited A-B multilayered films are then irradiated by inert gas ions (e.g., xenon ions) of high energy (typically, 200–300 keV), while the substrate is kept at low temperature (77 K) or ambient temperature (300 K).

The IBM process consists of two consecutive steps, that is, an atomic collision cascade and a relaxation (Liu et al. 2000; Russell 1984; Thompson 1969). In the first step, the irradiating ions are decelerated by elastic and inelastic collisions, transferring energy to the metal layers. The energy received by the atoms in the metal layers is very high and the atoms knocked out from original lattice positions. The knocked out atoms are able to generate a second generation of collisions, and so on so forth. This process is named as atomic collision cascade, resulting in intermixing of A and B metal layers. After receiving an adequate irradiation dose, the discrete layered structure of the A-B multilayered films is smeared out and a uniform A-B atomic mixture is obtained. The A-B atomic mixture obtained is highly energetic and in a disordered state.

After that, equilibrium thermodynamics comes into play to direct the disordered A-B atomic mixture to lower its energy towards equilibrium during relaxation. According to Atomic Collision Theory (Russell 1984; Thompson 1969), the time duration of relaxation is extremely short and the A-B atomic mixture is not allowed to proceed straightforward to its equilibrium state. It is estimated that an effective cooling speed could be as high as 10^{12}–10^{13} K/s (Averback and Rubia 1998; Sigmund 2006; Suryanarayana 1999; Thompson 1969). Naturally, during the relaxation, either some limited atomic rearrangements take place, leading the A-B atomic mixture to form an A-B metastable crystalline phase of simple structure, or the disordered state is preserved, thus forming a glassy phase. In fact, IBM experimental results show the nonequilibrium alloys produced by IBM are either metallic glasses or crystalline alloys mostly of simplest crystalline structure of bcc, fcc and hcp (Liu 1986).

An important fact is that IBM technique has produced a number of metallic glasses in many equilibrium immiscible systems with positive $\Delta H_{f'}$ indicating that such disorder state does exist in the immiscible systems. Table 2 lists some of the binary alloy systems, which have been studied by IBM.

Binary Metallic Formation by Solid-State-Amorphization

Solid-state-amorphization (SSA) was first observed in 1983 by Schwarz and Johnson in the Au-La system they attributed the SSA to a large negative ΔH_f and a very large atomic size difference (Schwarz and Johnson 1983).

From the mid 1990s, extensive experimental studies have shown that SSA could take place in some binary alloy systems having different characteristics. In fact, SSA has been observed, for example, in the Ni-Mo

Table 2: Glass-forming range (GFR, at% B) of the A-B binary metal systems obtained by ion beam mixing.

System	ΔH_f	GFR	References	System	ΔH_f	GFR	References
Ag-Mo	Positive	25–88	(Jin et al. 1995)	Fe-Zr	Negative	10–27	(Rao et al. 1987)
Ag-Nb	Positive	25–90	(Jin and Liu 1996)	Hf-Ta	Positive	26–90	(Liu et al. 1996a)
Ag-Ta	Positive	62–80	(Zhao et al. 2008)	Mo-Ni	Negative	35–65	(Li and Liu 1999)
Al-Pt	Negative	33–75	(Hung et al. 1983)	Mo-Y	Positive	22–78	(Zhang et al. 1995)
Al-Ti	Negative	25–75	(Liu and Cheng 1991)	Nb-Ni	Negative	20–85	(Zhang et al. 1993)
Au-Mo	Positive	75–85	(Pan and Liu 1996)	Ni-Sc	Negative	29–80	(Hu et al. 2003)
Co-Mo	Negative	35–65	(Lin et al. 2000b)	Ni-Ta	Negative	25–75	(Liu and Zhang 1994)
Co-Nb	Negative	20–77	(Zhang and Liu 1994b)	Ni-Ti	Negative	37.6–85	(Lai et al. 2001)
Co-Ta	Negative	25–75	(Zhang and Liu 1994a)	Ni-Zr	Negative	30–65	(Huang et al. 1988)
Fe-Nb	Negative	25–75	(Lin et al. 2000a)	Ta-Zr	Positive	31–80	(Jin and Liu 1997)
Fe-Ta	Negative	25–80	(Lin et al. 2000a)	Y-Zr	Positive	65–85	(Liu et al. 1998)

system with $\Delta H_f = -7$ kJ/mol (Zhang and Liu 1994c), in the immiscible Cu-Ta equilibrium system with $\Delta H_f = +3$ kJ/mol (Liu et al. 1996b) and in the Au-Ta system with two metals having almost identical atomic sizes (Pan et al. 1995). Up till date, SSA has been observed in 28 binary alloy systems (Dinda 2006), and it is also a technique to produce amorphous thin films and in SSA, the favoured composition range to form amorphous alloys is, in general, narrower than those in IBM.

Empirical Criteria/Rules to Predict Metallic Glass Formation

In past decades, researchers have striven to predict the glass-forming ability (GFA) or quantitatively estimate the glass formation range of a binary alloy system. For example, Turnbull proposed a deep eutectic criterion to predict the metallic glass formation upon LMQ (Turnbull 1969). Egami and Waseda predicted the metallic glass formation by considering the atomic size difference of the constituent metals (Egami and Waseda

1984). Based on IBM studies, Liu et al. proposed a structural difference rule (Liu et al. 1983b) and, a little later, further proposed that the total width of the two-phase region (or regions) observed from equilibrium phase diagram could approximately be the glass formation range of a binary alloy system (Liu 1986). Johnson (Johnson 1999) and Inoue (Inoue and Takeuchi 2002) have also predicted the element selection and composition range of glass forming alloys. In addition, thermodynamic approach, for example, by considering the formation enthalpy, formation energy, etc., has also been proposed to predict binary metallic glass formation (Bormann et al. 1988; Weeber and Bakker 1988).

Deep Eutectic Criterion

The deep eutectic criterion states that metallic glasses are likely to be formed by LMQ in a binary alloy system featuring a deep eutectic point (or points) in its equilibrium phase diagram and that the alloys locating in the vicinity of an eutectic point (or points) is amorphized much easier. Naturally, the eutectic composition is always located in a two-phase region in the phase diagram and is not good for crystallization of either crystalline phase of two sides. Besides, deep eutectic means a low melting point, which ensures that initial liquid melt could be maintained in over-heating state for much longer time than in under-cooling state.

As shown in Table 2, a number of metallic glasses have been obtained by IBM, not only in many systems classified as glass-forming systems, but also in many systems classified as non-glass-forming ones (based on LMQ data).

Size Difference Rule

Considering the atomic sizes of constituent metals, Egami argued that (Egami and Waseda 1984), due to the atomic size difference, the solute atoms would lead to an atomic level stress in solvent lattice and that once the solute concentration and atomic level stress achieve critical points, the solvent lattice would transform into an amorphous phase. For an A-B binary alloy system, the critical concentrations $C_{B\ in\ A}^{critical}$ and $C_{A\ in\ B}^{critical}$ for metallic glass formation could be estimated by

$$C_{A\ in\ B}^{critical} = \frac{\gamma^3}{10|\gamma^3 - 1|}, \tag{1a}$$

and

$$C_{B\ in\ A}^{critical} = \frac{1}{10|\gamma^3 - 1|}, \tag{1b}$$

where $\gamma = R_A/R_B$, and R_A and R_B are the radius of atoms A and B, respectively. The two critical concentrations $C_{B\ in\ A}^{critical}$ and $C_{A\ in\ B}^{critical}$ identify an alloy composition range, that is, the glass formation range of the A-B binary alloy system. It is noted that the prediction relevance is about 60% (Li et al. 2011).

Structural Difference Rule

In 1983, Liu et al. have proposed a structural difference rule, stating that an amorphous alloy would most likely be formed in a binary alloy system by IBM, when two constituent metals are of different crystalline structures and the alloy composition is chosen to be in middle of the two-phase region observed from the phase diagram of the system (Liu et al. 1983b).

Further studies show that when the overall composition is located at or in the vicinity of intermetallic compound having a complicated structure and a high melting point, an amorphous alloy could also be formed by IBM. Based on above results, Liu et al. have proposed an empirical model by using a single parameter, that is, the maximum possible amorphization range (MPAR) of a binary alloy system. The physical meaning of the proposed MPAR is that it represents the total width of the two-phase regions observed from the equilibrium phase diagram of the system. Accordingly, for a binary alloy system, the MPAR is defined to equal to 100%, as the whole composition range, minus two maximum terminal solid solubilities observed from its equilibrium phase diagram.

Interatomic Potential to Predict the Binary Metallic Glass Formation

Although the above criteria or rules have helped as guidelines for producing metallic glasses, they were empirical approaches, and if their predictions are not adequate, it is probably due to their starting bases, which could not reflect the intrinsic characteristics of the system.

In 2000, the present authors have chosen the interatomic potential of an alloy system as the starting base to develop related model/theory, as the interatomic potential of the system describes the major interactions among all the atoms involved in the system, governing the energetic states (of either crystalline or disordered structure) of all the alloy phases. The glass formation range (GFR) of the system could also be considered one of intrinsic characteristics and could also be deduced from the interatomic potential (Li et al. 2008; Li et al. 2011).

To proceed, it is necessary to have an appropriate simulation route. According to the experimental observations, formation of a metallic glass is always a non-equilibrium process and the available kinetic condition is extremely restricted. For example, in LMQ, a cooling speed of 10^6 K/s is frequently required (Wang et al. 2004b), and in IBM, the effective cooling speed can be up to 10^{13} K/s (Liu et al. 2000). Consequently, complicated structured phases can hardly nucleate and grow (Li et al. 2011), and thus the competing phase against metallic glass is the terminal solid solutions, which always have simple crystalline structures. Moreover, IBM results (Liu 1986, 1987) have shown that the maximum favoured composition range is approximately to be MPAR, that is, the total width of the two-phase regions observed from the equilibrium phase diagram of the system. In other words, the scientific issue of predicting the GFR of a binary alloy system can be converted into an issue of determining the two critical solid solubilities, which split the entire alloy composition range into two different categories, energetically favoring the formation of solid solutions and metallic glasses, respectively.

Consequently, to develop the atomistic theory, the first step is to construct a relevant interatomic potential of a binary alloy system and then to apply the constructed potential to carry out a series of molecular dynamics (MD) simulations, by using solid solution models to compare the relative stability of solid solution versus its disordered counterpart as a function of solute concentration.

Construction of n-body Potentials for the Binary Metal Systems

For transition metals and alloy systems (Daw et al. 1993; Heine et al. 1991; Li et al. 2008; Raeker and Depristo 1991; Voter 1996), there have been three major schemes for construction of n-body potentials: embedded atom method (EAM) potentials (Daw et al. 1993; Wu et al. 2006), second-moment approximation of tight-binding (TB-SMA) potentials (Cleri and Rosato 1993; Ducastelle and Cyrot-Lackmann 1970, 1971), and Finnis-Sinclair (F-S) potentials (Ackland et al. 1987; Finnis and Sinclair 1984; Rebonato et al. 1987; Sutton and Chen 1990). In these potentials, the energy is expressed by a pair term and n-body term (i.e., many-body term), and n-body term is a function of the electron density. By combining the unique features of the TB-SMA and F-S potentials, Li, Dai and Dai proposed a long-range n-body potential (abbreviated as the LDD potential) and it has been proven to be suitable for bcc, fcc and hcp structured metals and their alloys (Dai et al. 2007a; Dai et al. 2007b; Dai et al. 2009b; Li et al. 2007).

For an A-B binary alloy system, there are three atomic interactions, that is, A-A interaction, B-B interaction and A-B cross interaction, respectively.

The potential parameters are determined by fitting the coefficients to some physical properties of the system, such as cohesive energies, lattice constants, bulk modules and elastic constants obtained from experiments or theoretical calculations. In general, these physical properties of the constituent metals A and B and their compounds are used if available. In an equilibrium immiscible system, however, there is no any compound existing and no any datum available to fit the cross potential. This problem could also be encountered in some miscible binary metal systems, in which equilibrium compounds may exist, yet no adequate property data are available. A method named *ab initio* assisted construction of n-body potential has been proposed (Li et al. 2008; Voter 1996). In practice, *ab initio* calculations are employed to acquire some physical properties of some possible metastable compounds in the system, such as B2 and L1$_2$ structures (Brandes et al. 1992; Pearson 1958), etc. and acquired properties are used to fit the cross potential.

We have so far constructed about 20 n-body potentials for the binary alloy systems (He et al. 2007). For example, for the Ag-Mo system, its n-body potential was constructed under F-S formalism.

For Finnis–Sinclair (F-S) potential (Finnis and Sinclair 1984), total energy is given by

$$E_{tot} = \sum_{j \neq i} V(r_{ij}) - \sqrt{\sum_{j \neq i} A^2 \phi(r_{ij})}, \tag{2}$$

where the first term is a normal pair-wise energy consisting of a repulsive part. The parameter A in the second term is a constant in positive value (Dai et al. 2004; Tai et al. 2006). The function $V(r)$ was taken a polynomial form extending up to r^5, instead of a quadratic form in original F-S formalism. In deriving the Ag-Mo cross potential, *ab initio* calculations were carried out to acquire the cohesive energies and lattice constants of the possible metastable L1$_2$ Ag$_3$Mo and B2 AgMo compounds, and the calculated data were then used to fit the Ag-Mo cross potential. Table 3 lists the data to compare the cohesive energies and lattice constants of the Ag-Mo phases reproduced from constructed potential with those obtained from *ab initio* calculations. One sees the agreement well, confirming the constructed F-S Ag-Mo potential is relevant. Table 4 lists the fitted parameters for the F-S Ag-Mo potential.

Methods of Molecular Dynamics Simulation and Structural Characterization

We would briefly describe the methods of molecular dynamics simulation and structural characterization. For details, the readers are referred to some of the previous publications (Allen and Tildesley 1987; Frenkel and Simit 2002; Rapaport 2004).

Table 3: Comparison of the cohesive Energy Ec (eV) and lattice constant a (Å) derived from the constructed Ag-Mo potential vs. those from *ab initio* calculations.

System	Structure	*Ab initio* calculation		Present study	
		E_c	a	E_c	a
Ag_3Mo	$L1_2$	4.13	3.1166	4.13	3.1166
AgMo	B2	3.23	3.9161	3.23	3.9161
$AgMo_3$	$L1_2$	3.99	5.2072	3.99	5.4438

Table 4: Finnis-Sinclair potential parameters of the Ag-Mo system (Dai et al. 2004; Tai et al. 2006).

	Ag-Ag	Mo-Mo	Ag-Mo
A (eV/Å)	0.325514	1.848648	0.625749
d (Å)	4.41	4.1472	4.2
c (Å)	4.76	3.2572	4.5
c_0 (eV/Å²)	10.6812	47.98066	44.40681
c_1 (eV/Å³)	12.04517	34.09924	45.49026
c_2 (eV/Å⁴)	5.203072	5.832293	15.60511
c_3 (eV/Å⁵)	1.013304	0.101749	1.793704
c_4 (eV/Å⁶)	0.0742308	0.0203934	0.00
B (1/Å²)	1.293394	0.00	0.00

Parrinello and Rahman scheme for molecular dynamics simulation

Based on Anderson's work, Parrinello and Rahman have developed a scheme to simulate a condensed matter system under a constant pressure by introducing the volume and shape of the simulation box as the extended dynamical variables (Andersen 1980; Hoover 1986; Parrinello and Rahman 1981). The scheme allows both volume and shape of the simulation box to change or fluctuate depending on the imbalance between the external pressure and the internal instantaneous stress. In the scheme, the motion equations of atoms and simulation box could be written as

$$\ddot{\vec{s}}_i = \frac{1}{m_i} h^{-1} \vec{F}_i - G^{-1} G \dot{\vec{s}}_i, \tag{3}$$

$$\ddot{h} = \frac{\Omega}{W} (\Pi - p_{ext} I) h^{-T}. \tag{4}$$

Here, $\vec{F}_i = -\nabla_i E_{potential}$ is the force acting on atom i due to the interatomic interaction and can be computed from local gradient of potential energy.

$G = h^T h$ is for matrix tensor of simulation box, and $\Pi = \frac{1}{\Omega} \sum (m_i \dot{\vec{r}}_i \otimes \dot{\vec{r}}_i +$ $F_i \otimes \vec{r}_i)$ is for internal stress tensor. W is for fictional mass of extended system. p_{ext} is applied external pressure. I is unitary matrix. One dot and two dots above a variable denote the first- and second-derivatives respect to time, respectively. Superscripts -1 and $-T$ denote the reverse and transposed of a matrix or a tensor, respectively. The motion equations of atoms and simulation box are integrated using second-order 6-value predictor-corrector algorithm by Gear (Allen and Tildesley 1987; Ciccotti and Hoover 1985) with a time step of 5×10^{-15} second. The temperature of the system is regulated by scaling the velocities of atoms to corresponding value.

Solid solution simulation models

To compare relative stabilities of solid solution versus its disordered counterpart, bcc, fcc and hcp structured solid solution models were constructed. For bcc and fcc solid solution models, the [100], [010] and [001] crystalline directions were parallel to the x, y and z axes, respectively. For hcp solid solution model, the [100], [120] and [001] crystalline directions were parallel to the x, y and z axes, respectively. Along x, y and z directions, there were N_x, N_y and N_z conventional cells for the bcc/fcc solid solution models and equivalent orthorhombic cells for the hcp solid solution models. For bcc solid solution model, N_x, N_y and N_z were set to be 10 in the three directions, and there were $10 \times 10 \times 10 \times 2 = 2000$ atoms. For fcc solid solution model, $N_x = 8$, $N_y = 8$ and $N_z = 8$, and there were $11 \times 7 \times 7 \times 4 = 2156$ atoms. For hcp solid solution model, $N_x = 11$, $N_y = 7$ and $N_z = 7$, and there were $11 \times 7 \times 7 \times 4 = 2156$ atoms. The solvent atoms were substituted randomly by a certain number of the solvent atoms to form a uniform solid solution and the substitution is gradually increased to find out the critical solubility, thus splitting the composition range into favoring solid solution and disordered state, respectively. In simulations, the periodic boundary conditions were always adopted in x-, y- and z-directions.

Structural characterization

Several methods were employed to characterize structure changing during MD simulations. A major quantity was structure factor $S(k)$, which was obtained in the reciprocal space (Hansen and Mcdonald 1976; Leshansky and Brady 2005). To extract the topological characteristics of structure, planar structure factor could be computed by (Phillpot et al. 1989):

$$S(\vec{k},z) = \left\langle \frac{1}{N_z^2} \left| \sum_{j=1}^{N_z} \exp(i\vec{k} \cdot \vec{r}_j) \right|^2 \right\rangle, \tag{5}$$

where z labels the crystal plane to which the point $\vec{r}_j$ belonged at initial configuration, and $\vec{k} = 2\pi(1/N_x, 1/N_y)$. N_z is the number of atoms in atomic plane z. $S = 1$ refers to an entirely ordered crystal, while $S = 0$ stands for a completely disordered state (Phillpot et al. 1989). In practice, the atomic planes were defined as original crystal planes and $S(k)$ was used to indicate degrees of disordering in the atomic planes.

The second method is the pair-correlation function $g(r)$, also named as density autocorrelation in the real space. The pair-correlation function could be computed by (Zallen 1983)

$$g(r) = \frac{\Omega}{N^2} \sum_{i,j}^{N} \delta(r, r_{ij}), \tag{6}$$

where $\delta = 1$ for $r = r_{ij}$ and $\delta = 0$ for $r \neq r_{ij}$. r_{ij} is the distance between atoms i and j. The quantity r is correlation length or distance. For a crystalline structure, $g(r)$ exhibits a sequence of peaks that corresponding to the shells around an atom. For a liquid or disordered solid, $g(r)$ exhibits a few major peaks and oscillates with less pronounced peaks at long distance.

In addition, a plane density profile function, reflecting an amount of ordering in z-direction, is defined and could be computed by

$$\rho_z(l) = \frac{1}{N} \sum_{i}^{N} \delta(l, \vec{r}_{i,x}), \tag{7}$$

Prediction of the Glass Formation Ranges of the Binary Alloy Systems

Take the Ag-Mo as an example, MD simulations have been carried out by using the above proven realistic F-S potential (Tai et al. 2006). Figure 1 displays the projections of atomic positions for the $Ag_{95}Mo_5$, $Ag_{90}Mo_{10}$, and $Ag_{50}Mo_{50}$ fcc solid solution models upon annealing at 300 K for 120,000 simulation time steps. One sees that for the $Ag_{95}Mo_5$ solid solution model, the original fcc structure still remains and that for the $Ag_{90}Mo_{10}$ and $Ag_{50}Mo_{50}$ solid solution models, the projections of atomic positions are obviously disordered, indicating that the crystalline lattices have collapsed and turned into their respective disordered states.

Figure 2 shows corresponding density profiles $\rho_z(l)$ for $Ag_{95}Mo_5$, $Ag_{90}Mo_{10}$, and $Ag_{50}Mo_{50}$ fcc solid solution models.

One sees that, when the solute concentration is less than 10 at% Mo, the atomic planes can clearly be distinguished from the density profiles and that once Mo concentration equals to or exceeds 10 at%, a relatively uniform distributed density profile appears. To have firm evidence, total and partial pair correlation functions $g(r)$ for fcc Ag-based solid solution models are shown in Fig. 3.

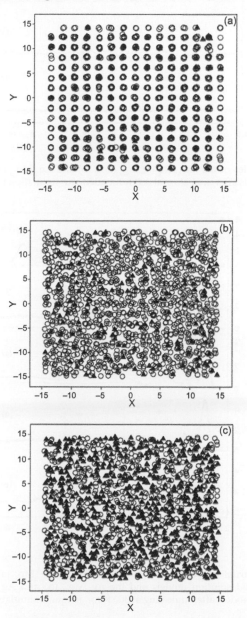

Fig. 1: Projections of atomic positions for the $Ag_{95}Mo_5$ (a), $Ag_{90}Mo_{10}$ (b), and $Ag_{50}Mo_{50}$ (c) fcc solid solution models upon annealing at 300 K for adequate simulation times.

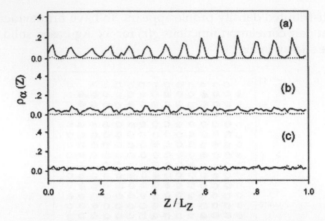

Fig. 2: Density profiles of the $Ag_{95}Mo_5$ (a), $Ag_{90}Mo_{10}$ (b) and $Ag_{50}Mo_{50}$ (c) fcc solid solution models after annealing at 300 K for adequate simulation times.

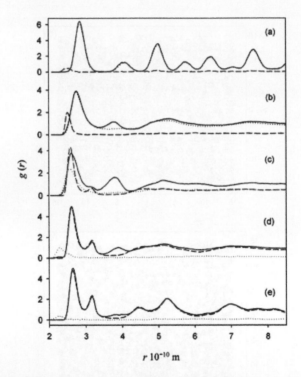

Fig. 3: Total and partial pair-correlation functions for the $Ag_{95}Mo_5$ (a), $Ag_{90}Mo_{10}$ (b) and $Ag_{50}Mo_{50}$ (c) fcc solid solution models and the $Ag_{15}Mo_{85}$ (d) and $Ag_{10}Mo_{90}$ (e) bcc solid solution models after annealing at 300 K for adequate simulation times.

Because the $g(r)$ curve of the $Ag_{95}Mo_5$ solid solution model exhibits apparent sharp peaks even at a large distance r, the $Ag_{95}Mo_5$ solid solution is still be in a crystalline structure. For the $Ag_{90}Mo_{10}$ and $Ag_{50}Mo_{50}$ solid solution models, the first and second peaks of the $g(r)$ curves are still clear, yet there are no discernible peaks beyond the third-nearest neighbors. Following Zallen's criterion (Zallen 1983), it is concluded that a crystal-amorphous transition has indeed taken place in both $Ag_{90}Mo_{10}$ and $Ag_{50}Mo_{50}$ solid solution models. These results suggest that for fcc Ag-based solid solution, the critical solid solubility is 10 at% Mo. Similar simulations and characterization have been carried out for a number of bcc Mo-based solid solution modes and the critical solid solubility was determined to be 12 at% Ag. Consequently, the two critical solid solubilities of the Ag-Mo system are 10 at% Mo in fcc Ag and 12 at% Ag in bcc Mo, respectively, thus predicting the glass formation range (GFR) of the Ag-Mo system is 10–88 at% Mo (or 12–90 at% Ag).

The present authors have worked out in eight selected binary alloy systems: the Ni-Nb, Ni-Mo, Ag-Mo, Cu-W, Co-Nb, Sc-W, Ni-Hf and Cu-Ru systems, covering various structural combinations of fcc, bcc and hcp, as well as various thermodynamic characteristics from large negative to large positive ΔH_f. First, their n-body potentials were constructed. Series MD simulations, using solid solution models, were then carried out. The simulation results reveal that when the solute concentration is less than the critical solid solubility, the crystalline solid solution is energetically favoured comparing to amorphous phase and that whenever the solute concentration reaches and/or exceeds the critical solid solubility, the crystalline solid solution becomes unstable and collapses into the disordered state, giving rise to the formation of a metallic glass. It follows that the physical origin of metallic glass formation, or of the crystal-to-amorphous transition, is the collapsing of the crystalline lattice, while the solute atoms are exceeding the critical solid solubility.

As the MD simulations determine two critical solubilities for each binary alloy system, the composition range bounded by two determined critical solubilities is defined as the glass formation range (GFR) of the system and within the GFR, metallic glass formation is energetically favoured. Table 5 lists the glass formation ranges (GFRs) of eight binary alloy systems predicted from their interatomic potentials through MD simulations. For comparison, the experimental data obtained by various glass-producing techniques are also included. In short, the predicted GFRs of the binary alloy systems from n-body potential are reasonable.

Table 5: Glass-forming ranges (GFRs, at% B) of the eight representative binary metal systems determined by their respective interatomic potentials and predicted by other proposed criteria or rules as well as measured in experiments.

A-B system	Interatomic potentials approach	Thermodynamics	Size difference rule	Experiments
Ni-Nb	13–81	20–69 (Miedema)		37.5–60 [a] 15–80 [b]
Ni-Mo	21–75			35–65 [c] 25–75 [d]
Ag-Mo	10–88			25–88 [e]
Cu-W	20–65			25–55 [f]
Co-Nb	18–84	15–74 (Calphad)	22–64	20–77 [g]
Sc-W	15–50	12–58 (Miedema)		25–40 [h]
Ni-Hf	18–75	16–75 (Miedema)	10–81	30–70 [i] 20–75 [j]
Cu-Ru	10–80			25–75 [k]

a) Liquid melt quenching (Skakov et al. 1991; Xia et al. 2006b)
b) Ion beam mixing (Zhang et al. 1993)
c) Ion beam mixing (Li and Liu 1999)
d) Solid state amorphization (Zhang and Liu 1994c)
e) Ion beam mixing (Jin and Liu 1996; Jin et al. 1995; Tai et al. 2006)
f) Vapor-deposition (Rizzo et al. 1993)
g) Ion beam mixing (Zhang and Liu 1994b)
h) Ion beam mixing (Zhang et al. 2005)
i) Liquid melt quenching (Zhang et al. 1994)
j) Solid state amorphization (Boyer and Atzmon 1993)
k) Ion beam mixing (He et al. 2007; He et al. 2003)

Interatomic Potential to Predict the Ternary Metallic Glass Formation

In the later 1980s, researchers have found a number of ternary and multi-component alloy systems, in which LMQ could readily produce the so-called bulk metallic glasses (BMGs) with a geometrical size of up to a magnitude of centimeter (Chen 1974; Inoue et al. 1989; Peker and Johnson 1993). From then on, the pursuit of BMGs in various alloy systems has become a hot topic due to the fundamental scientific importance and great potential for engineering applications (Inoue 2000; Miller and Liaw 2007; Schroers 2013; Wang et al. 2004b). In theoretical studies, one of the fundamental scientific issues is to establish a reliable model to predict the favoured composition region of a ternary alloy system for metallic glass formation.

After verifying the relevance of the atomistic theory in predicting the binary metallic glass, we take a step forward and proceed to investigate the ternary alloy systems, as a number of ternary alloy systems have been recognized as promising candidates in forming the BMGs.

Long-Range n-body Potentials Proposed for the Transition Metal Systems

In general, EAM and F-S potentials are applicable to fcc and bcc metals, whereas the TB-SMA potential is suitable to fcc and hcp metals. Therefore, the challenging problem is how to describe the interatomic interactions in *hcp-bcc* binary alloy systems and how to deal with a ternary alloy system with three components with fcc, bcc and hcp structures. The present authors' group has proposed a new long-range n-body potential (Dai et al. 2007a; Dai et al. 2007b; Dai et al. 2009b). In the following the newly proposed potential is abbreviated later by LDD potential for convenience.

The physical argument of the LDD potential is also based on the second moment approximation of the tight-binding theory. Accordingly, the potential energy of atom i in a condensed system consisting of N atoms, E_i, can be expressed by,

$$E_i = \frac{1}{2}\sum_{j\neq i}\phi(r_{ij}) - \sqrt{\sum_{j\neq i}\psi\left(r_{ij}\right)}, \tag{8}$$

here, r_{ij} is the distance between atoms i and j of the system. For sake of convenience, the physical argument can be ignored here and therefore ϕ and ψ are named as the pair term and n-body part, respectively. They can be computed by:

$$\phi(r) = (r_{c1} - r)^m (c_0 + c_1 r + c_2 r^2 + c_3 r^3 + c_4 r^4), \quad 0 < r \leq r_{c1}, \tag{9}$$

$$\psi(r) = \alpha^2 (r_{c2} - r)^n, \qquad\qquad 0 < r \leq r_{c2}, \tag{10}$$

where m and n are two adjustable parameters. r_{c1} and r_{c2} are cutoff radii of pair term and n-body part, respectively. It can be shown that ϕ, ψ and their first derivatives smoothly converge to zero at cutoff radii if m and n are both greater than 3. m, n, c_0, c_1, c_2, c_3, c_4, and α are potential parameters that are determined by fitting to the experimental or *ab initio* calculated results. Since the cutoff radius of this empirical potential is greater than 6 Å, which is greater than the sixth-neighbor distance, it can be considered a long-range empirical potential.

The pair term in the LDD potential inherits the feature of the original F-S potential, while the n-body part exhibits some characteristics of the traditional TB-SMA potential. In addition, the LDD potential has been proven to be suitable for bcc, fcc and hcp transition metals and their alloys.

Construction of n-body Potentials for the Ternary Alloy Systems

The Ni-Nb-Zr system, consisting of fcc, bcc and hcp metals, is a typical example to show the advantage of the LDD potential. There should be six sets of potential parameters in the: three sets for pure elements Ni-Ni, Nb-Nb and Zr-Zr and another three sets for dissimilar metals of Ni-Nb, Ni-Zr and Nb-Zr. Generally, the potential parameters of pure elements are determined by fitting to the basic physical properties such as the cohesive energies, lattice constants, bulk moduli and elastic constants of pure metals (Lide 2002). The cross potentials are determined by fitting to the basic physical properties of the intermetallic compounds. Since there is a lack of available physical properties of compounds in the system, *ab initio* calculations were carried out using CASTEP code (Clark et al. 2005; Segall et al. 2002). The geometry optimization was first performed to determine the lattice constant and the total energies of the concerned compounds, and then the elastic constants and bulk modulus were calculated. The cohesive energies of compounds were then easily derived from the total energies.

Table 6 shows the constructed n-body potentials for the Ni-Nb-Zr system. The interatomic potentials of three pure metals have been proven relevant to identify the relative stabilities between stable and their hypothetic structures (Dai et al. 2009a; Dai et al. 2009b). While in the Ni-Nb, Ni-Zr and Nb-Zr cross potentials fitting process, compounds

Table 6: Parameters of the constructed long-range LDD potentials for the Ni-Nb-Zr system.

	Ni-Ni	Nb-Nb	Zr-Zr	Ni-Nb	Ni-Zr	Zr-Nb
m	4	4	4	4	4	4
n	6	8	4	4	5	4
r_{c1} (Å)	5.749526	4.851379	6.437549	5.25695	4.783337	7.040277
r_{c2} (Å)	7.171908	7.117045	7.222787	9.864774	7.385404	7.2984
c_0 (eV/Å^m)	0.294724	10.53168	0.788297	0.616469	0.633714	0.371087
c_1 (eV/Å^{m+1})	−0.35871	−14.4959	−0.84608	−0.65777	−0.82273	−0.39984
c_2 (eV/Å^{m+2})	0.163193	7.514826	0.341507	0.278365	0.485298	0.160857
c_3 (eV/Å^{m+3})	−0.03303	−1.72564	−0.0611	−0.05807	−0.14258	−0.02815
c_4 (eV/Å^{m+4})	0.002535	0.147777	0.004081	0.005397	0.016805	0.001819
α (eV/Å$^{n/2}$)	0.014612	0.011261	0.150093	0.04074	0.056287	−0.26622

with different structures and compositions are involved to ensure that the constructed potential parameters can describe the interatomic interactions in the whole composition range. After careful computation, the constructed Ni-Nb-Zr potential was proven to be of relevance (Luo et al. 2014).

Predicting the Glass Formation Regions of the Ternary Alloy Systems

Based on the constructed LDD Ni-Nb-Zr interatomic potential (Luo et al. 2014), Monte Carlo (MC) simulations were carried out using solid solution models over the entire composition triangle of the system. Three types of solid solution models, that is, the *fcc*, *bcc* and *hcp* solid solution models were constructed according to which type of atoms is dominant in the alloy composition. For fcc and bcc models, the [100], [010] and [001] crystalline directions are parallel to the x, y and z axes, respectively, while for hcp model, the [100], [120] and [001] crystalline directions are parallel to the x, y and z axes respectively. Periodic boundary conditions were adopted in the three directions. The fcc, bcc and hcp solid solution models consists of $10 \times 10 \times 10 \times 4 = 4000$ atoms, $12 \times 12 \times 12 \times 4 = 3456$ atoms and $12 \times 10 \times 8 \times 4 = 4000$ atoms, respectively.

In the simulations, the initial solid solution $Ni_xNb_yZr_{1-x-y}$ was evolved in isothermal-isobaric MC simulations at 300 K and 0 Pa for sufficient simulation time until all the related dynamics variables showed no secular variation. Inspecting the results of MC simulations, according to the 3D atomic configurations and pair correlation functions $g(r)$, we classified all the $Ni_xNb_yZr_{1-x-y}$ alloys after equilibrating adequate MC time steps into three structural categories: the crystalline state (CS), the transitional state (TS) and the amorphous state (AS), and then constructed the glass formation composition diagram in Fig. 4.

One sees that the Ni-Nb-Zr composition triangle is divided into four regions by three critical solubility lines AB, CD and EF. When an alloy composition locates beyond the lines moving towards one of three corners, the alloy tends to keep its crystalline structure, and three corner regions are classified as crystalline regions. When an alloy composition falls into the central polygon region enclosed by the bold blue line, the crystalline solid solution becomes unstable and collapses into disordered state, giving rise to forming an amorphous phase (metallic glass). This region is defined as glass formation region (GFR) of the Ni-Nb-Zr system. Within the GFR, formation of metallic glasses is energetically favoured. It is concluded that the underlying physics of metallic glass formation is the spontaneous collapse of crystalline lattice while the solute concentration exceeds a critical value, thus the glassy phase possesses higher stability than its competing crystalline counterparts within the GFR.

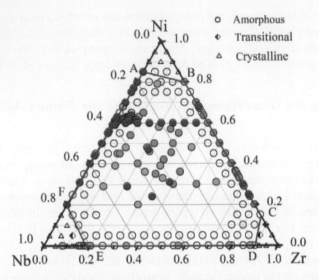

Fig. 4: Glass formation composition diagram derived from MC simulations at 300 K for the Ni-Nb-Zr system. Colored dots represent the experimental data obtained from different glass-producing techniques.

To validate the predicted GFR of Ni-Nb-Zr system, experimental results obtained by various glass-producing techniques (Altounian et al. 1983; Chang et al. 2007; Chen et al. 2007; Enayati et al. 2002; Kimura et al. 2003; Li et al. 2013; Paglieri et al. 2011; Sakurai et al. 2011; Xia et al. 2006b; Yamaura et al. 2006; Yamaura et al. 2005; Yuan et al. 2010; Zhu et al. 2006; Zhu et al. 2008; Zhu et al. 2007) were collected and shown in Fig. 4. The experimental data obtained are depicted in different colors. The red dots are for copper mold casting; the orange dots for melt-spinning technique; while the green ones are for other techniques, such as IBM, mechanical alloying, etc. One sees that all the experimental observations fall into the amorphous region. Besides, there are several other studies for the Ni-Nb-Zr system. For instance, Junpei Sakurai et al. have fabricated Ni-Nb-Zr thin films with 974 different compositions and 703 were confirmed to be single amorphous phases (Sakurai et al. 2011). Kimura et al. reported that the rapidly solidified Ni-Nb-Zr amorphous alloys were formed in the composition range of 20 to 75 at.% Ni, 0 to 60 at.% Nb and 0 to 80 at.% Zr (Kimura et al. 2003). Zhu et al. discovered that BMGs can be formed in quite a wide composition range within the ternary Ni-Nb-Zr system (Zhu et al. 2008). These compositions, which were not marked in Fig. 4, are also located in the predicted GFR of the Ni-Nb-Zr system.

Interatomic Potential to Calculate Amorphization Driving Force for Metallic Glass Formation

In this section, the atomistic approach will further be expanded to calculate the amorphization driving force (ADF) for the glassy alloy located in the predicted GFR of a ternary alloy system.

As discussion before, in the process of metallic formation, the amorphous phase competes with the fcc, bcc or hcp simple structured terminal solid solution. From a thermodynamic point of view, the energy difference between solid solution and amorphous phase can be considered as the amorphization driving force (ADF) for solid solution transforming into amorphous phase. Naturally, the larger the ADF, the easier the amorphous alloy forms (Lu et al. 2007; Wang et al. 2009; Xia et al. 2006a). Once the interatomic potential has been determined, the ADF for the metallic glass can be derived.

Take the Ni-Nb-Zr system as an example, the total energies of $Ni_xNb_yZr_{1-x-y}$ solid solutions and real alloys were calculated, respectively, and the energy differences between the two states (crystalline and amorphous) were defined as the amorphization driving forces (ADF). The ideal solid solution was obtained by annealing the simulation model with optimizing its lattice constant and simultaneously keeping the crystal symmetry unchanged; while for the real Ni-Nb-Zr alloy, the initial solid solution was evolved and equilibrated at 300 K in MC simulations for sufficient simulation time.

After MC simulations and calculations, the contour map of the ADF was plotted in Fig. 5.

One sees that ΔE is negative over entire GFR and that the larger the energy difference, the larger the ADF. One notes in Fig. 5 that the small composition sub-region marked by red dots have a lower ΔE than other area, implying that there exists a stronger driving force for the Ni-Nb-Zr alloys to synthesize metallic glasses. Consequently, within this composition sub-region, the alloys have greater GFA than those alloys located outside the sub-region. Interestingly, the collected experimental data in Fig. 4 were mainly distributed in this sub-region shown in Fig. 5.

The present authors have conducted similar studies for a number of other ternary alloy systems, that is, Cu-based Cu-Zr-Ti/Cu-Hf-Al/Cu-Hf-Ni, Zr-based Zr-Cu-Al/Zr-Ni-Al/Zr-Cu-Ni, Ni-based Ni-Nb-Ta/Ni-Hf-Ti/Ni-Zr-Ag, and Mg-based Mg-Cu-Ni/Mg-Cu-Y. These systems cover various structural combinations among fcc, bcc and hcp, and series MD or MC simulations, using solid solution models, were carried out as well. The GFRs derived from the n-body potentials through atomistic simulations are all well compatible with experimental observations obtained so far.

In addition, within the predicted GFR, the enthalpy of each glassy alloy could be computed. For instance, in the Cu-Hf-Al ternary system, the

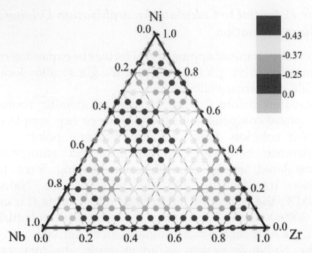

Fig. 5: Amorphization driving forces for the crystalline-to-amorphous transition calculated from MC simulations for the Ni-Nb-Zr system.

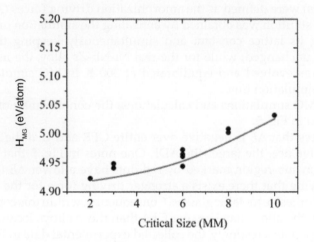

Fig. 6: Comparison between the enthalpies of glassy alloys calculated from MD simulations for the Cu-Hf-Al system and critical size of the metallic glass obtained by the typical Cu-mold casting technique.

enthalpies of glassy alloys were calculated in the predicted GFR. Figure 6 shows the comparison between calculated enthalpies with the critical size of the metallic glasses obtained by Cu-mold casting technique (Jia and Xu 2009). One sees, the enthalpy is in positive correlation to the GFA that is, the larger the enthalpy, the larger the size of the metallic glass obtained by the typical Cu-mold casting technique.

Concluding Remarks

(1) The interatomic potential of an A-B binary alloy system was taken as a starting base to develop a theory for metallic glass formation, as it governs the major interactions of the system.

(2) The experimental observations indicated that in a non-equilibrium processing with a cooling speed of 10^6–10^{13} K/sec, formation of binary metallic glass is a result of competition between solid solution and disordered glassy phase, which a complicated structured phase could not participate in.

(3) In a dozen binary alloy systems, their n-body potentials were constructed and applied to run MD and MC simulations, using solid solution models to compare the relative stability of solid solution vs. disordered state, over the entire composition range. The simulation results led to propose an atomistic theory, capable of predicting the favoured glass formation range (GFR) of the system.

(4) Some 20 ternary alloy systems were studied under similar methods. It turned out that an interatomic potential of a ternary alloy system could not only predict an energetically favoured glass formation region (GFR) in its composition triangle, but also calculate, for each possible glassy alloy, an amorphization driving force (ADF), which is somehow in positive correlation with the so-called glass-forming ability (GFA), which was defined by critical size or cooling rate.

Keywords: Metallic glass, Atomistic modeling, Interatomic potential, Molecular dynamic simulation, Monte Carlo simulation, Ab initial calculation, Favoured compositions, Glass formation range, Glass formation region, Amorphization driving force, Glass-forming ability

References

Ackland, G.J., G. Tichy, V. Vitek and M.W. Finnis. 1987. Simple N-body potentials for the noble metals and nickel. Philos. Mag. A 56: 735–756.

Akihisa, I., O. Katsumasa and M. Tsuyoshi. 1988. New amorphous Al-Y, Al-La and Al-Ce alloys prepared by melt spinning. Jpn. J. of Appl. Phys. 27: L736–L739.

Akiyama, E., H. Yoshioka, J.H. Kim, H. Habazaki, A. Kawashima, K. Asami et al. 1993. The effect of magnesium on the corrosion behavior of sputter-deposited amorphous AlMgTi ternary alloys in a neutral chloride solution. Corrosion Sci. 34: 27–40.

Allen, M.P. and D.J. Tildesley. 1987. Computer Simulation of Liquids. Clarendon Press, Oxford.

Altounian, Z., G.H. Tu and J.O. Strom-Olsen. 1983. Crystallization characteristics of Ni-Zr metallic glasses from $Ni_{20}Zr_{80}$ to $Ni_{70}Zr_{30}$. J. Appl. Phys. 54: 3111–3116.

Altounian, Z., C.A. Volkert and J.O. Strom-Olsen. 1985. Crystallization characteristics of Fe-Zr metallic glasses from $Fe_{43}Zr_{57}$ to $Fe_{20}Zr_{80}$. J. Appl. Phys. 57: 1777–1782.

Anantharaman, T.R. 1984. Metallic glasses: production, properties and applications. Trans Tech Publications, Switzerland.

Andersen, H.C. 1980. Molecular dynamics simulations at constant pressure and/or temperature. J. Chem. Phys. 72: 2384–2393.

Averback, R.S. and T.D. de la Rubia. 1998. Displacement damage in irradiated metals and semiconductors. pp. 281–402. *In*: H. Ehrenreich and F. Spaepen [eds.]. Solid State Physics, Vol. 51. Academic Press, New York.

Bacewicz, R. and J. Antonowicz. 2006. XAFS study of amorphous Al-RE alloys. Scr. Mater 54: 1187–1191.

Bormann, R., F. Gärtner and K. Zöltzer. 1988. Application of the CALPHAD method for the prediction of amorphous phase formation. J. Less-Common Met. 145: 19–29.

Boyer, W.S.L. and M. Atzmon. 1993. The effect of non-linear diffusion on solid state amorphization in Ni-Hf: experiment and simulation. J. Alloys Compd. 194: 213–220.

Brandes, E.A., G.B. Brook and P. Paufler. 1992. Smithells Metals Reference Book. Butterworth-Heinemann Ltd. Oxford.

Brunelli, K., M. Dabalà, R. Frattini, G. Sandonà and I. Calliari. 2001. Electrochemical behaviour of Cu-Zr and Cu-Ti glassy alloys. J. Alloys Compd. 317-318: 595–602.

Buschow, K.H.J. and N.M. Beekmans. 1979. Formation, decomposition, and electrical transport properties of amorphous Hf-Ni and Hf-Co alloys. J. Appl. Phys. 50: 6348–6352.

Buschow, K.H.J. 1984. Short-range order and thermal stability in amorphous alloys. J. Phys. F: Met. Phys. 14: 593–607.

Chang, H.J., E.S. Park, Y.S. Jung, M.K. Kim and D.H. Kim. 2007. The effect of Zr addition in glass forming ability of Ni-Nb alloy system. J. Alloys Compd. 434-435: 156–159.

Chen, H.S. 1974. Thermodynamic considerations on the formation and stability of metallic glasses. Acta Metall. 22: 1505–1511.

Chen, L.Y., H.T. Hu, G.Q. Zhang and J.Z. Jiang. 2007. Catching the Ni-based ternary metallic glasses with critical diameter up to 3 mm in Ni-Nb-Zr system. J. Alloys Compd. 443: 109–113.

Ciccotti, G. and W.G. Hoover. 1985. Molecular-Dynamics Simulation of Statistical - Mechanical System. North-Holland, New York.

Clark, S.J., M.D. Segall, C.J. Pickard, P.J. Hasnip, M.I.J. Probert, K. Refson et al. 2005. First principles methods using CASTEP. Zeit. für Kristallog.-Cryst. Mater 220: 567–570.

Clemens, B.M., W.L. Johnson and R.B. Schwarz. 1984. Proceedings of the Fifth International Conference on Liquid and Amorphous Metals Amorphous zirconium-nickel films formed by solid state reactions. J. Non-Cryst. Solids 61: 817–822.

Cleri, F. and V. Rosato. 1993. Tight-binding potentials for transition metals and alloys. Phys. Rev. B 48: 22–33.

Croat, J.J. 1981. Magnetic properties of melt-spun Pr-Fe alloys. J. Appl. Phys. 52: 2509–2511.

Dai, X.D., H.R. Gong and B.X. Liu. 2004. Structural stability of the metastable solid solution in the equilibrium immiscible Ag-Mo system predicted by an ab Initio derived potential. J. Phys Soc. Jpn. 73: 1222–1227.

Dai, X.D., Y. Kong and J.H. Li. 2007a. Long-range empirical potential model: Application to fcc transition metals and alloys. Phys. Rev. B 75: 104101.

Dai, X.D., J.H. Li and Y. Kong. 2007b. Long-range empirical potential for the bcc structured transition metals. Phys. Rev. B 75: 052102.

Dai, Y., J.H. Li, X.L. Che and B.X. Liu. 2009a. Proposed long-range empirical potential to study the metallic glasses in the Ni-Nb-Ta system. J. Phys. Chem. B 113: 7282–7290.

Dai, Y., J.H. Li and B.X. Liu. 2009b. Long-range empirical potential model: extension to hexagonal close-packed metals. J. Phys.: Condens. Matter 21: 385402.

Daw, M.S., S.M. Foiles and M.I. Baskes. 1993. The embedded-atom method: a review of theory and applications. Mater. Sci. Rep. 9: 251–310.

Dinda, G.P. 2006. Nonequilibrium Processing of Amorphous and Nanostructured Materials. Universität des Saarlandes, Karlsruhe.

Ducastelle, F. and F. Cyrot-Lackmann. 1970. Moments developments and their application to the electronic charge distribution of d bands. J. Phys. Chem. Solids 31: 1295–1306.

Ducastelle, F. and F. Cyrot-Lackmann. 1971. Moments developments: II. Application to the crystalline structures and the stacking fault energies of transition metals. J. Phys. Chem. Solids 32: 285–301.

Duwez, P., R.H. Willens and W. Klement. 1960. Continuous series of metastable solid solutions in silver-copper alloys. J. Appl. Phys. 31: 1136–1137.

Egami, T. and Y. Waseda. 1984. Atomic size effect on the formability of metallic glasses. J. Non-Cryst. Solids 64: 113–134.

Eifert, H.J., B. Elschner and K.H.J. Buschow. 1984. Electronic properties of amorphous $Zr_{(x)}$ $T_{(1-x)}$ alloys (T = Cu, Ni, Pd, Pt, Co, or Rh). Phys. Rev. B 29: 2905–2911.

Enayati, M.H., P. Schumacher and B. Cantor. 2002. The structure and thermal stability of mechanically alloyed Ni-Nb-Zr amorphous alloys. J. Mater. Sci. 37: 5255–5259.

Finnis, M.W. and J.E. Sinclair. 1984. A simple empirical N-body potential for transition metals. Philos. Mag. A 50: 45–55.

Frenkel, D. and B. Simit. 2002. Understanding Molecular Simulation: from Algorithms to Applications. Academic Press, San Diego.

Fukamichi, K. and T. Goto. 1991. Magnetic moment and spin glass behavior of AlCuMn and AlPdMn quasicrystalline and amorphous alloys. Science reports of the Research Institutes, Tohoku University Series A 36: 143–158.

Güntherodt, H.J. and H. Beck. 1981. Glassy Metals. Springer-Verlag, New York.

Gudzenko, V. N. and A.F. Polesya. 1975. Structure of splat cooled from liquid-state zirconium-aluminium alloys. Fizika Metallovi Metallovedenie 39: 1313–1315.

Guo, F.Q., S.J. Poon and G.J. Shiflet. 2004. CaAl-based bulk metallic glasses with high thermal stability. Appl. Phys. Lett. 84: 37–39.

Hansen, J.P. and I.R. Mcdonald. 1976. Theory of Simple Liquids. Academic Press, London.

Hasegawa, R. and L.E. Tanner. 1978. Superconducting transition temperatures of glassy and partially crystalline Be-Nb-Zr alloys. J. Appl. Phys. 49: 1196–1199.

Hassanain, N., A. Berrada, H. Lassri and R. Krishnan. 1995. Random anisotropy studies in amorphous Co-Tb ribbons. J. Magn. Magn. Mater 140-144: 337–338.

He, X., X.Y. Li and B.X. Liu. 2003. Metastable alloy formation in the immiscible Cu-Ru system by ion beam manipulation. J. Phys. Soc. Jpn. 72: 3032–3034.

He, X., S.H. Liang, J.H. Li and B.X. Liu. 2007. Atomistic mechanism of interfacial reaction and asymmetric growth kinetics in an immiscible Cu-Ru system at equilibrium. Phys. Rev. B 75: 045431.

Heine, V., I.J. Robertson and M.C. Payne. 1991. Many-atom interactions in solids. Philos. Trans. R. Soc. London A 334: 393–405.

Hoover, W.G. 1986. Constant-pressure equations of motion. Phys. Rev. A 34: 2499–2500.

Horikiri, H., A. Kato, A. Inoue and T. Masumoto. 1994. Proceedings of the Eighth International Conference on Rapidly Quenched and Metastable Materials New Mg-based amorphous alloys in Mg • Y-misch metal systems. Mater. Sci. Eng. A 179: 702–706.

Hu, L., Z.F. Li, W.S. Lai and B.X. Liu. 2003. Physical origin of ion mixing induced amorphization and associated asymmetric growth in the binary metal systems. Nucl. Instrum. Methods in Phys. Res., Sect. B 206: 127–131.

Huang, L.J., J.R. Ding, H.-D. Li and B.X. Liu. 1988. Growth of the fractal patterns in Ni-Zr thin films during ion-solid interaction. J. Appl. Phys. 63: 2879–2881.

Hung, L.S., M. Nastasi, J. Gyulai and J.W. Mayer. 1983. Ion-induced amorphous and crystalline phase formation in Al/Ni, Al/Pd, and Al/Pt thin films. Appl. Phys. Lett. 42: 672–674.

Inoue, A., Y. Takahashi, N. Toyota, T. Fukase and T. Masumoto. 1982. Superconducting properties of amorphous Zr-Nb-Ge alloys. Trans. Jpn. Inst. Met. 23: 693–702.

Inoue, A., H.S. Chen, J.T. Krause and T. Masumoto. 1984. Young's modulus sound velocity and Young's modulus of Ti-, Zr- and Hf-based amorphous alloys. J. Non-Cryst. Solids 68: 63–73.

Inoue, A., T. Zhang and T. Masumoto. 1989. Al-La-Ni amorphous alloys with a wide supercooled liquid region. Mater. Trans., JIM 30: 965–972.

Inoue, A., T. Zhang and T. Masumoto. 1990. Zr-Al-Ni amorphous alloys with high glass transition temperature and significant supercooled liquid region. Mater. Trans., JIM 31: 177–183.

Inoue, A., N. Nishiyama, K. Hatakeyama and T. Masumoto. 1994. New amorphous alloys in Al-Ca and Al-Ca-M (M = Mg or Zn) systems. Mater Trans., JIM 35: 282–285.

Inoue, A. 1998a. Bulk amorphous alloys: preparation and fundamental characteristics. Trans Tech Publications, Uetikon-Zürich, Switzerland.

Inoue, A. 1998b. Amorphous, nanoquasicrystalline and nanocrystalline alloys in Al-based systems. Prog. Mater. Sci. 43: 365–520.

Inoue, A. and A. Takeuchi. 2002. Recent progress in bulk glassy alloys. Mater. Trans. 43: 1892–1906.

Inoue, A. and W. Zhang. 2004. Formation, thermal stability and mechanical properties of Cu-Zr and Cu-Hf binary glassy alloy rods. Mater Trans. 45: 584–587.

Inoue, A. 2000. Stabilization of metallic supercooled liquid and bulk amorphous alloys. Acta Mater 48: 279–306.

Jastrow, L., U. Köster and M. Meuris. 2004. Catastrophic oxidation of Zr-TM (noble metals) glasses. Mater. Sci. Eng. A 375–377: 440–443.

Jia, P. and J. Xu. 2009. Comparison of bulk metallic glass formation between Cu-Hf binary and Cu-Hf-Al ternary alloys. J. Mater. Res. 24: 96–106.

Jin, O., Z.J. Zhang and B.X. Liu. 1995. Role of interface in ion mixing induced amorphization in the Ag-Mo system with very positive heat of formation. Appl. Phys. Lett. 67: 1524–1526.

Jin, O. and B.X. Liu. 1996. Unusual alloying behavior observed in the immiscible Ag-Mo and Ag-Nb systems under ion irradiation. Nucl. Instrum. Methods Phys. Res., Sect. B 114: 56–63.

Jin, O. and B.X. Liu. 1997. Chemical and interface effects on glass forming ability under ion mixing in Zr-Ta system of positive heat of formation. J. Non-Cryst. Solids 211: 180–186.

Johnson, W.L. 1999. Bulk glass-forming metallic alloys: Science and technology. MRS Bull. 24: 42–56.

Kawazoe, Y., J.-Z. Yu, A.-P. Tsai and T. Masumoto. 1997. Nonequilibrium Phase Diagrams of Ternary Amorphous Alloys. Springer-Verlag, Sendai.

Kimura, H., A. Inoue, S. Yamaura, K. Sasamori, M. Nishida, Y. Shinpo et al. 2003. Thermal stability and mechanical properties of glassy and amorphous Ni-Nb-Zr alloys produced by rapid solidification. Mater Trans. 44: 1167–1171.

Klement, W., R.H. Willens and P. Duwez. 1960. Non-crystalline structure in solidified gold-silicon alloys. Nature 187: 869–870.

Koster, U., J. Meinhardt, A. Aronin and Y. Birol. 1995. Crystallization of $Cu_{50}Ti_{50}$ glasses and undercooled melts. Z. Metallkd. 86: 171–175.

Lai, W.S., Q. Li, C. Lin and B.X. Liu. 2001. Critical solid solubility of the Ni-Ti system determined by molecular dynamics simulation and ion mixing. Phys. Status Solidi B 227: 503–514.

Lee, M.H., X. Yang, M.J. Kramer and D.J. Sordelet. 2006. Glass formation and crystallization in binary Zr-Pt systems. Philos. Mag. 86: 443–449.

Leonhardt, M., W. Löser and H.G. Lindenkreuz. 1999. Solidification kinetics and phase formation of undercooled eutectic Ni-Nb melts. Acta Mater. 47: 2961–2968.

Leshansky, A.M. and J.F. Brady. 2005. Dynamic structure factor study of diffusion in strongly sheared suspensions. J. Fluid Mech. 527: 141–169.

Li, J.H., X.D. Dai, T.L. Wang and B.X. Liu. 2007. A binomial truncation function proposed for the second-moment approximation of tight-binding potential and application in the ternary Ni-Hf-Ti system. J. Phys.: Condens. Matter 19: 086228.

Li, J.H., X.D. Dai, S.H. Liang, K.P. Tai, Y. Kong and B.X. Liu. 2008. Interatomic potentials of the binary transition metal systems and some applications in materials physics. Phys. Rep. 455: 1–134.

Li, J.H., Y. Dai, Y.Y. Cui and B.X. Liu. 2011. Atomistic theory for predicting the binary metallic glass formation. Mater. Sci. Eng. R: Reports 72: 1–28.

Li, Y., S.H. Liang, N. Li, J.H. Li and B.X. Liu. 2013. Synthesis of amorphous alloys and amorphous-crystalline composites in ternary Ni-Nb-Zr system by ion beam mixing. Mater. Chem. Phys. 141: 960–966.

Li, Z.C. and B.X. Liu. 1999. Experimental and theoretical studies on composition limits of metallic glass formation in the Ni-Mo system. Chin. Phys. Lett. 16: 667–669.

Liaw, P.K. and R.A. Buchanan. 2006. Bulk Metallic Glasses. The Minerals, Metals & Materials Society, Warrendale, Pennsylvania.

Lide, D.R. 2002. CRC Handbook of chemistry and physics: a ready-reference book of chemical and physical data. 83rd ed. CRC Press, Boca Raton, FL.

Lin, C., G.W. Yang and B.X. Liu. 2000a. Sequential formation of quasicrystalline and amorphous phases in the Fe-Nb and Fe-Ta multilayers upon ion irradiation. J. Appl. Phys. 87: 2821–2824.

Lin, C., G.W. Yang and B.X. Liu. 2000b. Different alloying behaviors of Co-Mo multilayered films upon ion irradiation and thermal annealing. Appl. Phys. A 70: 469–473.

Liu, B.X., B.M. Clemens, R. Gaboriaud, W.L. Johnson and M.-A. Nicolet. 1983a. Ion mixing to produce amorphous Mo-Ru superconducting films. Appl. Phys. Lett. 42: 624–626.

Liu, B.X., W.L. Johnson, M.-A. Nicolet and S.S. Lau. 1983b. Structural difference rule for amorphous alloy formation by ion mixing. Appl. Phys. Lett. 42: 45–47.

Liu, B.X. 1986. Ion mixing and metallic alloy phase formation. Phys. Status Solidi A 94: 11–34.

Liu, B.X. 1987. Prediction of metallic glass formation by ion mixing. Mater Lett. 5: 322–327.

Liu, B.X. and G.A. Cheng. 1991. Glass forming ability of the Al-Ti system under ion beam mixing. Phys. Status Solidi A 125: K93–K96.

Liu, B.X. and Z.J. Zhang. 1994. Formation of nonequilibrium solid phases by ion irradiation in the Ni-Ta system and their thermodynamic and growth-kinetics interpretations. Phys. Rev. B 49: 12519–12527.

Liu, B.X., O. Jin and Y. Ye. 1996a. Interface enhanced glass forming ability under ion mixing/solid-state reaction in an immiscible Hf-Ta system. J. Phys.: Condens. Matter 8: L79–L82.

Liu, B.X., Z.J. Zhang, O. Jin and F. Pan. 1996b. Role of interface in ion mixing or thermal annealing induced amorphization in multilayers in some immiscible systems. Materials Research Society Symposium-Proceedings 396: 71–82.

Liu, B.X. and O. Jin. 1997. Formation and theoretical modeling of non-equilibrium alloy phases by ion mixing. Phys. Status Solidi A 161: 3–33.

Liu, B.X., Z.J. Zhang and O. Jin. 1998. A comparative study of metastable alloy formation by ion mixing and thermal annealing of multilayers in the immiscible Y-Zr system. J. Alloys Compd. 270: 186–193.

Liu, B.X., W.S. Lai and Q. Zhang. 2000. Irradiation induced amorphization in metallic multilayers and calculation of glass-forming ability from atomistic potential in the binary metal systems. Mater. Sci. Eng. R: Reports 29: 1–48.

Liu, B.X., W.S. Lai and Z.J. Zhang. 2001. Solid-state crystal-to-amorphous transition in metal-metal multilayers and its thermodynamic and atomistic modelling. Adv. Phys. 50: 367–429.

Lu, Z.P., H. Bei and C.T. Liu. 2007. Recent progress in quantifying glass-forming ability of bulk metallic glasses. Intermetallics 15: 618–624.

Luborsky, F.E. 1983. Amorphous metallic alloys. Butterworths, London.

Luo, S.Y., J.H. Li, J.B. Liu and B.X. Liu. 2014. Atomic modeling to design favored compositions for the ternary Ni-Nb-Zr metallic glass formation. Acta Mater 76: 482–492.

Martin, G. and P. Bellon. 1997. Driven alloys. Solid State Phys. 50: 189–331.

Massalski, T.B., Joanne L. Murray, L.H. Bennett and Hugh Baker. 1986. Binary alloy phase diagrams. American Society for Metals, Metals Park, Ohio.

Miller, M. and P. Liaw. 2007. Bulk Metallic Glasses: An Overview. Springer-Verlag, Boston.

Paglieri, S.N., N.K. Pal, M.D. Dolan, S.-M. Kim, W.-M. Chien, J. Lamb et al. 2011. Hydrogen permeability, thermal stability and hydrogen embrittlement of Ni-Nb-Zr and Ni-Nb-Ta-Zr amorphous alloy membranes. J. Membr. Sci. 378: 42–50.

Pan, F., Y.G. Chen and B.X. Liu. 1995. Spontaneous vitrification in the Au-Ta system with a small size difference. Appl. Phys. Lett. 67: 780–782.

Pan, F. and B.X. Liu. 1996. Ion-mixing-induced amorphization in an immiscible Au-Mo system with a small size difference. J. Phys.: Condens. Matter 8: 383–388.

Parrinello, M. and A. Rahman. 1981. Polymorphic transitions in single crystals: A new molecular dynamics method. J. Appl. Phys. 52: 7182–7190.

Pearson, W.B. 1958. A Handbook of Lattice Spacings and Structures of Metals and Alloys. Pergamon Press, London.

Peker, A. and W.L. Johnson. 1993. A highly processable metallic glass: $Zr_{41.2}Ti_{13.8}Cu_{12.5}Ni_{10.0}Be_{22.5}$. Appl. Phys. Lett. 63: 2342–2344.

Phillpot, S.R., S. Yip and D. Wolf. 1989. How do crystals melt? Comput. Phys. 3: 20–31.

Raeker, T.J. and A.E. Depristo. 1991. Theory of chemical bonding based on the atom-homogeneous electron gas system. International Rev. Phys. Chem. 10: 1–54.

Rao, K.V., N. Karpe, D.-X. Chen, R. Goldfarb, J. Bottiger, K. Pampus et al. 1987. Magnetic and transport properties of ion-beam-mixed amorphous Fe-Zr alloys (abstract). J. Appl. Phys. 61: 3249.

Rapaport, D.C. 2004. The Art of Molecular Dynamics Simulation. Cambridge University Press, New York.

Rebonato, R., D.O. Welch, R.D. Hatcher and J.C. Bilello. 1987. A modification of the Finnis-Sinclair potentials for highly deformed and defective transition metals. Philos. Mag. A 55: 655–667.

Rizzo, H.F., T.B. Massalski and M. Nastasi. 1993. Metastable crystalline and amorphous structures formed in the Cu-W system by vapor deposition. Metall. Trans. A 24: 1027–1037.

Russell, K.C. 1984. Phase stability under irradiation. Prog. Mater Sci. 28: 229–434.

Ryan, D.H., J.M.D. Coey and J.O. Ström-Olsen. 1987. Magnetic properties of iron-rich Fe-Hf glasses. J. Magn. Magn. Mater 67: 148–154.

Saida, J., T. Sanada, S. Sato, M. Imafuku, C. Li and A. Inoue. 2008. Nano quasicrystal formation and local atomic structure in Zr-Pd and Zr-Pt binary metallic glasses. Z. Kristallogr. 223: 726–730.

Sakurai, J., M. Abe, M. Ando, Y. Aono and S. Hata. 2011. Searching for noble Ni-Nb-Zr thin film amorphous alloys for optical glass device molding die materials. Precis. Eng. 35: 537–546.

Schroers, J. 2013. Bulk metallic glasses. Phys. Today 66: 32–37.

Schwarz, R.B. and W.L. Johnson. 1983. Formation of an amorphous alloy by solid-state reaction of the pure polycrystalline metals. Phys. Rev. Lett. 51: 415–418.

Segall, M.D., J.D. Lindan Philip, M.J. Probert, C.J. Pickard, P.J. Hasnip, S.J. Clark et al. 2002. First-principles simulation: ideas, illustrations and the CASTEP code. J. Phys.: Condens. Matter 14: 2717–2744.

Sigmund, P. 2006. Particle Penetration and Radiation Effects. Springer-Verlag, Berlin, Heidelberg.

Skakov, Y.A., N.P. Djakonova, N.V. Edneral, M.R. Koknaeva and V.K. Semina. 1991. Some peculiarities of the atomic structure of metallic phases formed during liquid quenching and solid state reactions. Mater Sci. Eng. A 133: 560–564.

Suryanarayana, C. 1999. Non-Equilibrium Processing of Materials. Pergamon Press, Oxford, UK.

Sutton, A.P. and J. Chen. 1990. Long-range Finnis-Sinclair potentials. Philos. Mag. Lett. 61: 139–146.

Tai, K.P., X.D. Dai, Y.X. Shen and B.X. Liu. 2006. Formation and structural anomaly of the metastable phases in an immiscible Ag-Mo system studied by ion beam mixing and molecular dynamics simulation. J. Phys. Chem. B 110: 595–606.

Tanner, L.E. and R. Ray. 1979. Metallic glass formation and properties in Zr and Ti alloyed with Be-I the binary Zr-Be and Ti-Be systems. Acta Metall. 27: 1727–1747.

Thompson, M.W. 1969. Defects and Radiation Damage in Metals. Cambridge University Press, Cambridge.

Tsaur, B.Y., S.S. Lau and J.W. Mayer. 1980. Continuous series of metastable Ag-Cu solid solutions formed by ion-beam mixing. Appl. Phys. Lett. 36: 823–826.

Turnbull, D. 1969. Under what conditions can a glass be formed? Contemp. Phys. 10: 473–488.

Uhlig, H., L. Rohr, H.J. Güntherodt, P. Fischer, P. Lamparter and S. Steeb. 1993. Short range order in amorphous $Ni_{50}Ta_{50}$ alloys by means of X-ray and neutron diffraction. J. Non-Cryst. Solids 156–158, Part 1: 165–168.

Voter, A.F. 1996. Interatomic potentials for atomistic simulations. MRS Bull. 21: 17–19.

Wang, D., Y. Li, B.B. Sun, M.L. Sui, K. Lu and E. Ma. 2004a. Bulk metallic glass formation in the binary Cu-Zr system. Appl. Phys. Lett. 84: 4029–4031.

Wang, T.L., J.H. Li and B.X. Liu. 2009. Proposed thermodynamic method to predict the glass formation of the ternary transition metal systems. PCCP 11: 2371–2373.

Wang, W.H., C. Dong and C.H. Shek. 2004b. Bulk metallic glasses. Mater. Sci. Eng. R: Reports 44: 45–89.

Weeber, A.W. and H. Bakker. 1988. Extension of the glass-forming range of Ni-Zr by mechanical alloying. J. Phys. F: Met. Phys. 18: 1359–1369.

Whang, S.H. 1984. Glass forming property of a TiZrSi system determined by temperature-composition map. Scr. Metall. 18: 309–312.

Wu, Y.R., W.Y. Hu and S.C. Han. 2006. Theoretical calculation of thermodynamic data for gold-rare earth alloys with the embedded-atom method. J. Alloys Compd. 420: 83–93.

Xia, L., S.S. Fang, Q. Wang, Y.D. Dong and C.T. Liu. 2006a. Thermodynamic modeling of glass formation in metallic glasses. Appl. Phys. Lett. 88: 171905.

Xia, L., W.H. Li, S.S. Fang, B.C. Wei and Y.D. Dong. 2006b. Binary Ni–Nb bulk metallic glasses. J. Appl. Phys. 99: 026103.

Xu, D.H., B. Lohwongwatana, G. Duan, W.L. Johnson and C. Garland. 2004. Bulk metallic glass formation in binary Cu-rich alloy series -$Cu_{100-x}Zr_x$ (x = 34, 36, 38.2, 40 at.%) and mechanical properties of bulk $Cu_{64}Zr_{36}$ glass. Acta Mater 52: 2621–2624.

Yamaura, S., M. Sakurai, M. Hasegawa, K. Wakoh, Y. Shimpo, M. Nishida et al. 2005. Hydrogen permeation and structural features of melt-spun Ni-Nb-Zr amorphous alloys. Acta Mater 53: 3703–3711.

Yamaura, S., S. Nakata, H. Kimura and A. Inoue. 2006. Hydrogen permeation of Ni-Nb-Zr metallic glasses in a supercooled liquid state. Mater Trans. 47: 2991–2996.

Yang, W., F. Liu, H. Liu, H.F. Wang, Z. Chen and G.C. Yang. 2009. Glass forming ability in Cu-Zr binary alloy: Effect of nucleation mode. J. Alloys Compd. 484: 702–707.

Yuan, L., C. Pang, Y.M. Wang, Q. Wang, J.B. Qiang and C. Dong. 2010. Understanding the Ni-Nb-Zr BMG composition from a binary eutectic Ni-Nb icosahedral cluster. Intermetallics 18: 1800–1802.

Zallen, R. 1983. The Physics of Amorphous Solids. John Wiley & Sons, Inc, New York.

Zhang, C.L., D.H. Wang, Z.D. Han, H.C. Xuan, B.X. Gu and Y.W. Du. 2009. Large magnetic entropy changes in Gd-Co amorphous ribbons. J. Appl. Phys. 105: 013912.

Zhang, R.F., Y.X. Shen, H.F. Yan and B.X. Liu. 2005. Formation of Amorphous Alloys by Ion Beam Mixing and Its Multiscale Theoretical Modeling in the Equilibrium Immiscible Sc-W System. J. Phys. Chem. B 109: 4391–4397.

Zhang, T., A. Inoue and T. Masumoto. 1994. Amorphous (Ti,Zr, Hf)-Ni-Cu ternary alloys with a wide supercooled liquid region. Mater. Sci. Eng. A 181–182: 1423–1426.

Zhang, W., K. Arai, C.L. Qin, F. Jia and A. Inoue. 2008. Formation and properties of new Ni-Ta-based bulk glassy alloys with large supercooled liquid region. Mater. Sci. Eng. A 485: 690–694.

Zhang, Z.J., H.Y. Bai, Q.L. Qiu, T. Yang, K. Tao and B.X. Liu. 1993. Phase evolution upon ion mixing and solid-state reaction and thermodynamic interpretation in the Ni-Nb system. J. Appl. Phys. 73: 1702–1710.

Zhang, Z.J. and B.X. Liu. 1994a. Non-equilibrium crystalline phases formed in the Co-Ta system by ion beam mixing. J. Phys.: Condens. Matter 6: 9065–9074.

Zhang, Z.J. and B.X. Liu. 1994b. Solid state phase transition induced by ion irradiation in the Co-Nb system: experiments and thermodynamics. J. Phys.: Condens. Matter 6: 2647–2658.

Zhang, Z.J. and B.X. Liu. 1994c. Solid-state reaction to synthesize Ni-Mo metastable alloys. J. Appl. Phys. 76: 3351–3356.

Zhang, Z.J., O. Jin and B.X. Liu. 1995. Anomalous alloying behavior induced by ion irradiation in a system with a large positive heat of mixing. Phys. Rev. B 51: 8076–8085.

Zhao, M., X.D. Dai, Y.X. Shen and B.X. Liu. 2008. Glass forming ability in the equilibrium immiscible Ag-Ta system studied by molecular dynamics simulation and ion beam mixing. J. Phys. Soc. Jpn. 77: 074601.

Zhu, Z., H. Zhang, D. Pan, W. Sun and Z. Hu. 2006. Fabrication of binary Ni-Nb bulk metallic glass with high strength and compressive plasticity. Adv. Eng. Mater 8: 953–957.

Zhu, Z.W., H.F. Zhang, B.Z. Ding and Z.Q. Hu. 2008. Synthesis and properties of bulk metallic glasses in the ternary Ni-Nb-Zr alloy system. Mater. Sci. Eng. A 492: 221–229.

Zhu, Z.W., H.F. Zhang, W.S. Sun and Z.Q. Hu. 2007. Effect of Zr addition on the glass-forming ability and mechanical properties of Ni-Nb alloy. J. Mater. Res. 22: 453–459.

2.6

Shape Memory Alloys
Constitutive Modeling and Engineering Simulations

G. Scalet* and F. Auricchio

INTRODUCTION

Shape memory alloys (SMAs) were discovered around the 1960s, when Buehler et al. (1963) observed that a series of nickel-titanium (NiTi) alloys, composed of 53–57% nickel by weight, possessed an unusual property, successively named as shape memory effect (SME): after being deformed up to 8–15% strain, these alloys were able to recover the original shape with a thermal cycle. Later, it was observed that such alloys were also able to recover large deformations during mechanical loading-unloading cycles performed at constant high temperature; such a property became known as pseudoelasticity (PE).

NiTi alloys were then commercialized under the trade name NiTiNOL, in honour of their discovery at the Naval Ordinance Laboratory (NOL). The first commercial application appeared between the late 1960s and the early 1970s and consisted of pipe couplings (known as Cryofit) in F-14 fighter aircrafts. During the 1970s, NiTi alloys started to be exploited in

Dipartimento di Ingegneria Civile ed Architettura (DICAr), University of Pavia, Via Ferrata 3, 27100 Pavia, Italy.
* Corresponding author: giulia.scalet@unipv.it

biomedical applications and NiTi stents appeared in the market around the 1990s.

Thanks to their unique properties, NiTi alloys represent an attractive choice for applications in numerous commercial fields, for example, automotive, aerospace, robotics, civil engineering, biomedical, and fashion (see Jani et al. (2014) and references therein). The PE effect is employed in, for example, medical devices, components for vibration isolation and dampening, while the SME is exploited in, for example, actuators for micro-electromechanical systems (MEMS), automotive, robotic, and aerospace applications. To date, more than 10,000 patents have been published on SMA materials and components and the number of novel applications is still growing (Jani et al. 2014). If we type the keywords "shape memory alloy" or "nitinol" in Scopus, we obtain 12,052 patent results.

At present, NiTi-based alloys (e.g., NiTi, NiTiCu, NiTiPd) are the most common commercially available SMAs, thanks to their high ductility, stability in cyclic applications, and electrical resistivity, as well as good biocompatibility. However, other SMA families possess the SME and PE properties, such as Cu-based (e.g., CuZn, CuAlNi) or Fe-based (e.g., FeMnSi) SMAs. Alternative alloys, such as magnetic SMAs (e.g., FePd, NiMnGa), porous SMAs, high-temperature SMAs (e.g., TiNiPd), and hybrid or composite shape memory materials, are also widely investigated.

From 1980 up to now, the increasing interest in SMA materials and applications has motivated theoretical, numerical, and experimental researches. Noteworthy research has been dedicated to the formulation of constitutive models in order to describe the specificities of the complex thermomechanical behaviour of SMAs and to understand the physical mechanisms underlying material behaviour. In parallel, several works have focused on the development of robust computational tools to assist both the design and production of SMA-based devices.

The goal of the present chapter is to provide an overview about SMA constitutive models and their exploitation in the computer-based design of real-life applications. We will discuss the available modeling approaches to highlight their main features, differences, strengths, and weaknesses; then, we will show how such models have been employed in advanced computational simulations using structural finite element analysis (FEA) tools. The effectiveness of macroscopic phenomenological models to analyse and optimize SMA-based devices or components will be demonstrated through several significant examples from the literature. Given the large body of literature on the topic and the constant growth of the number of models and applications, our review is not exhaustive, but aims to discuss the main modeling approaches and the role of modeling and engineering simulations in the design of SMA applications.

Material Behaviour

This section presents a brief review on the macroscopic and microstructural aspects of the transformation behaviour in SMAs. It will be shown that SMA behaviour at the macroscopic scale is strongly connected to the evolution of material phases at the microscopic level. The reader is referred to, for example, Otsuka and Ren (2005) and Lagoudas (2008), for details on metallurgical, crystallographic, and phenomenological aspects of SMA response.

Macroscopic Properties

As anticipated above, SMAs possess the macroscopic properties known as PE and SME, which can be schematically represented in terms of stress, strain and temperature, as shown respectively in Figs. 1 and 2.

The PE response of the material subject to an isothermal tensile test is considered in Fig. 1. The material starts from a reference configuration at high temperature (point A). Upon mechanical loading, the behaviour is first linear elastic (path A-B) and, then, apparently plastic (path B-C). Differently from plastic deformations, the deformation reaches the saturation (point C) and an additional increase of the applied load leads to a linear elastic behaviour (path C-D). Upon unloading, the behaviour is first linear elastic (path D-E), then, again apparently plastic (path E-F) and,

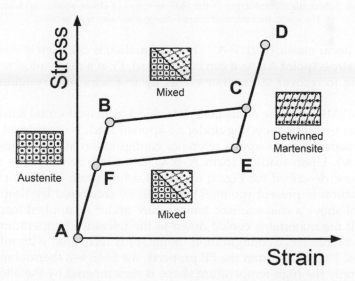

Fig. 1: Tensile test at fixed high temperature on a SMA. Schematic response in terms of stress and strain, reproducing the PE property of SMAs. The associated microstructural changes are also represented.

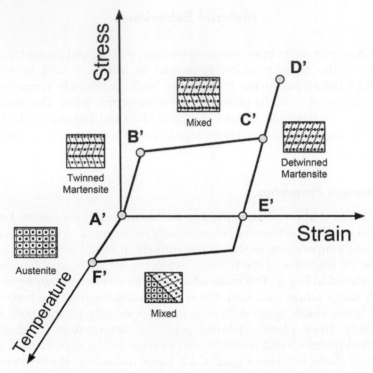

Fig. 2: Tensile test at fixed low temperature on a SMA, followed by thermal-cycling under zero stress. Schematic reproduction of the SME in terms of stress, strain and temperature. The associated microstructural changes are also represented.

finally, linear elastic (path F-A). The deformation is completely recovered at zero stress (point A). As it can be observed, PE is a mechanical recovery allowing to recover large strains (e.g., up to 8% in polycrystalline NiTi alloys).

The SME response of the material subject to an isothermal tensile test, followed by thermal-cycling under no applied load, is considered in Fig. 2. The material starts from a reference configuration at low temperature (point A'). Upon loading (path A'-B'-C'-D'), SMA behaviour is similar to the one described for Fig. 1; upon unloading (path D'-E'), a residual deformation is present (point E') and can be recovered by heating the material above a characteristic temperature under no applied load (path E'-F'). If the material is cooled down to the reference temperature (path F'-A'), the reference configuration (point A') is recovered without shape changes. Differently from the PE property, the SME is a thermal recovery where only the high-temperature shape is remembered by the alloy. This property is better known as the one-way SME to distinguish it from the two-way SME, which is related to the ability of the material to remember two different shapes at low and high temperature. While the one-way SME is an intrinsic characteristic of the material, the two-way SME is induced by specific treatments (e.g., training).

We recall that PE and SME can be observed in the same SMA, if it is tested at different temperatures. The characteristic temperature, mentioned above, that separates the temperature ranges of PE and SME, depends on both type and chemical composition of the alloy.

Microstructure

The macroscopic properties of SMAs, described above, result from crystallographic reversible thermoelastic martensitic transformations between two solid phases with different degrees of symmetry (Lagoudas 2008): (i) the austenite (or parent phase) which has a high symmetry crystal structure (typically body-centered cubic) and is stable at high temperature and low stress; and (ii) the martensite (or product phase) which has a low symmetry (typically rhombohedral or monoclinic) and can appear in crystallographically equivalent variants differing in their orientation with respect to material axes (up to 24 for NiTi). Martensitic or phase transformations (PTs) can be induced by temperature and/or mechanical loading and they occur by a shear lattice distortion without volume change (Lagoudas 2008).

NiTi alloys with near-equiatomic composition are the most used SMAs; in these alloys, austenite has an ordered bcc structure (B2), while martensite has an ordered monoclinic structure (B19'). Refer to Otsuka and Ren (2005) for details on the effect of small changes in the composition of the alloy on its properties. We remark that some NiTi alloys transform from B2 cubic austenite into monoclinic B19' martensite via an intermediate phase with a rhombohedral structure (R) (Ling and Kaplow 1981). This two-stage transformation can be obtained by cold working, ageing of the alloys with higher Ni content, or adding of a third element (e.g., Fe). The transformation from austenite to R-phase usually shows a small thermal hysteresis of 1–5°C, compared to hysteresis of 30–50°C for the austenite-martensite transformation, and a significant change in electrical resistivity. Moreover, the related SME is associated with limited recoverable shape changes of about 1% elongation, compared to the 10% elongation associated to martensite.

Accordingly, the PE and SME paths presented in Figs. 1 and 2 can be reinterpreted in terms of microstructural changes inside the material. In Fig. 1, the material is austenitic at high temperature (point A) and the loading initially causes elastic distortions of the austenite lattice (path A-B). At a certain level (point B), austenite becomes unstable and detwinned (stress-induced) martensite starts to nucleate; the coexistence of the two lattices determines macroscopic deformation (path B-C). Once reached the saturation (point C), austenite has transformed into martensite; recall that residuals of austenite may remain in polycrystals, which transform at higher stress levels (Shaw and Kyriakides 1995). Then, the deformation mechanism is governed by the elastic distortion of martensite (path C-D); a similar mechanism is obtained upon unloading (path D-E). Then, like austenite during loading, martensite becomes unstable at a certain level (point E) and austenite start to nucleate (path E-F). At point F the transformation into austenite is complete and further unloading leads to the elastic path F-A.

In Fig. 2, the material is composed of twinned martensite at low temperature and zero stress (point A'). Differently from the stress-induced martensite that consists of a single preferential variant according to the applied stress, twinned martensite has a random mixture of several variants. Upon loading, only elastic distortion take place (path A'-B') up to a critical value (point B'), where the movement of twin boundaries caused by stress (i.e., detwinning) results in change of orientation from one variant to another which is more favourably oriented to the stress direction (Shaw and Kyriakides 1995). Therefore, this leads to the formation of detwinned martensite and macroscopic deformation. At the saturation level (point C'), martensite is completely detwinned and further loading causes elastic deformation (path C'-D'). Upon unloading (path D'-E'), the macroscopic deformation is kept, since all the martensite variants are equally stable. Upon heating above a characteristic temperature, the transformation from detwinned martensite to austenite takes place (path E'-F') and allows to recover the macroscopic deformation (point F'). During cooling (path F'-A'), the PT occurs under no load and no macroscopic change is observed.

The correlations of stress and temperature in SMAs can be also represented in a schematic one-dimensional diagram, called phase diagram. Figure 3 illustrates the crystal structures of a given SMA with fixed composition, along with the PT regions (hatched strips). The loading paths associated to PE and SME are also represented in the phase diagram (the paths are indicated with the same letters used in Figs. 1 and 2). A discussion on phase diagrams taking into account twinned and detwinned martensite's properties is provided in Popov and Lagoudas (2007).

Besides PE, SME, and variant reorientation, which are in general indicated as primary effects, SMAs show other features important from the modeling point of view, termed as secondary effects, and including, for example, tension-compression asymmetry, different elastic properties of the solid phases and accumulated strain under cycling. We will see that constitutive models should be able to describe, at least, the primary effects to guarantee an effective representation of SMA behaviour.

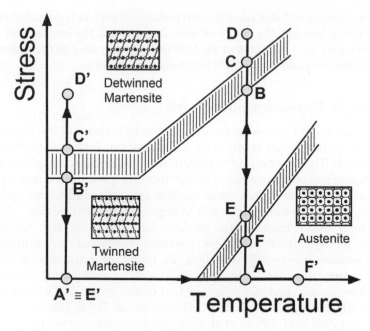

Fig. 3: One-dimensional phase diagram in terms of stress and temperature. The different crystal structures of a given SMA with fixed composition, along with the PT regions (hatched strips), are shown. The loading paths associated to the SME (i.e., A'-B'-C'-D'-E'-F') and PE (i.e., A-B-C-D-E-F) properties are also represented.

Modeling Approaches

Several modeling approaches have been proposed in the literature to describe the complex thermomechanical behaviour of SMAs. If we type the keywords "shape memory alloy" and "modeling" in Scopus, we obtain 1,669 results. Generally, SMA constitutive models are classified into the following categories:

1. Microscopic thermodynamics models
2. Micromechanics-based macroscopic models
3. Macroscopic phenomenological models

These categories have allowed the formulation of numerous one-dimensional and three-dimensional models. The first works on SMA modeling appeared in the early 1980s and were devoted to microscopic or simple phenomenological modeling (Tanaka and Nagaki 1982; Falk 1983). Currently, models allow reasonably accurate numerical simulations of complex SMA-based devices under general loading paths (Cisse et al. 2016).

In this section we will discuss the main features of each category. A detailed discussion is beyond the scope of this chapter and the reader can refer to the reviews by Lagoudas et al. (2006); Patoor et al. (2006); Lexcellent (2013); Concilio and Lecce (2015) and Cisse et al. (2016).

Microscopic Thermodynamics Models

Microscopic models focus on the description of features at the microscale (lattice or grain-crystal level) such as nucleation, interface motion and twin growth. They are helpful to understand the fundamental phenomena of SMA macroscopic behaviour, rather than its quantitative description. The main drawback is that these models are often difficult to apply to engineering applications and result in high computational costs.

Several SMA microscopic models are based on the Ginzburg-Landau theory which describes material behaviour by introducing a polynomial free-energy function in terms of variables as strain, stress, temperature and entropy (Falk 1983; Barsch and Krumhansl 1988; Abeyaratne and Knowles 1990; Levitas et al. 2003; Wang and Melnik 2007; Daghia et al. 2010; Grandi et al. 2012; She et al. 2013; Zhong and Zhu 2014; Dhote et al. 2015). Such models are simple, but they may be not able to accurately describe complex behaviours as well as evolutive processes.

Others SMA microscopic models adopt concepts of molecular dynamics, for example (Lai and Liu 2000; Suzuki et al. 2006; Sato et al. 2008; Mutter and Nielaba 2013).

Micromechanics-Based Macroscopic Models

Micro-macro models describe the polycrystalline response of SMAs by modeling a single grain and by averaging the results over a representative volume element. They use thermodynamics laws to describe PTs and micromechanics to estimate the interaction energy due to material transformation. These models require the definition of suitable observable variables (temperature, stress or strain) and internal variables (the volume fraction of martensite and/or a mean transformation strain).

Among the others, we cite two main micro-macro modeling routes for SMAs, namely the micromechanical models (Fischer and Tanaka 1992; Sun and Hwang 1993; Patoor et al. 1996; Gall et al. 2000; Huang et al. 2000; Thamburaja and Anand 2001; Lim and McDowell 2002; Blanc and Lexcellent 2004; Sadjadpour and Bhattacharya 2007; Levitas and Ozsoy 2009; Junker and Hackl 2011; Stupkiewicz and Gorzynska-Lengiewicz 2012; Yu et al. 2015) and the micro-plane/micro-sphere models (Brocca et al. 2002; Kadkhodaei et al. 2008; Mehrabi et al. 2014).

Micro-macro models have demonstrated good predictive results and, in some cases, successful descriptions of reorientation and detwinning of martensite variants (Gao and Brinson 2002). However, such models often employ a high number of internal variables and are difficult and expensive to use for engineering purposes.

Macroscopic Phenomenological Models

Macroscopic models are based on phenomenological and/or simplified thermodynamics, and/or on curves fitting experimental data. They are able to describe the global mechanical response of SMAs, but they neglect all the microscopic phenomena related to microstructural changes induced by PTs as well as their effects on macroscopic behaviour (Bernardini 2001). Moreover, since such models are calibrated on experimental measurements at the macroscopic scale, they may be also quite accurate.

Due to their features, phenomenological models become advantageous when studying a polycrystalline SMA with thousand grains having different crystallographic orientations that influence the PT. In such a case, the modeling of the macroscopic response by considering different PT conditions in each grain may become difficult. Therefore, macroscopic approaches are suitable for engineering applications thanks to their modeling simplicity, simple numerical implementation, and low computational times, compared to the other approaches, and they can be used for preliminary studies or design processes at the industrial level. Motivated by these advantages, in the following sections we will focus on this category that appears to be well-suited for the simulation of SMA-based applications.

Macroscopic Phenomenological Modeling

Macroscopic models generally represent each material point as a phase mixture, introduce energy potentials over homogenized material volumes and employ a method to describe material bulk behaviour.

The most common approach belongs to the theoretical framework of continuum thermomechanics with internal variables: the thermodynamic state of a SMA at a given material point and time instant is described by a set of state variables at that time instant, which depends only on the considered point. As discussed for the micromechanics-based macroscopic models, state variables can be divided into observable variables and internal variables, the latters introduced to describe microstructural features which are not directly appreciable, but important for the macroscopic modeling.

Another approach in the phenomenological framework is the hysteresis modeling which has been widely used in several fields, particularly for ferromagnetic materials. Hysteresis models are generally purely empirical and their implementation reduces to the identification of several mathematical parameters through numerous experiments that may be unavailable in practice; see Bo and Lagoudas (1999) for a review. These models constitute an important part of the SMA actuator modeling literature (Huo 1989; Likhacev and Koval 1992; Ortin 1992; Ivshin and Pence 1994). Huo (1989) introduced the macroscopic Landau-Ginzburg potential proposed by Falk (1983) to account for first-order martensitic PTs. Lagoudas and Bhattacharyya (1997) used Preisach weighting functions with distributions of single-crystal orientations in polycrystalline SMAs.

Finally, another macroscopic approach is based on statistical physics (Bhattacharya and Lagoudas 1997; Fischlschweiger and Oberaigner 2012). Models are obtained from equilibrium considerations for single-crystal SMAs and describe the kinetics and strain evolution for polycrystalline SMAs without assumptions on internal variables.

In the following subsection, we will focus on the first approach, that is, on models with internal variables within the infinitesimal strain framework. For a review and discussion on finite deformation and geometric non-linearity, refer to Cisse et al. (2016).

Models with Internal Variables

In the framework of continuum thermodynamics with internal variables the state of a macroscopic SMA material point can be described by observable variables (e.g., the total strain ε and temperature T), and a set of internal variables α (e.g., the martensite volume fraction ξ and/or the total transformation strain ε^{tr}). Assuming that the state is described by the set $\{\varepsilon, T, \alpha\}$, two key-ingredients characterize such an approach:

1. A free-energy function $\psi = \hat{\psi}\,(\varepsilon, T, \alpha)$ that enables the computation of the set $\{\varepsilon, T, \alpha\}$ and its conjugated set $\{\sigma, \eta, X\}$ by means of constitutive equations, where σ, η, and X are, respectively, the stress, entropy and thermodynamic force associated to the internal variables. The constitutive equations are obtained as partial derivatives of the free-energy function, after enforcing the Clausius-Duhem inequality for every process. If the heat conduction is considered, a constitutive equation (usually the Fourier equation) relating the temperature

gradient ∇T and the heat flux Q is also required in the form $Q = \hat{Q}(\varepsilon, T, \alpha, \nabla T)$.

2. A set of evolution equations for the internal variables, describing the microstructural changes induced by the PT in terms of material actual and past state, in the form $\dot{\alpha} = f(\varepsilon, T, \alpha, \dot{\varepsilon}, \dot{T})$. These equations are derived in a thermodynamically consistent framework.

Among the others, we recall models based on the theory of plasticity, for example (Boyd and Lagoudas 1996; Souza et al. 1998), where limit functions and evolution laws govern material processes. The evolution laws display a normality structure and obey conditions of consistency. Also, we recall models derived from thermodynamic potentials, for example (Lexcellent et al. 2000; Peultier et al. 2006; Zaki and Moumni 2007), which derive the energy potential from microscopic, phenomenological or physical considerations and, then, express the constitutive relations according to thermodynamic principles. A discussion on the equivalence between these two modeling formulations is proposed in the work by Cisse et al. (2016).

In all the models based on internal variables, the choice of an appropriate set of internal variables α represents an important step, since the model should appropriately describe at least SMA primary effects (i.e., PE, SME, and variant reorientation).

The first work seems to be due to Tanaka and Nagaki (1982) who introduced a scalar internal variable representing the martensitic volume fraction. Later, the three-dimensional models by Tanaka and Iwasaki (1985), Lin et al. (1994) and Tanaka et al. (1995) used exponential transformation hardening functions to describe martensite evolution. However, these models were implemented in a one-dimensional context. Liang and Rogers (1990) presented an evolution law based on cosine functions and Brinson (1993) split the martensite volume fraction into twinned and detwinned martensites. Boyd and Lagoudas (1994) formulated the Tanaka model in a three-dimensional framework, while Lagoudas et al. (1996) proposed a polynomial model and compared it to the exponential and cosine models cited above. The exponential model appeared less suitable in predicting the stress-strain behaviour and heat energy rate curves, compared to the other two models. Other three-dimensional models are those by Raniecki and Lexcellent (1994, 1998) who introduced the fractions of twinned and detwinned martensites as internal variables. Recall that the material properties of the two martensite phases are usually assumed to be the same, since the two phases are indistinguishable from a metallurgical point of view, while the different macroscopic mechanical behaviour is reflected in the kinematic considerations. A recent discussion on this topic is provided in Wang and Sehitoglu (2014).

Most of the above formulations are based on only a set of scalar variables, which is not complete due to the loss of directional information for the description of variant reorientation. The three-dimensional model by Frémond (2002) involves three scalar variables representing austenite and two martensite variants and assumes the transformation strain direction known. Successively, Savi and coworkers published several works based on the Frémond model, for example, Paiva et al. (Paiva et al. 2005).

Boyd and Lagoudas (1996) first proposed an inelastic strain tensor to describe martensite reorientation; the model has been recently generalized by Lagoudas et al. (2012) who proposed a model accounting for SMA smooth response.

Models with only tensorial internal variables, for example, (Alessi and Pham 2016), appear to be more successful, but may lead to constrained modeling approaches. The model by Souza et al. (1998), then generalized by Auricchio and Petrini (2004), Evangelista et al. (2009), Auricchio et al. (2016b), introduces the transformation strain tensor as tensorial internal variable, but it does not capture PTs for low levels of stress.

Therefore, several works have introduced scalar and tensorial internal variables to describe both martensite volume fraction and martensitic processes, for example, (Lexcellent et al. 2000; Peultier et al. 2006; Panico and Brinson 2007; Popov and Lagoudas 2007; Luig and Bruhns 2008; Arghavani et al. 2010; Saleeb et al. 2011). Recently, Auricchio and Bonetti (2013) and Auricchio et al. (2014a) generalized the works by Souza et al. (1998) and Frémond (2002) to account for martensite reorientation, tension-compression asymmetry, SMA smooth response, transformation-dependent elastic properties, and low-stress PTs.

Constitutive models have been enriched by including the description of secondary effects, for example, plastic strain (Zaki 2010), twins accommodation strain (Chemisky et al. 2011), cyclic behaviour (Mehrabi et al. 2015), R-phase and anisotropy (Sedlák et al. 2012) and viscoplasticity (Chemisky et al. 2014).

To date, SMA constitutive models have reached high levels of accuracy in the description of physical mechanisms; this, however, at the cost of complexity: models are often defined by several parameters (not always associated to physical quantities and also requiring costly calibrations) and they necessitate suitable numerical algorithms to treat non-smooth functions and/or constraints on internal variables and to guarantee numerical convergence. Therefore, a good balance between accuracy and efficiency has always to be taken into account.

Recall that most of the reviewed models are well-adapted to Cu- or NiTi-based SMAs, since they describe the PT effect. Only a few models, describing martensitic PT and plastic gliding, are suitable for Fe-based

SMAs, for example, (Nishimura and Tanaka 1998; Hartl and Lagoudas 2009; Jemal et al. 2009; Khalil et al. 2012), but they are mostly formulated in one-dimensional frameworks.

Engineering Simulations

Since the 1980s, numerous SMA constitutive models have been proposed and implemented into software analysis tools. Some of these models, for example (Auricchio and Taylor 1997), are available as built-in material models into commercial programs, as Abaqus or LS-Dyna; (Concilio and Lecce 2015). Moreover, the opportunity of integrating computer-based analysis tools, such as computer-aided design (CAD) and finite element analysis (FEA), in the design and optimization of SMA-based applications, has increased the attention on the formulation of novel constitutive models and robust numerical implementations.

Trial-and-error procedures are a standard design practice at the industrial level and they consist in testing many prototypes until the fulfilment of specific design criteria. These procedures may become problematic, expensive, and time-consuming under specific conditions (e.g., in case of complex devices, microscopic specimens or fatigue testing). In this context, the exploitation of engineering simulations is particularly useful for preliminary studies and/or validation stages to reduce material waste, time-to-market, and costs. Engineering simulations can play a valuable role by conducting realistic analyses of complicated device geometries with advanced constitutive models. Several analyses have been conducted in the literature; some of them have been also validated through a comparison with experimental data. In the following, we will show some examples of effective FEA approaches for a wide variety of SMA-based components, involving both PE and SME, in order to show the potential of the available computational framework in possibly offering a virtual engineering tool for the design, simulation and optimization of devices.

Biomedical Applications

FEA has been largely exploited in the last years to simulate or optimize SMA biomedical devices, due to the ever-increasing innovation in the medical field (Concilio and Lecce 2015).

Most of the available works focus on cardiovascular applications, specifically on self-expanding stents, for example, (Garcia et al. 2012; Brosse and Hartl 2015; Saleeb et al. 2015b). The aim of these studies is to investigate various aspects of stent design, such as radial force, flexibility, the effect of *in vivo* loading during crimping, pre- and post-deployment, fatigue-life, stent dimensions/geometry, vessel/stent interaction, and buckling behaviour. Therefore, the simulation of the stenting procedure generally involves the definition of a stent configuration, a vessel, a plaque, and contact to account for their interaction.

One of the pioneering works was published by Whitcher (1997) who adopted a von Mises elastoplastic model for the constitutive modeling of Nitinol. Later, Rebelo and Perry (2000) implemented the constitutive model by Auricchio and Taylor (1997) to simulate the expansion of a Nitinol stent. Theriault et al. (2006) proposed an effort to tailor the design of a novel stent. Recently, McGrath et al. (2014) investigated the buckling behaviour of tracheobronchial stents by using the Abaqus in-built superelastic model. They first simulated the buckling behaviour obtained experimentally to identify the main buckling mechanism (see Figs. 4a–d); then, they conducted a stability analysis, using the determined buckling mechanism; finally, the authors used the results to design a stent able to resist axial buckling, which was then tested experimentally (see Fig. 4e). Few studies have addressed the use of FEA for carotid stent design, for example (Conti et al. 2011), and for intracranial stents, for example (De Bock et al. 2012a).

Recently, the evidence of *in vivo* stent fractures has highlighted the problem of the fatigue resistance of stents whose rupture threatens the

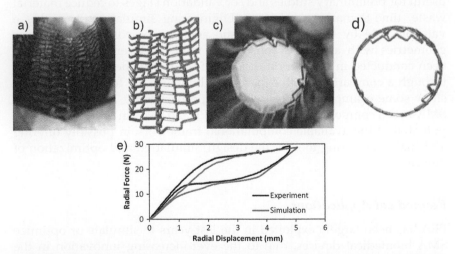

Fig. 4: Radial force test to investigate the buckling behaviour of tracheobronchial stents. Results for the A1 stent design (showing buckling): the (a) experimental and (b) numerical skewed axial views and the (c) experimental and (d) numerical radial views are shown after occurrence of buckling. (e) Results for the optimized B4 stent design (no indication of buckling): radial force vs. radial displacement curves. Reprinted from McGrath et al. (2014), with permission from Elsevier.

safety and life of the patient. The study of stent lifetime is difficult due to the complex fatigue behaviour of Nitinol and the lack of complete experimental investigations; the reader is referred to Auricchio et al. (2016a) for a review. FEA is now used to compute stent lifetime and to calculate design safety factors, especially considering the difficulties in conducting experiments on these devices; see, for example, Rebelo and Perry (2000), Pelton et al. (2008) and Azaouzi et al. (2012). The analyses are particularly useful in case of peripheral stents that present high fracture rates, for example, (Petrini et al. 2012).

In the cardiovascular field, the rapid evolution of self-expandable stent grafts in the last decade has also motivated the use of FEA in this sector, for example (Kleinstreuer et al. 2008), and its validation with experiments (De Bock et al. 2012b). Moreover, stent-mounted prosthetic valves have recently attracted the attention of researchers to study both the mechanical (Tzamtzis et al. 2013) and fatigue (Kumar et al. 2013) behaviours.

Beyond cardiovascular devices, other biomedical tools have been investigated, for example, spinal vertebrae spacer (Petrini et al. 2005), bone staples (Saleeb et al. 2015a), active catheter (Langelaar and van Keulen 2007) and pedicle screws (Tabesh et al. 2012).

Actuator Applications

SMA materials find successful applications in actuators, since they provide excellent advantages (e.g., design simplicity, low weight and cost, compact size) compared to classical devices; see Elahinia (2016) and references therein. SMA actuators are produced in several forms (e.g., wires, tubes, springs, or sheets) and are used in medium- and small-scale devices, as in biomimetic locomotion, miniature gripper, automotive tumble flaps or control valves for automobile engines or aerospace applications.

Despite the apparent simplicity, the design of SMA actuators requires close attention, since they operate under several thermomechanical cycles and complex multiaxial stress states. The literature presents numerous works on the modeling, design, simulation, and testing of SMA actuators. Actually, several works use FEA simulations for computer-based design, for example (Bhattacharyya et al. 1995; Huang et al. 2012; de Aguiar et al. 2013). As an example, Gravatt et al. (2010) experimentally and numerically investigated several SMA actuator beams. Langelaar et al. (2011) applied topology optimization to the design of thermal actuators in order to maximize the stroke. Several theoretical and numerical works focused on the analysis of SMA springs; generally, wire and spring diameters, number of active coils, and pitch angle represent design parameters to optimize the stroke and the hysteresis of the actuator response. Dumont and Kuhl (2005) proposed a FEA model of the Euler-Bernoulli beam for SMA springs, identified on experiments related to wires. Follador et al. (2012) presented a general model and the techniques for fabricating SMA spring actuators. The work by Saleeb et al. (2013) focuses on the characterization of the cyclic behaviour of 55NiTi and simulation of springs using the model in Saleeb et al. (2011). Model calibration was performed on a single thermal-cycling test at constant load on a SMA wire. Auricchio et al. (2014b) investigated the experimental behaviour of NiTi helical tension spring actuators through the numerical implementations of three phenomenological models (Auricchio and Petrini 2004; Auricchio et al. 2009; Auricchio et al. 2014a).

In the aerospace field, Hartl and Lagoudas (2007) performed an experimental characterization and FEA of commercial jet engine chevrons incorporating active SMA beam components. Active chevrons were analysed and tested to create the maximum deflection during take-off and landing and the minimum deflection into the flow during the remainder of flight. The model by Lagoudas et al. (1996) accurately predicted the experimental response. Oehler et al. (2011) presented a computational study of a morphing aerostructure with embedded thermally-actuating SMA ribbons and demonstrated the effective use of fluid-structure interaction modeling. Later, Oehler et al. (2012) proposed methods based on iterative analysis techniques to determine optimized designs for morphing aerostructures and considered the impact of uncertainty in model variables on the solution. The work by Hartl et al. (2011) proposes a general design optimization framework for the consideration of any SMA component by combining methods of design optimization, FEA, and constitutive modeling. Barbarino et al. (2011) compared experimental results with the FEA results of a SMA flap, obtained with the model by Liang and Rogers (1990).

Despite the numerous SMA aerostructural solutions, there are no industry- or government-accepted standardized test methods for SMAs when used as actuators (Hartl et al. 2015). The communication by Hartl et al. (2015) introduces a SMA specification and standardization effort to deliver the first ever-regulatory agency-accepted material specification and test standards for SMAs when used as actuators for commercial and military aviation applications.

The literature is very vast also in the automotive and robotic fields, but it focuses more on the experimental side, for example (Lan et al. 2009; Leary et al. 2013). In Suman et al. (2015) both experimental and numerical analyses on a morphing polymeric blade for an automotive axial fan are presented. Measurements of the fluid temperature, blade surface temperature pattern, and blade shape change, achieved by SMA strips, were performed, together with computational fluid dynamics simulations, in order to study the performance and fluid dynamic behaviour of the fan.

Recently, there has been a growing interest in using SMAs in micro and nanoscale systems, for example, as sensors and actuators in microelectromechanical (MEMS) and nanoelectromechanical (NEMS) systems or as microvalves (Choudhary and Kaur 2016). These applications motivate the use of micromechanical models to study phenomena in SMA polycrystals with a limited number of grains, such as grain size effect, the role of texture and intergranular stresses.

Self-Folding Applications

A recent field of investigation is constituted by the so-called self-folding structures. Its origin is attributed to the art of paper folding, worldwide known as origami, which has inspired the design of several engineering components and structures (hence the name origami-inspired engineering). Self-folding structures become advantageous for many applications as space systems, underwater robotics, small-scale devices and self-assembling systems (Peraza-Hernandez et al. 2014a).

Recently, researchers have started to investigate the way to obtain the folding behaviour by using SMAs, instead of applying external loads. Hartl, Lagoudas, Malak, and collaborators have treated the computational analysis of self-folding structures based on SMAs, for example, (Peraza-Hernandez et al. 2013b). Figure 5 shows an example of FEA simulation of a SMA-based sheet folding into a cube (Peraza-Hernandez et al. 2013a).

The work by Peraza-Hernandez et al. (2013c) studies a self-morphing laminate that includes two meshes of thermally-actuated SMA wire, separated by a compliant passive layer. The authors considered several design variables (e.g., mesh wire thickness and spacing, thickness of the insulating elastomer layer, heating power) as well as response parameters (e.g., folding angle, von Mises stress and temperature in the SMA, temperature in the elastomer, radius of curvature at the fold line) to maximize the folding capability under mechanical and thermal failure constraints. Peraza-Hernandez et al. (2013d) optimized the geometric and power input parameters of a self-folding sheet to achieve the tightest local fold possible subject to stress and temperature constraints. The authors investigated different angles relative to the orientation of the wire mesh. The results show that a relatively low elastomer thickness is preferable to generate the tightest fold possible and that the sheet does not require large power inputs to achieve an optimal folding performance. Recently, Peraza-Hernandez et al. (2014b) have considered self-folding laminate composites and proposed methods for designing folding patterns and determining temperature fields to obtain desired shapes and behaviours. Powledge et al. (2014) experimentally characterized a self-folding reconfigurable sheet and employed an FEA, based on the model by Lagoudas et al. (2012), to explore the effect of design parameters on the performance metrics of the sheet. Both experiment and simulation focused on the radius of curvature achieved by the sheet for a given set of design parameters and actuation path.

Despite these contributions, relatively little research exists on the topic and on the validation with experimental results (An and Rus 2014; Koh et al. 2014). This requires further investigations, especially considering that

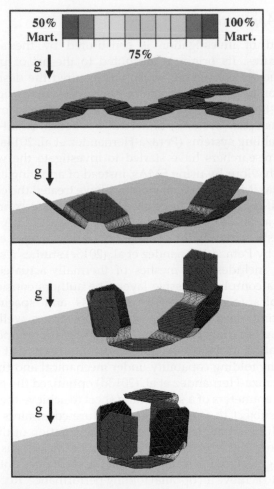

Fig. 5: FEA of a SMA-based laminate sheet that self-folds into a cube. Morphing is thermally-induced and the maximum temperature applied at the hinges is determined by means of trial-and-error simulations to set the desired final configuration at the prescribed final time. Contour plot indicates the distribution of martensite volume fraction. Reprinted from Peraza-Hernandez et al. (2013a), with permission from Elsevier.

self-folding origami sheets represent one of the most severe applications for SMAs, since they must be small enough to be inserted in the material layer and powerful to attain folding.

Civil Engineering Applications

The use of NiTi- and Cu-based SMAs has been always limited in the civil engineering field due to the structure size and the high costs in producing large quantities of these alloys. SMAs find applications in civil engineering structures for passive vibration damping and energy dissipation, active vibration control, tensioning applications, or as sensors and actuators (Ozbulut et al. 2011; Jani et al. 2014; Concilio and Lecce 2015).

Therefore, few works have been dedicated to the simulations of SMA components for civil engineering structures. As an example, the investigations of SMAs for earthquake applications aim at regularizing the structural response, shifting the fundamental period of vibration, and increasing the global energy dissipation (McCormick et al. 2006; Attanasi et al. 2009). We recall, also, the analysis of SMAs as expansion joints for bridges (Yang and Des Roches 2015).

The use of Fe-based SMAs in civil engineering applications, as tendons for repairing existing structures or reinforcing new structures, is becoming a promising field (Cladera et al. 2014). Among the others, FeMnSi alloys are the most promising candidates for the construction industry due to their lower cost and other advantages, for example, large hysteresis (Czaderski et al. 2014). Currently, Fe-based SMA applications are designed using empirical methods, mainly based on experimental results (Khalil et al. 2012). This highlights the need for suitable design numerical tools for the development of novel components based on these alloys.

Conclusions and Perspectives

The overview provided in the present chapter highlights that both the research field and technology based on SMAs have been extensively explored during the past decades. It is expected an ever-increasing exploitation of SMA materials at the three levels defined in Jani et al. (2014): (i) the *material level* through the development of novel or improved alloys, for example, FeMnSi SMAs (Dong et al. 2009), (ii) the *structure level* by combining the properties of SMAs and other materials, for example, active textiles, self-healing composite, hybrid or composite shape memory materials (Lester et al. 2015), and (iii) the *market level* by searching for new application fields, for example, thin-films, additive manufacturing

(Meisel et al. 2014), or 4D printing (Tibbits 2014). The growing utilization of SMA materials is actually demonstrated by the increasing number of worldwide publications and patents. This growth motivates the need for reliable design methodologies to use in the development of novel and sophisticated components or devices. In this context, constitutive models and computer-based design techniques can offer a valid support to speed-up the development process, especially for preliminary studies and validation of high-risk and/or expensive projects. As discussed, many potential constitutive models have been proposed in the literature. The choice of the material model to use for the simulation depends on the purpose of the analysis and on the considered problem under investigation; as shown, phenomenological models have been successfully exploited in advanced software simulations of real-life applications. In this direction, the main challenges for designing a SMA component should be: (i) the formulation and numerical implementation of a simple, but accurate, constitutive *model*; (ii) model *calibration* (generally on uniaxial tests); (iii) model *validation* through comparison with experimental data; (iv) *prediction* of the component response through the validated model; (v) component *optimization*; (vi) development of *guidelines/certifications/ standard* to homogenize methods.

Despite the listed successes, the SMA community has still to outline complete standards for SMAs (e.g., concerning testing, fabrication methods or treatments) to be used as guidelines in various fields. A lot of work has still to be done and a close collaboration between material scientists, software developers and engineering designers is needed to spread out the products over the market (Jani et al. 2014). Moreover, future methodologies can be established to control material variability during manufacturing and processing, which actually represents a substantial limit for SMA applications.

Acknowledgments

This work is partially funded by Fondazione Cariplo, grant number 2013-1779.

Keywords: Shape memory alloy, NiTi alloys, Constitutive modeling, Finite element analysis, Engineering simulation, SMA application

References

Abeyaratne, R. and J.K. Knowles. 1990. On the driving traction acting on a surface of strain discontinuity in a continuum. J. Mech. Phys. Solids 38: 345–360.

Alessi, R. and K. Pham. 2016. Variational formulation and stability analysis of a three dimensional superelastic model for shape memory alloys. J. Mech. Phys. Solids 87: 150–176.

An, B. and D. Rus. 2014. Designing and programming self-folding sheets. Rob. Auton. Syst. 62: 976–1001.

Arghavani, J., F. Auricchio, R. Naghdabadi, A. Reali and S. Sohrabpour. 2010. A 3D phenomenological constitutive model for shape memory alloys under multiaxial loadings. Int. J. Plast. 26: 976–991.

Attanasi, G., F. Auricchio and G.L. Fenves. 2009. Feasibility assessment of an innovative isolation bearing system with shape memory alloys. J. Earthquake Eng. 13: 18–39.

Auricchio, F., A. Coda, A. Reali and M. Urbano. 2009. SMA numerical modeling versus experimental results: Parameter identification and model prediction capabilities. J. Mater. Eng. Perform. 18: 649–654.

Auricchio, F., A. Constantinescu, M. Conti and G. Scalet. 2016a. Fatigue of metallic stents: From clinical evidence to computational analysis. Ann. Biomed. Eng. 44: 287–301.

Auricchio, F. and E. Bonetti. 2013. A new flexible 3D macroscopic model for shape memory alloys. Discret. Contin. Dyn. Syst. S. 6: 277–291.

Auricchio, F. and L. Petrini. 2004. A three-dimensional model describing stress-temperature induced solid phase transformations: solution algorithm and boundary value problems. Int. J. Numer. Methods Eng. 61: 807–836.

Auricchio, F. and R.L. Taylor. 1997. Shape-memory alloys: Modelling and numerical simulation of the finite-strain superelastic behavior. Comput. Methods Appl. Mech. Eng. 143: 175–194.

Auricchio, F., E. Boatti, A. Reali and U. Stefanelli. 2016b. Gradient structures for the thermomechanics of shape-memory materials. Comput. Methods Appl. Mech. Eng. 299: 440–469.

Auricchio, F., E. Bonetti, G. Scalet and F. Ubertini. 2014a. Theoretical and numerical modeling of shape memory alloys accounting for multiple phase transformations and martensite reorientation. Int. J. Plast. 59: 30–54.

Auricchio, F., G. Scalet and M. Urbano. 2014b. A numerical/experimental study of nitinol actuator springs. J. Mater. Eng. Perform. 23: 2420–2428.

Azaouzi, M., A. Makradi and S. Belouettar. 2012. Deployment of a self-expanding stent inside an artery: A finite element analysis. Mater Des. 41: 410–420.

Barbarino, S., R. Pecora, L. Lecce, A. Concilio, S. Ameduri and L. De Rosa. 2011. Airfoil structural morphing based on S.M.A. Actuator Series: Numerical and experimental studies. J. Intell. Mater Syst. Struct. 22: 987–1004.

Barsch, G. and J. Krumhansl. 1988. Nonlinear and nonlocal continuum model of transformation precursors in martensites. Metall. Trans. A. 19: 761–775.

Bernardini, D. 2001. On the macroscopic free energy functions for shape memory alloys. J. Mech. Phys. Solids. 49: 813–837.

Bhattacharya, A. and D. Lagoudas. 1997. A stochastic thermodynamic model for the gradual thermal transformation of SMA polycrystals. Smart Mater Struct. 6: 235–250.

Bhattacharyya, A., D.C. Lagoudas, Y. Wang and V.K. Kinra. 1995. On the role of thermoelectric heat transfer in the design of SMA actuators: Theoretical modeling and experiment. Smart Mater Struct. 4: 252–263.

Blanc, P. and C. Lexcellent. 2004. Micromechanical modelling of a CuAlNi shape memory alloy behaviour. Mater Sci. Eng. A. 378: 465–469.

Bo, Z. and D.J. Lagoudas. 1999. Thermomechanical modeling of polycrystalline SMAs under cyclic loading, Part IV: Modeling of minor hysteresis loops. Int. J. Eng. Sci. 37: 1205–1249.

Boyd, J. and D. Lagoudas. 1994. Thermomechanical response of shape memory composites. J. Intell. Mater Syst. Struct. 5: 333–346.

Boyd, J.G. and D.C. Lagoudas. 1996. A thermodynamical constitutive model for shape memory materials. Part I. The monolithic shape memory alloy. Int. J. Plast. 12: 805–842.

Brinson, L. 1993. One-dimensional constitutive behavior of shape memory alloys: thermomechanical derivation with non-constant material functions and redefined martensite internal variable. J. Intell. Mater Syst. Struct. 4: 229–242.

Brocca, M., L. Brinson and Z. Bazant. 2002. Three-dimensional constitutive model for shape memory alloys based on microplane model. J. Mech. Phys. Solids 50: 1051–1077.

Brosse, T. and D.J. Hartl. 2015. Design of a shape memory alloy self-expanding stent via open source optimization methods. Proc. Int. Conf. Shape Mem. Superelastic Technol. 23–24.

Buehler, W.J., J.V. Gilfrich and R.C. Wiley. 1963. Effect of low-temperature phase changes on the mechanical properties of alloys near composition TiNi. J. Appl. Phys. 34: 1475–1477.

Chemisky, Y., A. Duval, E. Patoor and T. Ben Zineb. 2011. Constitutive model for shape memory alloys including phase transformation, martensitic reorientation and twins accommodation. Mech. Mater 43: 361–376.

Chemisky, Y., G. Chatzigeorgiou, P. Kumar and D.C. Lagoudas. 2014. A constitutive model for cyclic actuation of high-temperature shape memory alloys. Mech. Mater 68: 120–136.

Choudhary, N. and D. Kaur. 2016. Shape memory alloy thin films and heterostructures for MEMS applications: A review. Sens. Actuators, A. 242: 162–181.

Cisse, C., W. Zaki and T.B. Zineb. 2016. A review of constitutive models and modeling techniques for shape memory alloys. Int. J. Plast. 76: 244–284.

Cladera, A., B. Weber, C. Leinenbach, C. Czaderski, M. Shahverdi and M. Motavalli. 2014. Iron-based shape memory alloys for civil engineering structures: An overview. Constr. Build. Mater 63: 281–293.

Concilio, L. and A. Lecce. 2015. Shape Memory Alloy Engineering: for Aerospace, Structural and Biomedical Applications. Elsevier Butterworth-Heinemann, Boston, USA.

Conti, M., D. Van Loo, F. Auricchio, M. De Beule, G. De Santis, B. Verhegghe et al. 2011. Impact of carotid stent cell design on vessel scaffolding: a case study comparing experimental investigation and numerical simulations. J. Endovasc. Ther. 18: 397–406.

Czaderski, C., M. Shahverdi, R. Brönnimann, C. Leinenbach and M. Motavalli. 2014. Feasibility of iron-based shape memory alloy strips for prestressed strengthening of concrete structures. Constr. Build. Mater 56: 94–105.

Daghia, F., M. Fabrizio and D. Grandi. 2010. A non isothermal Ginzburg-Landau model for phase transitions in shape memory alloys. Meccanica. 45: 797–807.

De Aguiar, A.A., W. de Castro Leo Neto, M. Savi and P. Calas Lopes Pacheco. 2013. Shape memory alloy helical springs performance: Modeling and experimental analysis. Mater Sci. Forum. 758: 147–156.

De Bock, S., F. Iannaccone, G. De Santis, M. De Beule, P. Mortier, B. Verhegghe et al. 2012a. Our capricious vessels: The influence of stent design and vessel geometry on the mechanics of intracranial aneurysm stent deployment. J. Biomech. 45: 1353–1359.

De Bock, S., F. Iannaccone, G. De Santis, M. De Beule, D. Van Loo, D. Devos et al. 2012b. Virtual evaluation of stent graft deployment: A validated modeling and simulation study. J. Mech. Behav. Biomed. Mater 13: 129–139.

Dhote, R.P., H. Gomez, R.N.V. Melnik and J. Zu. 2015. 3D coupled thermo-mechanical phase-field modeling of shape memory alloy dynamics via isogeometric analysis. Comput. Struct. 154: 48–58.

Dong, Z., U.E. Klotz, C. Leinenbach, A. Bergamini, C. Czaderski and M. Motavalli. 2009. A novel Fe-Mn-Si shape memory alloy with improved shape recovery properties by VC precipitation. Adv. Eng. Mater 11: 40–44.

Dumont, G. and C. Kuhl. 2005. Finite element simulation for design optimisation of shape memory alloy spring actuators. Eng. Comput. 22: 835–848.

Elahinia, M. 2016. Shape Memory Alloy Actuators: Design, Fabrication and Experimental Evaluation. John Wiley and Sons Ltd, Chichester, United Kingdom.

Evangelista, V., S. Marfia and E. Sacco. 2009. Phenomenological 3D and 1D consistent models for shape-memory alloy materials. Comput. Mech. 44: 405–421.

Falk, F. 1983. Ginzburg-Landau theory of static domain walls in shape-memory alloys. Zeitschrift Physik B Condens. Matter 51: 177–185.

Fischer, F. and K. Tanaka. 1992. A micromechanical model for the kinetics of martensitic transformation. Int. J. Solids Struct. 29: 1723–1728.

Fischlschweiger, M. and E.R. Oberaigner. 2012. Kinetics and rates of martensitic phase transformation based on statistical physics. Comput. Mater Sci. 52: 189–192.

Follador, M., M. Cianchetti, A. Arienti and C. Laschi. 2012. A general method for the design and fabrication of shape memory alloy active spring actuators. Smart Mater Struct. 21: 1–10.

Frémond, M. 2002. Non-smooth Thermomechanics. Springer-Verlag, Berlin.

Gall, K., T.J. Lim, D.L. McDowell, H. Sehitoglu and Y.I. Chumlyakov. 2000. The role of intergranular constraint on the stress-induced martensitic transformation in textured polycrystalline NiTi. Int. J. Plast. 16: 1189–1214.

Gao, X. and L.C. Brinson. 2002. A simplified multivariant SMA model based on invariant plane nature of martensitic transformation. J. Intell. Mater. Syst. Struct. 13: 795–810.

Garcia, A., E. Pena and M. Martinez. 2012. Influence of geometrical parameters on radial force during self-expanding stent deployment. Application for a variable radial stiffness stent. J. Mech. Behav. Biomed. Mater 10: 166–75.

Grandi, D., M. Maraldi and L. Molari. 2012. A macroscale phase-field model for shape memory alloys with non-isothermal effects: Influence of strain rate and environmental conditions on the mechanical response. Acta Mater 60: 179–191.

Gravatt, L.M., J.H. Mabe, F.T. Calkins and D.J. Hartl. 2010. Characterization of varied geometry shape memory alloy beams. Proc. SPIE Ind. Commer. App. Smart Struct. Tech. 7645.

Hartl, D.J. and D.C. Lagoudas. 2007. Characterization and 3-D modeling of Ni60Ti SMA for actuation of a variable geometry jet engine chevron. Proc. SPIE Int. Soc. Opt. Eng. 6529: 1–12.

Hartl, D.J. and D.C. Lagoudas. 2009. Constitutive modeling and structural analysis considering simultaneous phase transformation and plastic yield in shape memory alloys. Smart Mater Struct. 18: 1–17.

Hartl, D.J., D.C. Lagoudas and F.T. Calkins. 2011. Advanced methods for the analysis, design, and optimization of SMA-based aerostructures. Smart Mater Struct. 20: 094006.

Hartl, D.J., J.H. Mabe, O. Benafan, A. Coda, B. Conduit, R. Padan et al. 2015. Standardization of shape memory alloy test methods toward certification of aerospace applications. Smart Mater Struct. 24: 82001–82006.

Huang, M., X. Gao and L.C. Brinson. 2000. A multivariant micromechanical model for SMAs Part 2. polycrystal model. Int. J. Plast. 16: 1371–1390.

Huang, S., M. Leary, T. Ataalla, K. Probst and A. Subic. 2012. Optimisation of Ni-Ti shape memory alloy response time by transient heat transfer analysis. Mater Des. 35: 655–663.

Huo, Y. 1989. A mathematical model for the hysteresis in shape memory alloys. Continuum Mech. Thermodyn. 1: 283–303.

Ivshin, Y. and T. Pence. 1994. A constitutive model for hysteretic phase transition behavior. Int. J. Eng. Sci. 32: 681–704.

Jani, J.M., M. Leary, A. Subic and M. Gibson. 2014. A review of shape memory alloy research, applications and opportunities. Mater Des. 56: 1078–1113.

Jemal, F., T. Bouraoui, T. Ben Zineb, E. Patoor and C. Bradai. 2009. Modelling of martensitic transformation and plastic slip effects on the thermomechanical behaviour of Fe-based shape memory alloys. Mech. Mater 41: 849–856.

Junker, P. and K. Hackl. 2011. Finite element simulations of poly-crystalline shape memory alloys based on a micromechanical model. Comput. Mech. 47: 505–517.

Kadkhodaei, M., M. Salimi, R. Rajapakse and M. Mahzoon. 2008. Modeling of shape memory alloys based on microplane theory. J. Intell. Mater Syst. Struct. 19: 541–550.

Khalil, W., A. Mikolajczak, C. Bouby and T.B. Zineb. 2012. A constitutive model for Fe-based shape memory alloy considering martensitic transformation and plastic sliding coupling: Application to a finite element structural analysis. J. Intell. Mater Syst. Struct. 23: 1143–1160.

Kleinstreuer, C., Z. Li, C.A. Basciano, S. Seelecke and M.A. Farber. 2008. Computational mechanics of Nitinol stent grafts. J. Biomech. 41: 2370–2378.

Koh, J., S. Kim and K. Cho. 2014. Self-folding origami using torsion shape memory alloy wire actuators. Proc. ASME 38th Mechanisms Robotics Conf. 1–7.

Kumar, G.P., F. Cui, A. Danpinid, B. Su, J.K. Fatt Hon and H.L. Leo. 2013. Design and finite element-based fatigue prediction of a new self-expandable percutaneous mitral valve stent. Comput.-Aided Des. 45: 1153–1158.

Lagoudas, D., D. Hartl, Y. Chemisky, L. Machado and P. Popov. 2012. Constitutive model for the numerical analysis of phase transformation in polycrystalline shape memory alloys. Int. J. Plast. 32: 155–183.

Lagoudas, D., Z. Bo and M. Qidwai. 1996. A unified thermodynamic constitutive model for SMA and finite element analysis of active metal matrix composites. Mech. Com. Mat. Struct. 3: 153–179.

Lagoudas, D.C. 2008. Shape Memory Alloys—Modeling and Engineering Applications. Springer US, New York.

Lagoudas, D.C. and A. Bhattacharyya. 1997. On the correspondence between micromechanical models for isothermal pseudoelastic response of shape memory alloys and the Preisach model for hysteresis. Math. Mech. Solids 2: 405–440.

Lagoudas, D.C., P.B. Entchev, P. Popov, E. Patoor, L.C. Brinson and X. Gao. 2006. Shape memory alloys, part II: modeling of polycrystals. Mech. Mater 38: 430–462.

Lai, W. and B. Liu. 2000. Lattice stability of some Ni-Ti alloy phases versus their chemical composition and disordering. J. Phys.: Condens. Matter 12: L53.

Lan, C., J. Wang and C. Fan. 2009. Optimal design of rotary manipulators using shape memory alloy wire actuated flexures. Sens. Actuators, A. 153: 258–266.

Langelaar, M. and F. van Keulen. 2007. Design optimization of shape memory alloy active structures using the R-phase transformation. Proc. SPIE Act. Pass. Smart Struct. Int. Syst. 1–12.

Langelaar, M., G.H. Yoon, Y.Y. Kim and F. van Keulen. 2011. Topology optimization of planar shape memory alloy thermal actuators using element connectivity parameterization. Int. J. Numer. Meth. Eng. 88: 817–840.

Leary, M., S. Huang, T. Ataalla, A. Baxter and A. Subic. 2013. Design of shape memory alloy actuators for direct power by an automotive battery. Mater Des. 43: 460–466.

Lester, B.T., T. Baxevanis, Y. Chemisky and D.C. Lagoudas. 2015. Review and perspectives: shape memory alloy composite systems. Acta Mech. 226: 3907–3960.

Levitas, V.I. and I.B. Ozsoy. 2009. Micromechanical modeling of stress-induced phase transformations. Part 1. thermodynamics and kinetics of coupled interface propagation and reorientation. Int. J. Plast. 25: 239–280.

Levitas, V.I., D.L. Preston and D.W. Lee. 2003. Three-dimensional Landau theory for multivariant stress-induced martensitic phase transformations. III. Alternative potentials, critical nuclei, kink solutions and dislocation theory. Phys. Rev. B. 68: 134–201.

Lexcellent, C. 2013. Handbook of Shape Memory Alloys. Wiley-ISTE.

Lexcellent, C., S. Leclercq, B. Gabry and G. Bourbon. 2000. The two way shape memory effect of shape memory alloys: an experimental study and a phenomenological model. Int. J. Plast. 16: 1155–1168.

Liang, C. and C. Rogers. 1990. One-dimensional thermomechanical constitutive relations for shape memory materials. J. Intell. Mater Syst. Struct. 1: 207–234.

Likhacev, A. and Y. Koval. 1992. On the differential equation describing the hysteretic behavior of shape memory alloys. Scr. Metall. Mater 27: 223–227.

Lim, T. and D. McDowell. 2002. Cyclic thermomechanical behavior of a polycrystalline pseudoelastic shape memory alloy. J. Mech. Phys. Solids 50: 651–676.

Lin, P., H. Tobushi, K. Tanaka, T. Hattori and M. Makita. 1994. Pseudoelastic behaviour of TiNi shape memory alloy subjected to strain variations. Mech. Mater 5: 694–701.

Ling, H.C. and R. Kaplow. 1981. Stress-induced shape changes and shape memory in the R and martensite transformations in equiatomic NiTi. Metall. Trans. A. 12: 2101–2111.

Luig, P. and O. Bruhns. 2008. On the modeling of shape memory alloys using tensorial internal variables. Mat. Sci. Eng. A-Struct. 481-482: 379–383.

McCormick, J., R. DesRoches, D. Fugazza and F. Auricchio. 2006. Seismic vibration control using superelastic shape memory alloys. J. Eng. Mater Technol. 128: 294–301.

McGrath, D.J., B. O'Brien, M. Bruzzi and P.E. McHugh. 2014. Nitinol stent design -understanding axial buckling. J. Mech. Behav. Biomed. Mater 40: 252–263.

Mehrabi, R., M. Kadkhodaei and M. Elahinia. 2014. A thermodynamically-consistent microplane model for shape memory alloys. Int. J. Solids Struct. 51: 2666–2675.

Mehrabi, R., M. Shirani, M. Kadkhodaei and M. Elahinia. 2015. Constitutive modeling of cyclic behavior in shape memory alloys. Int. J. Mech. Sci. 103: 181–188.

Meisel, N.A., A.M. Elliott and C.B. Williams. 2014. A procedure for creating actuated joints via embedding shape memory alloys in PolyJet 3D printing. J. Intell. Mater Syst. Struct. 26: 1498–1512.

Mutter, D. and P. Nielaba. 2013. Simulation of the shape memory effect in a NiTi nano model system. J. Alloys Compd. 577: S83–S87.

Nishimura, F. and K. Tanaka. 1998. Phenomenological analysis of thermomechanical training in an Fe-based shape memory alloy. Comput. Mater Sci. 12: 26–38.

Oehler, S., D. Hartl, D. Lagoudas and R. Malak. 2011. Design optimization of a shape memory alloy actuated morphing aerostructure. Proc. ASME Conference on Smart Materials, Adaptive Structures and Intelligent Systems. 421–429.

Oehler, S., D.J. Hartl, R. Lopez, R.J. Malak and D.C. Lagoudas. 2012. Design optimization and uncertainty analysis of SMA morphing structures. Smart Mater Struct. 21: 1–16.

Ortin, J. 1992. Preisach modeling of hysteresis for a pseudoelastic Cu-Zn-Al single crystal. J. Appl. Phys. 71: 1454–1461.

Otsuka, K. and X. Ren. 2005. Physical metallurgy of Ti-Ni-based shape memory alloys. Prog. Mater Sci. 50: 511–678.

Ozbulut, O.E., S. Hurlebaus and R. DesRoches. 2011. Seismic response control using shape memory alloys: A review. J. Intell. Mater Syst. Struct. 22: 1531–1549.

Paiva, A., M. Savi, A. Braga and P. Pacheco. 2005. A constitutive model for shape memory alloys considering tensile-compressive asymmetry and plasticity. Int. J. Solids Struct. 42: 3439–3457.

Panico, M. and L. Brinson. 2007. A three-dimensional phenomenological model for martensite reorientation in shape memory alloys. J. Mech. Phys. Solids 55: 2491–2511.

Patoor, E., A. Eberhardt and M. Berveiller. 1996. Micromechanical modelling of superelasticity in shape memory alloys. J. Phys. IV. C1: 277–292.

Patoor, E., D.C. Lagoudas, P.B. Entchev, L.C. Brinson and X. Gao. 2006. Shape memory alloys, part I: General properties and modeling of single crystals. Mech. Mater 38: 391–429.

Pelton, A.R., V. Schroeder, M.R. Mitchell, X.-Y. Gong, M. Barney and S.W. Robertson. 2008. Fatigue and durability of Nitinol stents. J. Mech. Behav. Biomed. Mater 1: 153–164.

Peraza-Hernandez, E.A, S. Hu, H.W. Kung, D. Hartl and E. Akleman. 2013a. Towards building smart self-folding structures. Comput. Graphics 37: 730–742.

Peraza-Hernandez, E., D. Hartl and R. Malak. 2013b. Simulation-based design of a self-folding smart material system. Proc. ASME Des. Eng. Tech. Conf. 13439: 1–9.

Peraza-Hernandez, E., D. Hartl, E. Galvan and R. Malak. 2013c. Design and optimization of a shape memory alloy-based self-folding sheet. J. Mech. Des. 135: 1–11.

Peraza-Hernandez, E.A, D.J. Hartl and R.J. Malak. 2013d. Design and numerical analysis of an SMA mesh-based self-folding sheet. Smart Mater Struct. 22: 1–17.

Peraza-Hernandez, E.A., D.J. Hartl, R.J. Malak and D.C. Lagoudas. 2014a. Origami-inspired active structures: A synthesis and review. Smart Mater Struct. 23: 1–50.

Peraza-Hernandez, E.A., K.R. Frei, D.J. Hartl and D.C. Lagoudas. 2014b. Folding patterns and shape optimization using SMA-based self-folding laminates. Proc. SPIE Int. Soc. Optical Eng.

Petrini, L., F. Migliavacca, P. Massarotti, S. Schievano, G. Dubini and F. Auricchio. 2005. Computational studies of shape memory alloy behavior in biomedical applications. Trans. ASME 127: 716–725.

Petrini, L., W. Wu, E. Dordoni, A. Meoli, F. Migliavacca and G. Pennati. 2012. Fatigue behavior characterization of nitinol for peripheral stents. Funct. Mater Lett. 5: 1250012.

Peultier, B., T. Ben Zineb and E. Patoor. 2006. Macroscopic constitutive law for SMA: Application to structure analysis by FEM. Mech. Mater 38: 510–524.

Popov, P. and D.C. Lagoudas. 2007. A 3-D constitutive model for shape memory alloys incorporating pseudoelasticity and detwinning of self-accommodated martensite. Int. J. Plast. 23: 1679–1720.

Powledge, A.C., D.J. Hartl and R.J. Malak. 2014. Experimental analysis of self-folding SMA-based sheet design for simulation validation. Proc. ASME Conf. Smart Mater., Adapt. Struct. Intell. Syst. Rhode Island, USA. 1–9.

Raniecki, B. and C. Lexcellent. 1994. RL-models of pseudoelasticity and their specification for some shape memory solids. Eur. J. Mech. A. Solids 13: 21–50.

Raniecki, B. and C. Lexcellent. 1998. Thermodynamics of isotropic pseudoelasticity in shape memory alloys. Eur. J. Mech. A Solids 17: 185–205.

Rebelo, N. and M. Perry. 2000. Finite element analysis for the design of nitinol medical devices. Minim. Invasive Ther. Appl. Tech. 9: 75–80.

Sadjadpour, A. and K. Bhattacharya. 2007. A micromechanics inspired constitutive model for shape-memory alloys: The one-dimensional case. Smart Mater Struct. 16: S51–S62.

Saleeb, A., S. Padula II and A. Kumar. 2011. A multi-axial, multimechanism based constitutive model for the comprehensive representation of the evolutionary response of SMAs under general thermomechanical loading conditions. Int. J. Plast. 5: 655–687.

Saleeb, A., B. Dhakal, M. Hosseini and S. Padula II. 2013. Large scale simulation of NiTi helical spring actuators under repeated thermomechanical cycles. Smart Mater Struct. 22: 1–20.

Saleeb, A.F., B. Dhakal and J.S. Owusu-Danquah. 2015a. Assessing the performance characteristics and clinical forces in simulated shape memory bone staple surgical procedure: The significance of SMA material model. Comput. Biol. Med. 62: 185–195.

Saleeb, A.F., B. Dhakal and J.S. Owusu-Danquah. 2015b. On the role of SMA modeling in simulating NiTinol self-expanding stenting surgeries to assess the performance characteristics of mechanical and thermal activation schemes. J. Mech. Behav. Biomed. Mater 49: 43–60.

Sato, T., K. Saitoh and N. Shinke. 2008. Atomistic modelling of reversible phase transformations in Ni-Ti alloys: A molecular dynamics study. Mater Sci. Eng. A. 481: 250–253.

Sedlák, P., M. Frost, B. Benešová, T. Ben Zineb and P. Šittner. 2012. Thermomechanical model for NiTi-based shape memory alloys including R-phase and material anisotropy under multi-axial loadings. Int. J. Plast. 39: 132–151.

Shaw, J.A. and S. Kyriakides. 1995. Thermomechanical aspects of NiTi. J. Mech. Phys. Solids 43: 1243–1281.

She, H., Y. Liu, B. Wang and D. Ma. 2013. Finite element simulation of phase field model for nanoscale martensitic transformation. Comput. Mech. 52: 949–958.

Souza, A.C., E.N. Mamiya and N. Zouain. 1998. Three-dimensional model for solids undergoing stress-induced phase transformations. Eur. J. Mech. A Solids 17: 789–806.

Stupkiewicz, S. and A. Gorzynska-Lengiewicz. 2012. Almost compatible X-microstructures in CuAlNi shape memory alloy. Continuum Mech. Thermodyn. 24: 149–164.

Suman, A., A. Fortini, N. Aldi, M. Pinelli and M. Merlin. 2015. Using shape memory alloys for improving automotive fan blade performance: experimental and computational fluid dynamics analysis. Proc. Inst. Mech. Eng., Part A.

Sun, Q.P. and K.C. Hwang. 1993. Micromechanics modelling for the constitutive behavior of polycrystalline shape memory alloys I. derivation of general relations. J. Mech. Phys. Solids 41: 1–17.

Suzuki, T., M. Shimono, X. Ren, K. Otsuka and H. Onodera. 2006. Study of martensitic transformation by use of monte-carlo method and molecular dynamics. Mater Sci. Eng. A. 438: 95–98.

Tabesh, M., V. Goel and M.H. Elahinia. 2012. Shape memory alloy expandable pedicle screw to enhance fixation in osteoporotic bone: Primary design and finite element simulation. J. Med. Devices 6: 1–8.

Tanaka, K. and S. Nagaki. 1982. A thermomechanical description of materials with internal variables in the process of phase transitions. Ing. Archiv. 51: 287–299.

Tanaka, K. and R. Iwasaki. 1985. A phenomenological theory of transformation superplasticity. Eng. Fract. Mech. 4: 709–720.

Tanaka, K., F. Nishimura, T. Hayashi, H. Tobushi and C. Lexcellent. 1995. Phenomenological analysis on subloops and cyclic behavior in shape memory alloys under mechanical and/or thermal loads. Mech. Mater 19: 281–292.

Thamburaja, P. and L. Anand. 2001. Polycrystalline shape-memory materials: Effect of crystallographic texture. J. Mech. Phys. Solids 49: 709–737.

Theriault, P., P. Terriault, V. Brailovski and R. Gallo. 2006. Finite element modeling of a progressively expanding shape memory stent. J. Biomech. 39: 2837–44.

Tibbits, S. 2014. 4D Printing: Multi-Material Shape Change. Arch. Des. 84: 116–121.

Tzamtzis, S., J. Viquerat, J. Yap, M.J. Mullen and G. Burriesci. 2013. Numerical analysis of the radial force produced by the Medtronic-CoreValve and Edwards-SAPIEN after transcatheter aortic valve implantation (TAVI). Med. Eng. Phys. 35: 125–30.

Wang, J. and H. Sehitoglu. 2014. Martensite modulus dilemma in monoclinic NiTi-theory and experiments. Int. J. Plast. 61: 17–31.

Wang, L.X. and R.V. Melnik. 2007. Thermo-mechanical wave propagations in shape memory alloy rod with phase transformations. Mech. Adv. Mater Struct. 14: 665–676.

Whitcher, F. 1997. Simulation of *in vivo* loading conditions of nitinol vascular stent structures. Comput. Struct. 64: 1005–1011.

Yang, C. and R. Des Roches. 2015. Bridges with innovative buckling restrained SMA expansion joints having a high symmetrical tension/compression capacity. Proc. Struct. Congress 452–461.

Yu, C., G. Kang and Q. Kan. 2015. A micromechanical constitutive model for anisotropic cyclic deformation of super-elastic NiTi shape memory alloy single crystals. J. Mech. Phys. Solids 82: 97–136.

Zaki, W. and Z. Moumni. 2007. A three-dimensional model of the thermomechanical behavior of shape memory alloys. J. Mech. Phys. Solids 55: 2455–2490.

Zaki, W. 2010. An approach to modeling tensile-compressive asymmetry for martensitic shape memory alloys. Smart Mater Struct. 19: 025009.

Zhong, Y. and T. Zhu. 2014. Phase-field modeling of martensitic microstructure in NiTi shape memory alloys. Acta Mater 75: 337–347.

Prediction of the Thermoelectric Properties of Half-Heusler Phases from the Density Functional Theory

V.V. Romaka,[1,]* *P.F. Rogl,*[2] *R. Carlini*[3] *and C. Fanciulli*[4]

INTRODUCTION

Thermoelectric Performance of Materials

The global demand for energy has intensified research for more effective means of power generation. Among renewable power sources like solar, wind and biomass, an increasingly important role is played by the direct conversion of waste heat that can come from the combustion of fossil fuels, from sunlight, or as a byproduct of various processes (e.g.,

---

[1] Department of Materials Science and Engineering, Lviv Polytechnic National University, 79013 Lviv, Ustiyanovycha Str. 5, Ukraine.

[2] Institute of Material Chemistry and Research, University of Vienna, Währingerstrasse 42, A-1090 Wien, Austria.

[3] Chemistry and Industrial Chemistry Department, University of Genoa, Via Dodecaneso 31, 16146 Genova, Italy.

[4] CNR – ICMATE, Corso Promessi Sposi 29, 23900 Lecco, Italy.

* Corresponding author: romakav@lp.edu.ua

combustion, chemical reactions, and nuclear decay) into electricity using thermoelectric (TE) generators. Such scalability is a key advantage of TE generators. The effectiveness, reliability and potential of thermoelectric generators were proven along the recent 50 years in extreme terrestrial and extraterrestrial conditions (deep space probes such as Voyager). High-efficiency thermoelectric materials for power-generation devices can be used also in refrigeration devices. Peltier coolers provide precise thermal management for such solid-state devices like high-clocked CPUs, image sensors in professional digital cameras, optoelectronics and even for passenger seat cooling in automobiles. The process of direct conversion of heat energy into electrical energy is based on the Seebeck effect—the appearance of an electrical current caused by a temperature gradient in a material. The inverse phenomenon is known as the Peltier effect (Rowe 1995, 1999, 2006; Anatychuk 1998). The potential of a material for thermoelectric applications can be evaluated by the adimensional figure of merit, ZT, a parameter strictly related to material properties calculated by the relation:

$$ZT = (S^2 \sigma \kappa^{-1})T \tag{1}$$

where S is Seebeck coefficient, σ – electrical conductivity, κ – thermal conductivity, T – temperature.

According to Eq. 1 there is not an upper limit to the value of ZT that could be achieved in hypothetical material with high values of Seebeck coefficient and electrical conductivity and low thermal conductivity at a certain temperature. However, in a real thermoelectric material, the three properties are strongly correlated and it is impossible to reach such a ratio between these physical parameters.

In fact, all these parameters are defined by the electronic structure of material and scattering of charge carriers. Maximizing ZT of a thermoelectric material involves a compromise of thermal conductivity, electrical resistivity and Seebeck coefficient. Good thermoelectric materials are typically heavily doped semiconductors with a carrier concentration $n = 10^{19} \div 10^{20}$ carriers cm^{-3}. The term $S^2\sigma$ in Eq. 1 is called the power factor (PF) and it is usually optimized in narrow-gap semiconductors as a function of carrier concentration. A large PF means that a large voltage and a high current are generated. According to this consideration in many technological applications the most important parameter is PF which is directly related to the power produced by the material. The thermal conductivity (κ) consists of two terms—κ_L and κ_{el}:

$$\kappa = \kappa_L + \kappa_{el} \tag{2}$$

where κ_L is the lattice thermal conductivity and κ_{el}—the one due to the charge carriers. For semiconductors with $n \leq 10^{19}$ carriers/cm^3 dominates the lattice contribution into thermal conductivity, while in heavily doped semiconductors with Fermi level deep in conduction or valence bands and metals the thermal conductivity of charge carrier takes the lead (Sootsman 2009; Tritt 2006).

The importance of the ZT parameter is related to the possibility to describe the performances of TE devices in terms of the material's characteristic. In addition to ZT, the efficiency of a TE system depends on the operating conditions, meaning the effective temperature gradient applied, ΔT. The thermoelectric efficiency for power generation is limited by Carnot efficiency ($\Delta T / T_{hot}$) and adimensional figure of merit (ZT):

$$\eta = \frac{\Delta T \sqrt{1 + ZT_{avg}} - 1}{T_{hot} \sqrt{1 + ZT_{avg}} + \frac{T_{cold}}{T_{hot}}} \tag{3}$$

Where T_{hot} and T_{cold} are the temperature of the hot and cold ends in a thermoelectric device, ΔT is their difference, and T_{avg} the average temperature. According to Eq. 3, the larger the temperature gradient and higher the ZT values, the higher is the efficiency of the thermoelectric device (Sootsman 2009; Anatychuk 1998).

Ways of Improving ZT

According to Eq. 1, there are two main ways to improve ZT. The first one is based on increasing the power factor, and the second one—decreasing thermal conductivity (Sootsman 2009; Anatychuk 1998Anat).

The power factor depends only on two parameters, the Seebeck coefficient and the electrical conductivity. However, in materials it is almost impossible to maximize both parameters simultaneously as they change in opposite directions with the carrier concentration (Fig. 1). The aim of improving PF is keeping the high values of Seebeck coefficient while increasing the electrical conductivity. These electrical transport properties are generally described in the framework of Boltzmann transport theory with Seebeck coefficient expressed by Mott equation:

$$S = \frac{\pi^2 k^2 T}{3e} \cdot \frac{d \ln \sigma(E)}{dE} \bigg|_{E = E_F} \tag{4}$$

where $\sigma(E)$ is electronic conductivity. This relation is valid in most of the cases, whether the conduction is through band states, localized states, hopping or other mechanisms. From Mott's relation it is clear that any mechanism enhancing the energy-dependence of the conductivity, $d\sigma(E)/dE$, will increase the Seebeck coefficient. The $\sigma(E)$ in the case of band conduction can be written as:

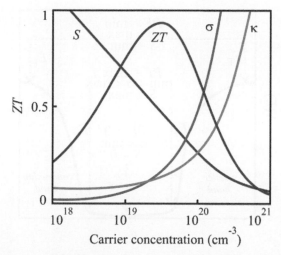

Fig. 1: Carrier concentration dependence of thermoelectric figure of merit (ZT) affected by Seebeck coefficient (S), electrical conductivity (σ), and thermal conductivity (κ).

$$\sigma(E) = \mu(E)en(E) \tag{5}$$

where $\mu(E)$ is the carrier mobility, e is the elementary charge, and $n(E)$ the carrier density, a function of the density of state distribution $N(E)$. If electronic scattering does not depend on energy, then $\sigma(E)$ is just proportional to the density of states at Fermi level.

As a consequence, to achieve maximum value of Seebeck coefficient the Fermi level should be at the inflection point on the $N(E)$ plot.

This means that materials with narrow bands near to the Fermi energy display a high Seebeck coefficient. However, at the same time, both the number and the mobility of the carriers are significantly affected by this factor improving with the broadening of the density of states distribution at the Fermi level.

According to Fig. 2 and Eq. 4 if the material is intrinsic (pure) or fully compensated semiconductor with Fermi level at the middle of the energy gap, then its Seebeck coefficient and power factor are equal to zero. To improve the value of Seebeck coefficient the Fermi level should be shifted towards the conduction (for n-type material) or valence (for p-type material) band. The easiest way to do that is to dope the materials with donor or acceptor impurities, respectively, to create substitutional solid solution. The key problem is how much of dopant is required to reach the maximum S value. If the dopant concentration is too high, then the Fermi level shifts too deeply into one of the bands and material shows the properties of a typical metal (very low Seebeck coefficient and high electrical conductivity).

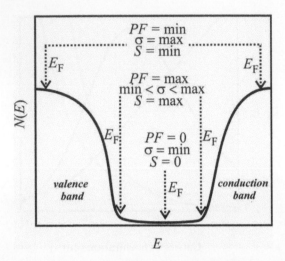

Fig. 2: Hypothetical distribution of the density of states with different cases of Fermi level position.

The doping of the n- or p-type material with dopants of the same type (sign) could increase the PF through increasing of electrical conductivity. However, if such material is doped with a dopant of opposite type, then the Seebeck coefficient will pass through zero and change the sign, and its absolute values will be lower in comparison to initial material.

Other ways for improving Seebeck are related to the activation of quantum effects when the size goes down enough. Looking at Eqs. (4) and (5), it is possible to see that to improve Seebeck there are two main possibilities: working on $dn(E)/dE$ or on $d\mu(E)/dE$. To increase $dn(E)/dE$ means to enhance the dependence of the density of state distribution on energy, target that can be achieved, for example, thanks to the size-quantization effects in the low-dimensional structures (quantum wires, quantum dots,...). In these system the carriers, due to the specific shape of the density of state function, have, as quantum numbers, a momentum k along 3, 2 or 1 axes of the crystal in 3, 2, or 1D respectively, and a set of fixed quantum numbers (1, 2, 3...) along directions in which the motion is constrained (1 direction in 2D systems, 2 directions in 1D systems, all 3 directions in 0D systems). This results in a large increase in the Seebeck coefficient. The consequent improving on the ZT parameter was deduced and discussed in a Hicks and Dresselhaus paper (Hicks and Dresselhaus 1993).

Increasing the energy dependence of the carrier mobility means to increase the scattering time's energy dependence, creating scattering mechanism acting differently on carriers depending on their energy: this is the principle of energy filtering. Several descriptions exist detailing how nanostructures can increase the Seebeck coefficient by filtering electrons according to their energy (Heremans 2005). An example can be the use of potential energy barriers in 2D structures: the bands produced by the barriers in the energy band structure can filter electrons according to their energy.

An additional possibility to change electrical conductivity and Seebeck coefficient is to introduce additional atoms inside the crystal structure (if it is possible) to create impurity zones (levels) inside the band-gap.

One more parameter that affects thermoelectric performance is thermal conductivity, which, however cannot be established directly from the electronic structure. Disorder of the crystal structure caused by formation of substitutional and interstitial solid solutions, nanostructuring of material, formation of precipitates (nanoinclusions) of other phases lead to higher phonon scattering in the material and thus, lower thermal conductivity. Among numerous thermoelectric materials the half-Heusler phases exhibit promising thermoelectric performance with $ZT > 1$. Their simple crystal structure, ability to obtain n- and p-type materials, large number of dopands, absence of precious components and high melting points makes them good candidates for high-temperature thermoelectric materials.

Application of Density Functional Theory to Half-Heusler Phases

The electronic structure study of half-Heusler phases within the Density Functional Theory (DFT) could be divided into three parts:

The first part consists of chemical bonding evaluation using semi-empirical extended Hückel tight-binding method (EHTB). Deeper understanding and sometimes misunderstanding of the chemical bonding in half-Heusler phases was brought by analysis of the charge density and electron localization function (ELF). In general the interpretation of the chemical bonding is made via the analysis of the distribution of the density of states (DOS) and defining the overlapping states (s, p, d, f) of each element.

The second part contains thermodynamic calculations using heat of formation (at $T = 0$ K) obtained from the total energy calculations and temperature dependences of the entropy, free energy and enthalpy from phonon calculations in quasiharmonic approximation. These methods are widely used to predict the formation of solid solutions, binodal and spinodal curves, and help to evaluate the phase diagram of the corresponding system by providing the enthalpy data for calculation of phase diagram (CALPHAD).

The third part includes physical properties calculation (mainly electrical resistivity and Seebeck coefficient) using the linear response techniques in Korringa-Kohn-Rostocker (KKR) method with coherent potential approximation (CPA) and analysis of DOS spectrum.

This classification is not strict and the most feedback from the theoretical methods is obtained when using all of them in a complex investigation of new material. The most widely methods used in DFT calculations are the KKR methods, especially with CPA, Tight-Binding Linear Muffin-Tin Orbitals (TB-LMTO) methods and a group of plane-wave methods with their pseudopotential and full potential realizations.

Structure and Representatives of Half-Heusler Phases

The half-Heusler phases belong to MgAgAs structure type (sometimes with reference to LiAlSi as prototype), an ordered derivative of the CaF_2 type, and crystallize in cubic $F\text{-}43m$ space group. Among almost 200 known ternary half-Heusler compounds only one of them is formed without metallic component—SiNC (Kawamura 1965). The rest of ternary compounds contain from two to three metals and their general composition could be presented as M'MX, where M' occupies $4a$ (0, 0, 0) crystallographic site, M – $4c$ (1/4, 1/4, 1/4), and X – $4b$ (1/2, 1/2, 1/2) as presented in Fig. 3.

Filling the vacancies in $4d$ position transforms the half-Heusler phase into full-Heusler (often referred to as Heusler) phase. However for the DFT calculations with KKR method the model with an origin shift is usually used: M' in $4d$ (3/4, 3/4, 3/4), M in $4a$ (0, 0, 0), and X in $4c$ (1/4, 1/4, 1/4) in order to introduce an empty Muffin-Tin (MT) sphere (vacancy) in $4b$ site and fill all the empty sites (voids) in the crystal structure and make it closed-packed.

The distribution of the constituent elements of the half-Heusler phases in the periodic table (Fig. 4) shows that X-elements are concentrated mainly in IIIA-VIA groups around the Zint line that congenitally separates metals from nonmetals. In the compounds with In and Ga, these elements play mainly the role of X-element, except several compounds (e.g., LiGaSi (Nowotny 1960)) where they represent the M-element.

To the M'-elements belong a part of alkali and alkaline earth metals, the major part of the rare earth metals, d-elements of the IVA (Ti, Zr, Hf), VA (V, Nb, Ta), and VIA (Cr) groups, some actinides (Th and U). The following elements: Mn, Zn, Cd, Mg, Ce, Nd, Sm, and Yb could play role of both M' or M-element. To the M-elements partially belong metals of

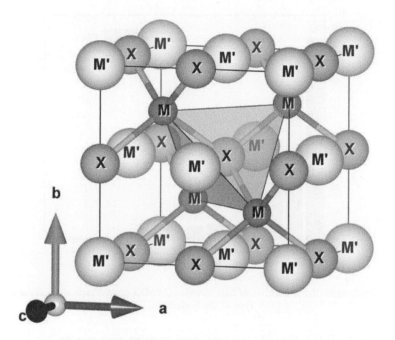

Fig. 3: The crystal structure model of the half-Heusler compounds with general composition M'MX.

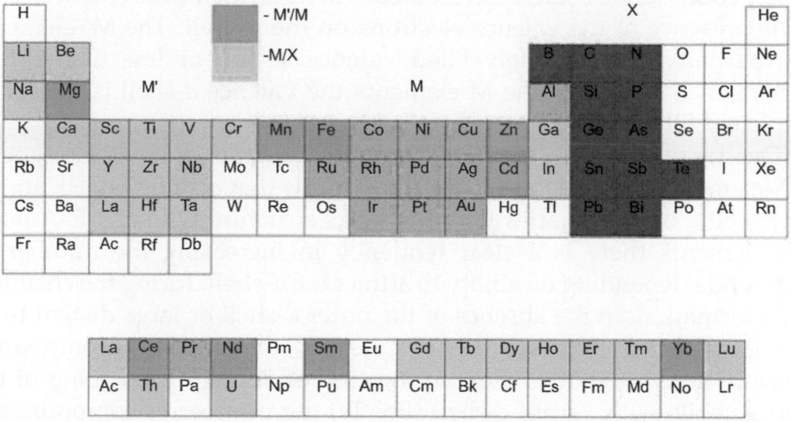

H																	He
Li	Be											B	C	N	O	F	Ne
Na	Mg		M'					M				Al	Si	P	S	Cl	Ar
K	Ca	Sc	Ti	V	Cr	Mn	Fe	Co	Ni	Cu	Zn	Ga	Ge	As	Se	Br	Kr
Rb	Sr	Y	Zr	Nb	Mo	Tc	Ru	Rh	Pd	Ag	Cd	In	Sn	Sb	Te	I	Xe
Cs	Ba	La	Hf	Ta	W	Re	Os	Ir	Pt	Au	Hg	Tl	Pb	Bi	Po	At	Rn
Fr	Ra	Ac	Rf	Db													

- M'/M
- M/X
X

La	Ce	Pr	Nd	Pm	Sm	Eu	Gd	Tb	Dy	Ho	Er	Tm	Yb	Lu
Ac	Th	Pa	U	Np	Pu	Am	Cm	Bk	Cf	Es	Fm	Md	No	Lr

Fig. 4: Distribution of the constituent elements of the half-Heusler phases with general formula M'MX in the periodic table.

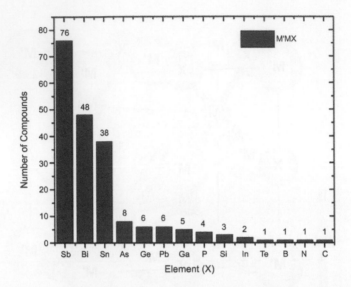

Fig. 5: Distribution of the number of half-Heusler (M'MX) compounds depending on the type of the X-element.

the VII-VIII, IB-IIB, IIA-IIIA groups, and cerium group with Yb. Despite it is impossible to definitely divide the constituent elements of the half-Heusler phases into groups; however several tendencies in the formation of these compounds could be evaluated. The X-elements are characterized by the presence of the valence electrons on the *p*-shell. The M'-elements have partially of completely filled valence *s*-shell or less than a half filled valence *d*-shell. In the M-elements the valence *d*-shell is almost or completely filled.

The distribution of the M'MX compounds depending on the type of the X-element is not balanced. The compounds that contain Sn, Sb, and Bi cover > 80% of all ternaries with MgAgAs structure type Fig. 5. Among all X-elements there is a clear tendency in increasing the number of compounds depending on ability to attract the *d*-shell during the chemical bonds formation. In the absence of the outer *d*-shell or large distant from the valence *s*- and *p*-shells (B, Si, Ge, P, As, Ga), the number of compounds is very poor. On the other side, in the case of strong overlapping of the valence shells with vacant *d*-shell (Pb, Te) the number of compounds is also rather low. The highest number of half-Heusler compounds forms with elements that exhibit both metallic and semi-metallic properties (Sn, Sb and Bi).

The ternary half-Heusler phases M'MSn and M'MSb are characterized by a systematically higher values of electronegativity (Pauling scale) of M and X components in comparison to M' (Fig. 6).

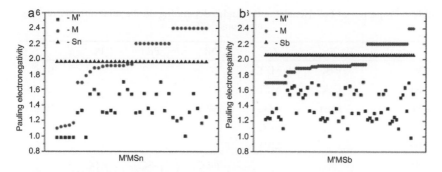

Fig. 6: The distribution of electronegativity (Pauling scale) of the constituent elements in M'MSn (a) and M'MSb (b) half-Heusler compounds.

Table 1: Properties of half-Heusler stannides depending on the total number of valence electrons (VEC).

	Ni	Pd	Pt	Co	Rh	Ir	Fe	
Ti	n-SC		p-SC	MET	MET	MET	MET	18 el.
Zr	n-SC	p-SC	p-SC			MET		17 el.
Hf	n-SC	p-SC	p-SC			MET		16 el.
Nb				n-SC	MET			20 el.
Mn			MET			MET		

Such a difference in electronegativity allows us to describe these compounds as polar intermetallics where the covalently bonded M and X elements form the cationic part and M' the anionic one.

The Zintl concept can be used for a simple prediction of properties of half-Heusler compounds (the total number of valence electrons, VEC). For values VEC = 8 (or 18)/f.u. the Fermi level will be in the bandgap and the compound will have semiconductor properties. When the VEC values deviate from 8 (18), a compound will behave like a typical metal, because the Fermi level will be in the valence (VEC < 8 (18)) or in the conduction (VEC > 8 (18)) band. Thus, for the practical application of the half-Heusler compounds, as a basis for high-performance thermoelectric materials, the compounds with the total number of valence electrons 8 (18) have the highest interest.

Despite numerous half-Heusler compounds with Sn, the physical properties were investigated only for the small part of them (Table 1) (Skolozdra 1993; Dudkin et al. 1993; Kuentzler et al. 1992; Aliev et al. 1989; Ahilan et al. 2004). It is easy to find that the compounds with VEC < > 18 exhibit metallic properties, while all stannides with VEC = 18 are semiconductors, except NbRhSn, which is characterized by the

Table 2: Properties of half-Heusler antimonides depending on the total number of valence electrons (VEC).

	Fe	Co	Ni	Pd	Pt	
V	n-SC	MET				18 el.
Nb	n-SC	MET				17 el.
Ti	MET	n-SC	MET			19 el.
Zr		n-SC				
Hf		MET				
Sc			p-SC	p-SC	p-SC	
Y			p-SC	p-SC	p-SC	
Tb			SC			
Dy			SC			
Ho			SC			
Gd					SC	
Mg			MET			

metallic type of conductivity and low (-10 μVK^{-1}) Seebeck coefficient at 300 K (Skolozdra 1993). Stannides with Ti, Zr, Hf, and Ni are n-type semiconductors, but substitution Ni for Pd and Pt changes the conductivity to p-type, which is strange, as Ni, Pd, and Pt have the same number of valence electrons. A similar situation is observed in half-Heusler antimonides (Table 2) (Kaczmarska et al. 1998; Dudkin et al. 1993; Young et al. 2000; Melnyk et al. 2000; Caruso et al. 2003; Terada et al. 1970, 1972; Evers et al. 1997; Morozkin and Nikiforov 2005; Xia et al. 2000; Stadnyk et al. 2001; Sekimoto et al. 2005; Kleinke 1998; Oestreich et al. 2003; Karla et al. 1998a,b, 1999; Harmening et al. 2009; Dhar et al. 1993; Hartjes and Jeitschko 1995; Suzuki et al. 1993). The NbFeSb compound appeared to be the n-type semiconductor with Seebeck coefficient -85 μVK^{-1} (280 K) (Melnyk et al. 2000) in one case and fully compensated semiconductor with almost zero value of Seebeck coefficient in another (Young et al. 2000). The deviation from the equiatomic composition was also observed in TiNiSb (VEC = 19 el./f.u.) (Terada et al. 1972), ScNiSb and ScPdSb (VEC = 18 el./f.u.) (Morozkin and Nikiforov 2005; Harmening et al. 2009). The absence of ZrNiSb and HfNiSb half-Heusler compounds with the same VEC as TiNiSb confirms the limitations of the Zintl concept in predicting the new compounds with MgAgAs structure type.

In contrast to TiCoSb and ZrCoSb, which exhibit semiconductor behavior (Xia et al. 2000; Stadnyk et al. 2001; Kleinke 1998; Melnyk et al. 2000), the HfCoSb antimonide is characterized by the metallic type of conductivity (Melnyk et al. 2000; Xia et al. 2000), and such a difference also could not be explained by the Zint concept. Similar differences were also observed in isovalent compounds VCoSb (Curie-Weiss paramagnet above 80 K) (Terada et al. 1970, 1972) and NbCoSb (Pauli paramagnet) (Kaczmarska et al. 1998). Such compounds as GdNiSb, TbNiSb, DyNiSb, HoNiSb, ErNiSb, and TmNiSb are Curie-Weiss paramagnets, while LuNiSb – Pauli paramagnet (Hartjes and Jeitschko 1995). The experimental values of the corresponding magnetic moments are in a good agreement with the theoretical one's for the R^{3+} ions, proving that Ni atoms are in nonmagnetic state. These facts prove the existence of structural defects of donor/acceptor nature in almost all "pure" ternary half-Heusler compounds. These defects are usually not taken into account during the electronic structure calculations and thus affect predicted properties of half-Heusler materials.

Electronic Structure of Thermoelectric Half-Heusler Phases

Evaluation of the Chemical Bonding and Band Gap Formation in Half-Heusler Phases

The electronic structure calculations of half-Heusler compounds at the early stage of their study were done mainly to find some property or physical parameter that is responsible for the phase formation/stability and band gap appearance. The band structure of thermoelectric ZrNiSn compound (Fig. 7) shows the presence of the energy gap at the Fermi level.

In the paper by Ögut and Rabe (Ögut and Rabe 1995) the authors using the pseudopotential method with a plane-wave basis set (GGA) concluded that the stability of ZrNiSn half-Heusler phase is based on the stability of ZrSn rocksalt substructure, while the formation of the band gap induced by the addition of the symmetry-breaking Ni sublattice has a minor contribution. The authors also showed that only one atomic configuration within the MgAgAs structure type has the lowest total energy and exhibits energy gap at Fermi level, while two others doesn't. Similar conclusions were made in (Ishida et al. 1997) for related systems M-Ni-X (M = Sc, Ti, Zr, Hf; X = Sn, Sb) by analyzing the total energy and band structure using the Local Spin Density (LSD) approximation and the LMTO method with Atomic Sphere Approximation (ASA). More precise DFT calculations using Full-Potential Lincrized Auguumented Plane Wave (FP-LAPW) method with Generalized Gradient Approximation (GGA) performed by Larson et al. (Larson et al. 1999) for Ni-containing half-Heusler phases revealed that the *f*-states of the rare earth atoms in RNiSb (R = Y, La, Lu) compounds lie far away from the Fermi level and the compounds are narrow-band semiconductors. The origin of the gap is a

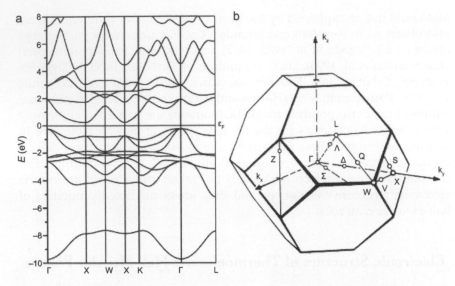

Fig. 7: Band structure of ZrNiSn (a) along the main symmetry points in Brillouin zone (b) calculated with FP-LMTO method (Savrasov and Savrasov 1992).

result of the interaction between Ni and rare earths d-states and somewhat differs from the origin of the gap formation in ZrNiSn (Ögut and Rabe 1995) where Sn p-states and Zr d-states also played a crucial role. Using the KKR method with Local Density Approximation (LDA) the authors (Tobola and Pierre 2000) came to the same conclusion analyzing M'MX (M = Fe, Co, Ni; M' = Ti, Zr, V, Nb, Mn; X = Sn, Sb) and noticed that the d-states of M component lie below Fermi level and have dominant bonding character, while the d-states of M' component lie above Fermi level and exhibit antibonding character. The authors also mentioned that half-Heusler compounds are stable only when the Fermi level falls rather close to the band gap (17 ≤ VEC ≤ 19) and with VEC = 20 the Fermi level enters the antibonding region, decreasing the stability of the lattice. However, the chemical bonding in half-Heusler phases was evaluated not solely by *ab initio* methods. Semi-empirical methods like extended Huckel tigh-binding (EHTB) appeared to be very useful in chemical bonding disclosure. In Jung et al. (Jung et al. 2000), the authors analysed the bonding in VCoSb (VEC = 19) compound by means of EHTB method utilizing the total and projected DOS and comparing it with appropriate crystal orbital overlap population (COOP) for atom pairs. It was also mentioned that for the formation of the band-gap the interaction between the two transition metal atoms (M and M') in M'MX half-Heusler phases with difference in electronegativity is essential. In simplified approximation, the electronic structure of M'MX compounds with VEC > = 18 is described in terms of

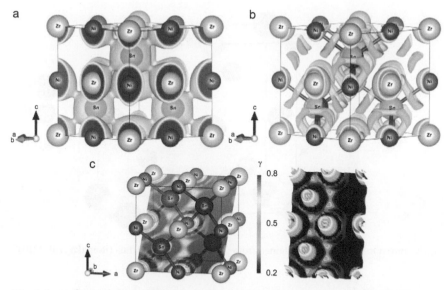

Fig. 8: Isosurfaces of charge density at 0.053 e/(10³ nm³) (a), electron localization function (ELF) at Υ = 0,40 (b) and distribution of ELF in Zr-Ni-Sn lattice plane (c) in ZrNiSn.

d^{10} state for M, s^2p^6 state for X and d^n state for M'. Even in the case of VEC = 17 the M component has electronic configuration close to d^{10}. Wide utilization of the electron localization function (ELF) and crystal orbital Hamiltonian population (COHP) for chemical bonding evaluation and interpretation (Kandpal 2006; Nanda and Dasgupta 2003; Offernes et al. 2007) using TB-LMTO-ASA and FP-LMTO methods finally revealed strong covalent bonding between M and X components. The charge density between M and M' (e.g., ZrNiSn, TiNiSn) appeared to be shifted to the more electronegative M atom (Romaka et al. 2010a). The calculation of the ELF distribution using FP-LAPW (GGA) method for HfNiSn (Romaka et al. 2014a) and ZrCoSb (Romaka et al. 2014b) showed its strong localization between Ni-Sn and Co-Sb atoms. A typical charge density and ELF distribution in ZrNiSn are presented in Fig. 8.

The electron density localization between Ni and Sn atoms was confirmed by X-ray diffraction (XRD) study of single crystals and polycrystalline samples (Romaka et al. 2009a, 2010a). It was also shown in (Romaka et al. 2010a) that the electron density around Ni atom has not spherical but more tetrahedral like shape with maxima toward the Zr atoms (Fig. 9).

In the work of Skolozdra (Skolozdra 1993) the high energy shift for the $K\alpha_{1,2}$ lines of Ni was observed which confirms the increasing of d-electrons density of Ni atoms.

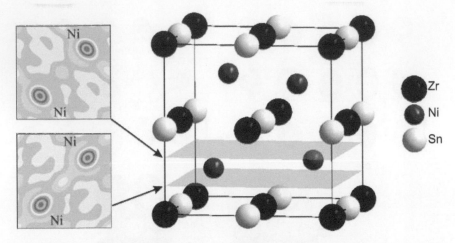

Fig. 9: Fourier maps of ZrNiSn obtained from the analysis of XRD data (Romaka et al. 2010a).

Disorder Effects in Half-Heusler Phases

Despite a very good qualitative explanation of the origin of the band gap and chemical bonding in half-Heusler phases with different theoretical methods, the quantitative characteristics like energy gap and activation energy values and the presence of impurity bands that define type of conductivity were far from being experimentally observed. First disorder effects in ZrNiSn were found in (Aliev et al. 1989) where the authors assumed that Ni atoms ($4c$ site) and vacancies ($4d$ site) occupy their usual positions, but Zr ($4a$) and Sn ($4b$) atoms substitute each other giving the general formula $Zr_{1-x}Sn_xNi(Vac)Sn_{1-x}Zr_x$ ($0.1 < x < 0.3$). Using this model (Ögut and Rabe 1995) and (Larson et al. 2000) performed electronic structure calculations and found that such a disorder could take place. Unfortunately this model was not able to explain the excess of electrons in n-ZrNiSn and related n-HfNiSn and n-TiNiSn compounds (Hohl et al. 1999; Aliev 1989, 1990, 1991, 1994; Käfer et al. 1997) because the chemical composition of compound remained the same and so did the number of electrons. Significant deviations from the equiatomic compositions were also observed in thin films of n-HfNiSn ($Hf_{29(1)}Ni_{34}Sn_{37(2)}$) (Wang et al. 2010). The authors (Hazama et al. 2010) explained the deviation in composition of TiNiSn with partial filling of vacancies in $4d$ site. In works (Uher et al. 1997, 1999; Skolozdra 1993) the authors showed that without the annealing the compounds are characterized by the highest disorder due to the lowest lattice thermal conductivity. From the other side, the disorder initiates the appearance of impurity states in the energy gap that significantly affect (decrease) the resistivity and Seebeck coefficient. In

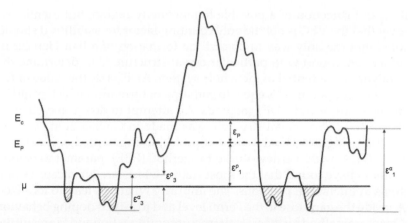

Fig. 10: Energy diagram of compensated semiconductor (Shklovskii and Efros 1984). E_c—energy of conduction band, E_p—energy of percolation level, μ—Fermi energy, ε—activation energy.

general the approach of (Ögut and Rabe 1995) and (Larson et al. 2000) is counterproductive because it is practically impossible to precisely determine the value of the energy gap of semiconductor with impurity concentration $\sim 10^{20}$ cm^{-3} (Shklovskii and Efros 1984). The authors showed that taking into account the influence of electrostatic interaction between differently charged complexes that are randomly distributed significantly affects the band structure of semiconductor (Fig. 10) and is accompanied by fluctuation of potential relief and modulation of continuous energy bands.

In such a case, the approaches for description of traditional semiconductors and heavily-doped semiconductors have fundamental differences. For the heavily-doped semiconductors, the electron is considered to move not in the periodic field of the crystal, but in the chaotic field of impurities with potential energy that could not be neglected. At low temperatures the doped crystalline semiconductor is a disordered system that reminds amorphous. The higher is the doping of such a system, the higher is the temperature to exhibit these properties.

Another attempt to predict the deviation in stoichiometry of half-Heusler phases was made by (Offernes et al. 2008). The authors showed that VEC is an important parameter for composition considerations on half-Heusler phases. The optimum VEC found by combining information from DOS and COHP profiles for the equiatomic half-Heusler phase is used to predict its composition or possible homogeneity region. The authors assumed that the knowledge of the constituents of the half-Heusler phase is enough to predict whether the composition of the stable phase is likely to be shifted away from the equiatomic stoichiometry or not and also the position and direction of a possible homogeneity region, but mentioned however that the VEC is not the only counting factor for stability. It should be noted that the only way to predict the formation of a half-Heusler or any other compound with particular crysal structure is to determine the thermodynamic potential of the whole system. At T = 0 K the value of the thermodynamic potential is equal to enthalpy of formation which could be calculated using *ab initio* DFT methods. An attempt to determine the type of defects in TiNiSn crystal structure was made in Colinet et al. (Colinet et al. 2014) using the pseudopotential projector augmented wave (PAW) method (GGA with Perdew-Burke-Ernzerhof (PBE) parameterization). The authors have found that the most stable defects are Ni vacancies and Ni atoms in interstitial positions. The impurity Ni atoms lead to localized states inside the gap close to the Fermi level and to *n*-type doping behavior.

One more method for the crystal structure optimization and as a result the composition is based on the comparison of the experimentally determined activation energies, type of conductivity, Seebeck coefficient, magnetic susceptibility with parameters that could be extracted from the DOS profiles (DOS at Fermi level, activation energy from Fermi level to percolation, Seebeck coefficient, etc.). The aim of the method is to get the calculated properties that fit the best with experimental by tuning the crystal structure of half-Heusler compound. For the optimization the KKR-CPA-LDA (Akai 1989) method is used with different types of parameterization. The benefit of the method is coherent potential approximation that is used for alloys with random distribution of components. It allows calculating the electronic structure of material with very low (0.1% and even less) concentration of impurities. The application of method will be demonstrated by several examples.

Crystal Structure Optimization of ZrNiSn Compound

According to Romaka et al. (Romaka et al. 2010a) the crystal structure of ZrNiSn is disordered—the position of Zr atoms (4*a*) is occupied by the statistical mixture of ~ 90% Zr and ~ 10% Ni, and two other sites are fully occupied by Ni and Sn atoms, which was partially confirmed by the energy dispersive X-ray (EDX) data of single crystal. The resistivity and Seebeck coefficient measurements showed that ZrNiSn is an *n*-type semiconductor with low- and high-temperature activation regions. The calculated value of the activation energy from Fermi level to the conduction band (ε_1^ρ) equals to 97.6 meV. The value of the magnetic susceptibility χ(300 K) = –0.07 × 10^{-6} emu/g indicates that ZrNiSn is a weak diamagnet with a Fermi level inside the band gap. These experimental results confirm that the structure of ZrNiSn is disordered.

To check the results of crystal structure refinement and to find a structural model that fits the best to the investigated physical properties the models illustrated in Fig. 11 were used. The electronic structure was calculated using KKR-CPA-LDA method. The energy gap for the ordered ZrNiSn ε_g = 514 meV and the Fermi level (ε_F) appears to lie slightly inside the conduction band (Fig. 12a). This result confirms the *n*-type of conductivity of semiconductor, but the absence of activation energy (ε_1) from ε_F to conduction band indicates the wrong model of the ordered crystal structure.

For the disordered model of the crystal structure that is described by the following formula $(Zr_{0.99}Ni_{0.01})NiSn$, ε_g = 306 meV and ε_F lies inside the gap near the conduction band (Fig. 12b). Such position of the Fermi level confirms the n-type of conductivity and the existence of the activation energy ε_1. The most valuable argument for this model is the results of electrical properties investigation of the $ZrNiSn_{0.98}In_{0.02}$ alloy (Romaka et al. 2007a, 2010a) which appeared to be fully compensated semiconductor with ε_1 = 180 meV. In the heavily doped and fully compensated semiconductor, the Fermi level lies exactly at the middle of the band gap as in intrinsic semiconductor and thus in $ZrNiSn_{0.98}In_{0.02}$, the band gap is doubled ε_1 activation energy and ε_g = 360 meV, which is close to the theoretical value for $(Zr_{0.99}Ni_{0.01})NiSn$.

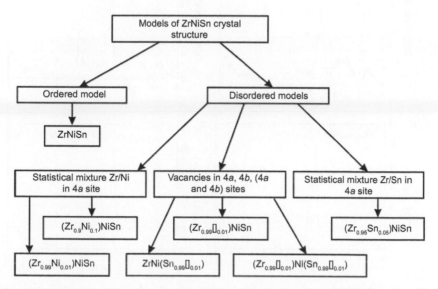

Fig. 11: Ordered and disordered models of ZrNiSn crystal structure used in DFT calculations

In the case of the $(Zr_{0.9}Ni_{0.1})NiSn$ model, the energy gap is absent and calculations predict the metallic type of conductivity (Fig. 12c). For the $(Zr_{0.95}Sn_{0.05})NiSn$ model the $\varepsilon_g = 282$ meV, but the Fermi level like in ZrNiSn is inside the conduction band (Fig. 12d). The calculations for the models with vacancies in $4a$, $4b$, and simultaneously in $4a$ and $4b$ sites also do not fit the experimental results. The vacancies in Sn ($4b$) site are equal to donor impurities in the semiconductor and that is why the Fermi level shifts deeper into the conduction band. Similar theoretical and experimental results were obtained for n-HfNiSn and n-TiNiSn semiconductors (Romaka et al. 2010a, 2011, 2013a).

Crystal Structure Optimization of LuNiSb Compound

The half-Heusler LuNiSb antimonide is characterized by some deviation from the stoichiometry 1:1:1 and corresponds to the composition $Lu_{34.68}Ni_{30.57}Sb_{34.75}$ with lower Ni content (electron probe micro-analysis data). The sample at composition $Lu_{34}Ni_{33}Sb_{33}$ is located in the two-phase

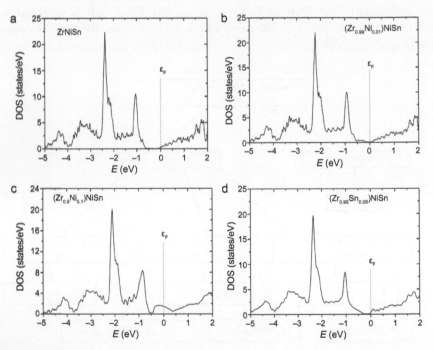

Fig. 12: (a–d) Distribution of the density of states in ordered and disordered models of the ZrNiSn crystal structure.

region and contains the main phase LuNiSb in equilibrium with the LuNi$_5$ binary. The crystal structure refinement of LuNiSb indicated a presence of the statistical mixture of Ni and Lu in Ni (4c) crystallographic site. More precise refinement with fixed (according to the EPMA) Lu content in 4c site also revealed the presence of vacancies, and thus the statistical mixture in 4c position consists of 92.8% Ni + 1.4% Lu + 5.8% Vac.

Temperature dependences of electrical resistivity $\ln\rho(1/T)$ and differential thermopower $\alpha(1/T)$ are shown in Fig. 13 and are typical for doped and compensated p-type semiconductors with high- and low-temperature activation regions. At $T > 100$ K the activation of halls from Fermi level to the valence band takes place. The activation energy for p-LuNiSb $\varepsilon_1^\rho = 23.1$ meV. Thus, this result is possible only if the crystal is characterized by structural defects of acceptor nature that was partially evaluated by crystal structure investigation.

To predict energy and kinetic characteristics of LuNiSb the calculation of their electronic structure was carried out using the KKR-CPA-LDA method (Fig. 14). The calculation of the distribution of the density of electronic states for the ordered crystal structure with 100% filling of crystallographic sites in accordance to MgAgAs structure type revealed that the Fermi level is located in the energy gap near the bottom of the conduction band. This result conflicts with the experimental data and confirms the presence of structural defects in LuNiSb. For the refinement of the crystal structure model of LuNiSb, the optimization method based on the results of the electronic structure calculation and physical properties was used. Using the experimental value of the activation energy ε_1^ρ from

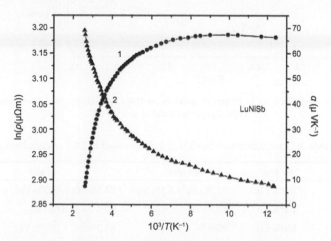

Fig. 13: Temperature dependences of electrical resistivity (1) and Seebeck coefficient (2) in LuNiSb compound.

Fermi level to the percolation level of the valence band, several models (Table 3) of the crystal structure were simulated.

The most appropriate and the best option is the arrangement of atoms, which involves filling the following crystallographic position in the LuNiSb (Model 5): Ni (4c) = 92.65% Ni + 1.35% Lu + 6% Vac; Lu (4a) = 100% Lu; Sb (4b) = 100% Sb. The obtained results confirm the defects nature in RNiSb compounds and show advantages of the optimization method in the crystal structure evaluation. The calculated activation energy from Fermi level to the percolation level of the valence band is 28(7) meV which is close to the experimental value (ε_1^ρ = 23.1 meV). Thus, the composition of the LuNiSb obtained from the DFT optimization

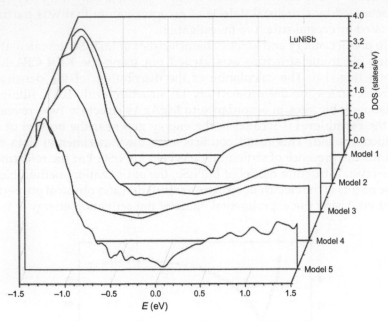

Fig. 14: DOS distribution in different models of the crystallographic sites occupancy in LuNiSb (Fermi level at E = 0 eV).

Table 3: Crystal structure models of LuNiSb used in DFT calculations.

LuNiSb	4*a* site	4*b* site	4*c* site
Model 1	100% Lu	91.7% Sb + 8.3% Vac	83.3% Ni + 16.7% Vac
Model 2	100% Lu	100% Sb	90% Ni + 10% Lu
Model 3	100% Lu	95% Sb + 5% Lu	83.3% Ni + 16.7% Vac
Model 4	100% Lu	90% Sb + 10% Lu	90% Ni + 10% Sb
Model 5	100% Lu	100% Sb	92.65% Ni + 1.35% Lu + 6% Vac

procedure $Lu_{34.48}Ni_{31.51}Sb_{34.01}$ is very close to those from Ritveld refinement $Lu_{34.46}Ni_{31.54}Sb_{34.00}$ and EPMA data $Lu_{34.68}Ni_{30.57}Sb_{34.75}$.

It is important to note that similar procedure was used for n-ZrCoSb compound, which is characterized by the homogeneity region along the isoconcentrate of antimony (Romaka et al. 2014b,c) and revealed the presense of Zr, Co and vacancies in Co $4c$ crystallographic site. Shifting the composition along the homogeneity region leads to decreasing of Co content and increasing of Zr and number of vacancies in $4c$ site. Both p-LuNiSb and n-ZrCoSb semiconductors have the same mechanism of disorder, which, however, causes different physical properties.

Doping of Half-Heusler Phases

Doping is one of the most widely used methods to improve thermoelectric figure of merit of a half-Heusler materials. Despite numerous experimental results on half-Heusler solid solutions, the theoretical calculations were performed only for some of them. The major part of DFT calculations of solid solutions was performed using the KKR-CPA method. These solid solutions can be divided into several groups depending on the doping site.

The first group includes M'MX compounds doped in M' ($4a$) crystallographic site with rare earth elements (R): $Zr_{1-x}R_xNiSn$, (Romaka et al. 2005, 2009a,b, 2010a,b; Hlil et al. 2010; Stadnyk et al. 2011), $Ti_{1-x}R_xNiSn$ (Romaka et al. 2005, 2010a), and $Hf_{1-x}R_xNiSn$ (Romaka et al. 2010a), with V: $Ti_{1-x}V_xNiSn$ (Romaka et al. 2010a; Stadnyk et al. 2010), $Ti_{1-x}V_xCoSb$ (Romaka et al. 2008a, 2010a), with Nb and Mo: $Zr_{1-x}Nb_xNiSn$ (Romaka et al. 2010a), $Zr_{1-x}Mo_xNiSn$ (Romaka et al. 2010a), with Cu, Mn, and Ti: $Pd_{1-x}Cu_xMnSb$ (Wu et al. 2013), $Zr_{1-x}Ti_xNiSn$ (Wunderlich et al. 2011), $Ti_{1-x}Mn_xCoSb$ (Mena et al. 2015); and with Li: $Sr_{1-x}Li_xAlSi$ (Kunkel et al. 2015).

To the second group belongs solid solutions that are characterized by In, Sb, and Bi doping in X ($4b$) position: $ZrNiSn_{1-x}In_x$ (Romaka et al. 2007b, 2010a), $ZrNiSn_{1-x}Sb_x$ (Romaka et al. 2010a; Kawaharada et al. 2004), $ZrNiSn_{1-x}Bi_x$ (Romaka et al. 2012a), $TiNiSn_{1-x}In_x$ (Romaka et al. 2010a), $HfNiSn_{1-x}Sb_x$ (Romaka et al. 2010a, 2014a), and $LuPdSb_{1-x}Bi_x$ (Nourbakhsh 2013). The third group is characterized by doping of M'MX compounds with d-elements in M ($4c$) crystallographic site: $ZrNi_{1-x}M_xSn$ (M = Cr, Mn, Fe, Co, Cu) (Fruchart et al. 2007; Romaka et al. 2009c, 2010a), $TiNi_{1-x}V_xSn$ (Romaka et al. 2010a), $TiNi_{1-x}Co_xSn$ (Romaka et al. 2009d, 2010a), $HfNi_{1-x}Co_xSn$ (Romaka et al. 2010a, 2012b), $TiCo_{1-x}Ni_xSb$ (Stadnyk et al. 2006; Romaka et al. 2006, 2010a), $VFe_{1-x}Cu_xSb$ (Stadnyk et al. 2005),

HfNi$_{1-x}$Ru$_x$Sn (Romaka et al. 2014d), and HfNi$_{1-x}$Rh$_x$Sn (Romaka et al. 2013b).

And the last group covers interstitial solid solutions ZrNi$_{1+x}$Sn (Romaka et al. 2013a,c), HfNi$_{1+x}$Sn (Romaka et al. 2013a), and TiNi$_{1+x}$Sn (Romaka et al. 2013d) where additional Ni atoms fill the vacancies in the 4d crystallographic site. The first three groups keep the equiatomic composition constant, while the last – shifts the composition toward the M'Ni$_2$Sn Heusler phase.

In M'$_{1-x}$R$_x$NiSn solid solutions the substitution of Ti, Zr or Hf on rare earth atoms shifts the Fermi level from the edge of conduction band (n-type material), through the middle of the gap (compensated material) toward the valence band (p-type material) (Fig. 15).

It is equivalent to the doping of n-type semiconductor with acceptor impurity and allows obtaining n- and p-type materials within one solid solution. The density of states at Fermi level passes through the minimum at the bottom of the band gap (Fig. 16).

As the DOS at Fermi level is proportional to the number of free charge carriers in the semiconductor, it is also connected with electrical conductivity of material. The minimum DOS value at Fermi level corresponds to the maximum resistivity of material and almost zero value of Seebeck coefficient. For Zr$_{1-x}$Y$_x$NiSn the calculated Seebeck coefficient values at different concentrations of x(Y) (Fig. 17) are in a good agreement with experimental and the maximum of resistivity (Fig. 18) corresponds with minimum of DOS at Fermi level (Fig. 15). Complete substitution of Zr doesn't change the band structure of material (Fig. 18c), it is still a semiconductor but with metallic type of conductivity. However the complete substitution of Zr within one solid solution is not possible as YNiSn crystallizes not in the MgAgAs structure type, but in the TiNiSi type (Romaka et al. 2012c) and is characterized by the metallic type of conductivity (Romaka et al. 2008b) and the absence of the band gap (Fig. 18d). Thermodynamic calculations of the enthalpy of formation with configuration entropy (Fig. 19a) showed that maximum solubility of Y in ZrNiSn is ~ 0.3, which correlates with experimental dependence of the lattice parameter (Fig. 19b).

The presence of Y atoms in the crystallographic site of Zr leads to lower ELF values between Y and Ni (Fig. 20).

The initial disordered crystal structure of ZrNiSn completely orders at x(Y) = 0.02. The correctness of the model is also proved by comparing the magnetic susceptibility of solid solution at 300 K with the density of states at Fermi level (Fig. 21).

Practically the same results were obtained for the rest of rare earth metals including magnetic (Romaka 2010b). Similar to the described above solid solutions is Ti$_{1-x}$V$_x$CoSb, but in this case the p-type material

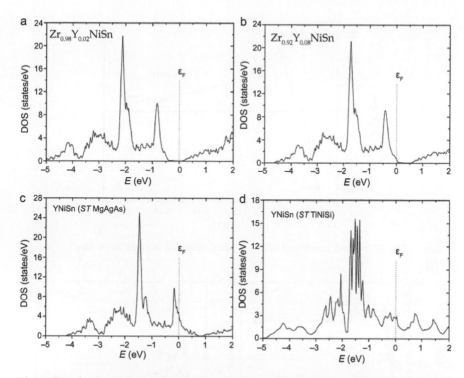

Fig. 15: Distribution of the density of states in $Zr_{0.98}Y_{0.02}NiSn$ (a), $Zr_{0.92}Y_{0.08}NiSn$ (b), hypothetical YNiSn (*ST* MgAgAs) (c), real YNiSn (*ST* TiNiSi) (d).

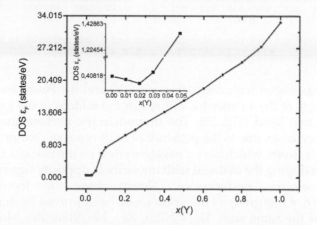

Fig. 16: Concentration dependence of the density of states at Fermi level in $Zr_{1-x}Y_xNiSn$ solid solution.

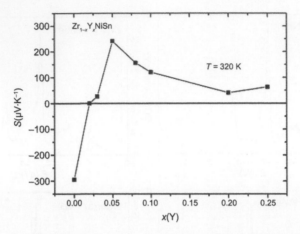

Fig. 17: Seebeck coefficient of $Zr_{1-x}Y_xNiSn$ extracted from DOS profiles.

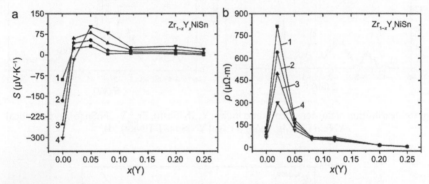

Fig. 18: Concentration dependence of Seebeck coefficient (a) and resistivity (b) at $1 - 1 - T = 80$ K; $2 - T = 160$ K; $3 - T = 250$ K; $4 - T = 370$ K.

(TiCoSb) was doped with donor impurity (V) and the Fermi level shifted from the edge of the valance band through the middle of the gap toward the conduction band (Fig. 22). The thermoelectric performance of such materials decreases due to the presence of both types of charge carriers— holes and electrons, which have a negative effect on the Seebeck coefficient value. Thus doping the material with impurity of opposite signs decreases its thermoelectric performance. Nevertheless, the thermoelectric performance of this group of materials could be improved by doping with impurity of the same sign: $Ti_{1-x}V_xNiSn$, $Zr_{1-x}Nb_xNiSn$, $Zr_{1-x}Mo_xNiSn$; or neutral: $Zr_{1-x}Ti_xNiSn$. In the first case the number of electrons increases and the Fermi level shifts deeper into the conduction band.

At concentration of dopant when the Fermi level just enters the conduction band the density of states and electrical conductivity begin

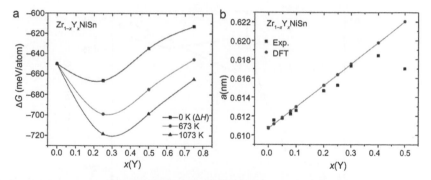

Fig. 19: Concentration dependence of thermodynamic potential (a) and lattice parameter (b) of $Zr_{1-x}Y_xNiSn$ solid solution.

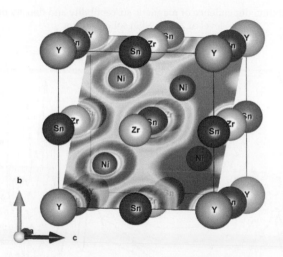

Fig. 20: ELF distribution in $Zr_{0.75}Y_{0.25}NiSn$ alloy due to the acceptor nature of Y in comparison to Zr.

to increase and the Seebeck coefficient value pass through the maximum according to Eq. 4. Such effect positively impacts the value of the power factor. In this case the initial disordered crystal structure of TiNiSn also orders but at $x(V) = 0.005$.

In the case of doping with impurity of neutral sign $Zr_{1-x}Ti_xNiSn$ the Fermi level should stay at the same position (Fig. 23), but according to (Wunderlich et al. 2011) the band gap in this material is absent (pseudopotential PAW method). This result looks strange as according to our calculations with KKR-CPA-LDA method the band gap is present at all concentrations and could be explained by the wrong atomic configuration like it was showed above. The isovalent substitution has practically no

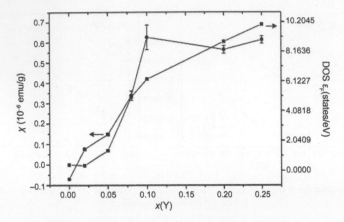

Fig. 21: Concentration dependence of magnetic susceptibility and density of states at Fermi level of $Zr_{1-x}Y_xNiSn$ solid solution.

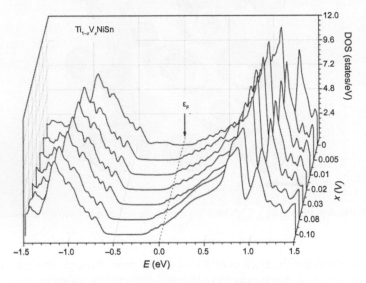

Fig. 22: Distribution of DOS in $Ti_{1-x}V_xNiSn$ solid solution.

influence on the electronic structure and power factor. Improving of ZT values is achieved by decreasing of the lattice thermal conductivity through the disordering of M′ (4a) crystallographic site.

Significant improvement of the power factor could be achieved in solid solutions from the second group except with In. Doping with Sb and Bi has the same effect as in $Ti_{1-x}V_xNiSn$ solid solution. In the case of $ZrNiSn_{1-x}In_x$ and $TiNiSn_{1-x}In_x$ using the crystal structure optimization

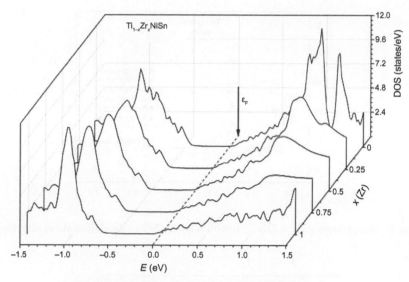

Fig. 23: Distribution of DOS in $Ti_{1-x}Zr_xNiSn$ solid solution.

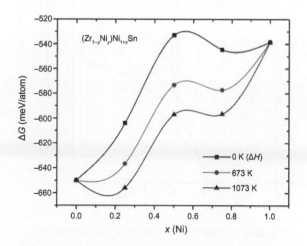

Fig. 24: Concentration dependence of thermodynamic potential of $ZrNi_{1+x}Sn$ solid solution.

method described above the In atoms fill both Sn (4*b*) and Ni (4*c*) atomic positions increasing the disorder of the crystal structure. Solid solutions from the third group like *n*-type $ZrNi_{1-x}Cu_xSn$ and $VFe_{1-x}Cu_xSb$ improve the power factor, because Cu in these cases acts like donor of electrons and the rest behaves like $M'_{1-x}R_xNiSn$.

The fourth group of solid solutions is a subject of great interest as it gives an additional degree of freedom for improving *ZT*. Thermodynamic

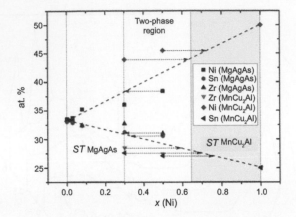

Fig. 25: Phase composition of ZrNi$_{1+x}$Sn solid solution (EPMA data) (Romaka et al. 2013a).

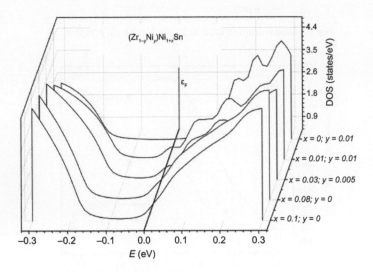

Fig. 26: Distribution of DOS in ZrNi$_{1+x}$Sn solid solution.

calculation of ZrNi$_{1+x}$Sn solid solution (Fig. 24) shows that the solubility is limited and is in a good agreement with EPMA data (Fig. 25).

Complete filling of the vacant 4d site leads to the formation of ZrNi$_2$Sn compound with MnCu$_2$Al structure type, which is a typical metal. Optimization of the crystal structure revealed it's ordering (Fig. 26) that was proved by comparison of theoretical values of activation energy from Fermi level to the conduction band with experimental (Fig. 27).

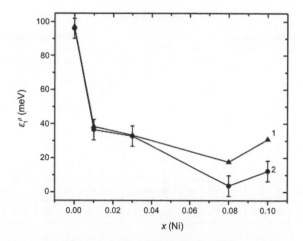

Fig. 27: Concentration dependence of the calculated (2) and (1) experimental values of activation energy ε_1 in $ZrNi_{1+x}Sn$ solid solution.

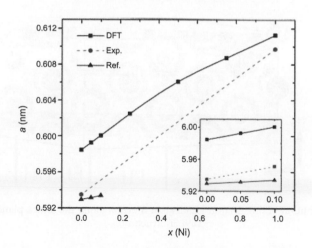

Fig. 28: Concentration dependences of optimized (DFT), experimental and reference (Douglas et al. 2014) values of lattice parameter a of $TiNi_{1+x}Sn$ solid solution.

The density of states at Fermi level is affected by the tails of the conduction band due to the two concurrent types of hybridization of Ni and Zr atoms.

The interstitial solid solution based on half-Heusler TiNiSn phase modeled by reducing the symmetry of the crystal to $P1$ and successively filling of four available voids ($4d$ site) with Ni atoms. The comparison of optimized (FP-LAPW-GGA) and experimental values of lattice parameter (Fig. 28) with a set from Ref. (Douglas et al. 2014) shows that the theoretical

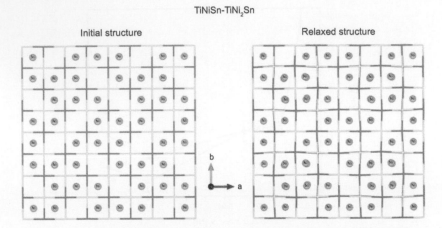

Fig. 29: The initial and relaxed geometry of the hypothetical TiNiSn-TiNi₂Sn structure.

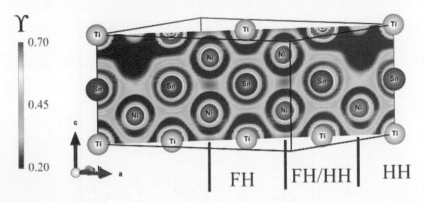

Fig. 30: Distribution of the electron localization function along the lattice plane in TiNiSn-TiNi₂Sn structure.

and interpolated experimental dependencies are almost parallel, while the slope of the reference dependence is more horizontal. Such deviation in slope means that the corresponding to the lattice parameters x(Ni) values in Ref. (Douglas et al. 2014) might be significantly lower. To explain the increasing of the lattice parameter we have modeled a structure that consists of four TiNiSn unit cells where one Ni atom from each of them diffused to form one full-Heusler TiNi₂Sn unit cell. This is a model of the coherent grain boundary between TiNiSn and TiNi₂Sn phases (Fig. 29). The crystal structure geometry optimization for this model shows that the lattice parameters of TiNiSn and TiNi₂Sn unit cells vary in the ranges 0.58693 ÷ 0.58686 and 0.60766 ÷ 0.67062 nm, respectively.

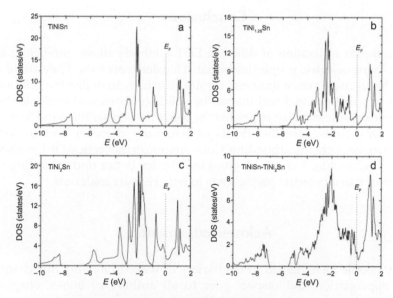

Fig. 31: The distribution of the total density of states for TiNiSn, TiNi$_{1.25}$Sn, TiNi$_2$Sn, and grain boundary TiNiSn-TiNi$_2$Sn.

The projection of the electron localization function (Fig. 30) shows localized maxima between Ni and Sn atoms within the half-Heusler (HH) cell and somewhat lower in full-Heusler (FH) cell. The calculations predict that the presence of the TiNi$_2$Sn phase strongly affects the TiNiSn lattice parameter value due to the stress on the grain boundary between Heusler and half-Heusler phases. The distribution of the total density of states for TiNiSn, TiNi$_{1.25}$Sn, TiNi$_2$Sn, and TiNiSn-TiNi$_2$Sn grain boundary (Fig. 31) confirmed that TiNiSn is a semiconductor and TiNi$_2$Sn is characterized by the metallic type of conductivity. In the case of TiNi$_{1.25}$Sn (ST MgAgAs) the additional Ni atoms in $4d$ position generate a new band inside the energy gap and eliminate it. Thus the material predicted to has metallic type of conductivity similar to TiNi$_2$Sn. The DOS distribution at the grain boundary of TiNiSn-TiNi$_2$Sn significantly differs from those in pure TiNiSn and TiNi$_2$Sn. The main difference from TiNiSn and TiNi$_2$Sn is that the energy bands are strongly delocalized. Like in the case of TiNi$_{1.25}$Sn and TiNi$_2$Sn the electrical conductivity at the grain boundary predicted to be metallic. The deformation of the crystal structure at the TiNiSn-TiNi$_2$Sn grain boundary lead to additional phonon scattering and decreasing of the thermal conductivity of the material. The increased electrical conductivity and decreased thermal conductivity should have a positive effect on the value of the thermoelectric figure of merit (ZT).

Conclusions

The complex utilization of different DFT methods allows predicting the main thermoelectric properties of half-Heusler materials. They could be used to precisely refine their real crystal structure, to define the solubility limits of substitutional and interstitial solid solutions and evaluate their thermodynamic stability, to predict the behavior of resistivity and Seebeck coefficient, to explore the effects of structural disorder and to show peculiarities in chemical bonding. Combination of theoretical DFT methods with experimental techniques gives the key for better understanding the nature of thermoelectric phenomena in half-Heusler materials.

Acknowledgements

Many people have helped in the variety of ways in making this chapter. Our appreciation and respect goes to all authors of books, chapters, monographs, research articles, computer programs cited and used throughout this text. We are grateful to all our colleges from University of Vienna, Lviv Polytechnic National University and Ivan Franko National University for their willingness to share the results of their unpublished work. We would like to acknowledge financial support of the Ministry of Education and Science of Ukraine under Grant 0116U004142. Finally yet importantly, we extend our gratitude to our families for their exceptional patience and support.

Keywords: Thermoelectric materials, Half-Heusler phases, DFT calculations, Seebeck coefficient, Resistivity, Crystal structure, Disorder, Chemical bonding

References

Ahilan, K., M.C. Bennett, M.C. Aronson, N.E. Anderson, P.C. Canfield et al. 2004. Magnetotransport in single-crystal half-Heusler compounds. Phys. Rev. B. 69: 245116.

Akai, H. 1989. Fast Korringa-Kohn-Rostoker coherent potential approximation and its application to FCC Ni-Fe systems. J. Phys.: Condens. Matter 1: 8045–8063.

Aliev, F.G., N.B. Brandt, V.V. Moshchalkov, V.V. Kozyrkov, R.V. Skolozdra and A.I. Belogorokhov. 1989. Gap at the Fermi level in the intermetallic vacancy system RNiSn (R = Ti, Zr, Hf). Z. Phys. B—Condensed Matter 75: 167–171.

Aliev, F.G., V.V. Kozyrkov, V.V. Moshchalkov, R.V. Scolozdra and K. Durczewski. 1990. Narrow band in the intermetallic compounds MNiSn (M = Ti, Zr, Hf). Z. Phys. B—Condensed Matter 80: 353–357.

Aliev, F.G. 1991. Gap at Fermi level in some new d- and f-electron intermetallic compounds. Physica B. 171: 199–205.

Aliev, F.G., J.M. Calleja, N. Mestres, R. Villar, S. Vieira, V.V. Pryadun et al. 1994. Localization induced transformation of the lattice modes of MNiSn (M = Zr, Hf, Ti) compounds. Physica B. 194-196: 1089–1090.

Anatychuk, L.I. 1998. Thermoelectricity, Volume I. Physics of Thermoelectrivity. Institute of Thermoelectricity, Chernivtsi, Ukraine.

Caruso, A.N., C.N. Borca, D. Ristoiu, J.P. Nozieres and P.A. Dowben. 2003. A comparison of surface segregation for two semi-Heusler alloys: TiCoSb and NiMnSb. Surf. Sci. 525: L109–L115.

Colinet, C., P. Jund and J.-C. Tédenac. 2014. NiTiSn a material of technological interest: *Ab initio* calculations of phase stability and defects. Intermetallics 46: 103–110.

Dhar, S.K., K.A. Gschneidner, Jr. and R. Vijayaraghavan. 1993. YbNiSn—a possible heavy-fermion compound. Physica B. 186-188: 463–465.

Douglas, J.E., C.S. Birkel, N. Verma, V.M. Miller, Mao-Sheng Miao, G.D. Stucky et al. 2014. Phase stability and property evolution of biphasic Ti–Ni–Sn alloys for use in thermoelectric applications. J. Appl. Phys. 115: 043720.

Dudkin, L.D., Z.M. Dashevskii and R.V. Skolozdra. 1993. On crystal chemical nature of ternary compounds $M^{VIII}M^{IV(V)}Sn(Sb)$ with MgAgAs structure type (M–transition metal). Inorganic Materials (Russian). 29: 313–318.

Evers, C.B.H., C.G. Richter, K. Hartjes and W. Jeitschko. 1997. Ternary transition metal antimonides and bismuthides with MgAgAs-type and filled NiAs-type structure. J. Alloy Compd. 252: 93–97.

Fruchart, D., V.A. Romaka, Yu.V. Stadnyk, L.P. Romaka, Yu.K. Gorelenko, M.G. Shelyapina et al. 2007. Conductivity mechanisms in heavy-doped n-ZrNiSn intermetallic semiconductor. J. Alloy Compd. 438: 8–14.

Harmening, T., H. Eckert and R. Pöttgen. 2009. Defects in half-Heusler type antimonides ScTSb (T = Ni, Pd, Pt). Solid State Sci. 11: 900–906.

Hartjes, K. and W. Jeitschko. 1995. Crystal structures and magnetic properties of the lanthanoid nickel antimonides LnNiSb (Ln = La-Nd, Sm, Gd-Tm, Lu). J. Alloy Compd. 226: 81–86.

Hazama, H., R. Asahi, M. Matsubara and T. Takeuchi. 2010. Study of electronic structure and defect formation in $Ti_{1-x}Ni_{1+x}Sn$ half-Heusler alloys. J. Electr. Mater 39: 1549–1553.

Heremans, J.P. 2005. Low-dimensional thermoelectricity. Acta Phys. Polon. A. 108: 609–634.

Hicks, L.D. and M.S. Dresselhaus. 1993. Thermoelectric figure of merit of a one-dimensional conductor. Phys. Rev. B 47: 16631.

Hlil, E.K., Yu. Stadnyk, Yu.K. Gorelenko, L. Romaka, A. Horyn and D. Fruchart. 2010. Synthesis, electronic transport and magnetic properties of $Zr_{1-x}Y_xNiSn$ ($x = 0 \div 0.25$), solid solutions. J. Solid State Chem. 183: 521–526.

Hohl, H., A.P. Ramirez, C. Goldmann, G. Ernst, B. Wolfing and E. Bucher. 1999. Efficient dopants for ZrNiSn-based thermoelectric materials. J. Phys. Condens. Matter 11: 1697–1709.

Ishida, S., T. Masaki, S. Fujii and S. Asano. 1997. Effects on electronic structures of atomic configurations in ternary compounds Ni-M-Z (M = Sc, Ti, Zr, Hf; Z = Sn, Sb). Physica B. 237-238: 363–364.

Jung, D., H.-J. Koo and M.-H. Whangbo. 2000. Study of the 18-electron band gap and ferromagnetism in semi-Heusler compounds by non-spin-polarized electronic band structure calculations. Journal of Molecular Structure (Theochem) 527: 113–119.

Kaczmarska, K., J. Pierre, J. Beille, J. Tobola, R.V. Skolozdra and G.A. Melnik. 1998. Physical properties of the weak itinerant ferromagnet CoVSb and related semi-Heusler compounds. J. Magn. Magn. Mater 187: 210–220.

Käfer, W., K. Fess, Ch. Kloc, K. Friemelt and E. Bucher. 1997. Thermoelectric porperties of MnNiSn (M = Ti, Zr, Hf) single crystals and related alloys. Proc. 16th Int. Conf. Thermoelectrics 489–492.

Kandpal, H.C., C. Felser and R. Seshadri. 2006. Covalent bonding and the nature of band gaps in some half-Heusler compounds. J. Phys. D: Appl. Phys. 39: 776–785.

Karla, I., J. Pierre and R.V. Skolozdra. 1998a. Physical properties and giant magnetoresistance in RNiSb compounds. J. Alloy Compd. 265: 42–48.

Karla, I., J. Pierre and B. Ouladdiaf. 1998b. Magnetic structures of RNiSb compounds (R = rare earths) investigated by neutron diffraction. Physica B 253: 215–221.

Karla, I., J. Pierre, A.P. Murani and M. Neumann. 1999. Crystalline electric field in RNiSb compounds investigated by inelastic neutron scattering. Physica B 271: 294–303.

Kawaharada, Y., H. Uneda, H. Muta, K. Kurosaki and S. Yamanaka. 2004. High temperature thermoelectric properties of NiZrSn half-Heusler compounds. J. Alloy Compd. 364: 59–63.

Kawamura, T. 1965. Silicon carbide crystals grown in nitrogen atmosphere. Mineralogical Journal (Japan) 4: 333–355.

Kleinke, H. 1998. Die neuen Antimonide ZrNiSb and HfNiSb: Synthese, Struktur und Eigenschaften im Vergleich zu ZrCoSb und HfCoSb. Z. Anorg. Allg. Chem. 624: 1272–1278.

Kuentzler, R., R. Clad, G. Schmerber and Y. Dossmann. 1992. Gap at the Fermi level and magnetism in RMSn ternary compounds (R = Ti, Zr, Hf and M = Fe, Co, Ni). J. Magn. Magn. Mater 104: 1976–1978.

Kunkel, N., C. Reichert, M. Springborg, D. Wallacher and H. Kohlmann. 2015. Hydrogenation properties of $Li_xSr_{1-x}AlSi$ studied by quantum-chemical methods ($0 \leq x \leq 1$) and *in-situ* neutron powder diffraction ($x = 1$). J. Solid State Chem. 221: 318–324.

Larson, P., S.D. Mahanti, S. Sportouch and M.G. Kanatzidis. 1999. Electronic structure of rare-earth nickel pnictides: Narrow-gap thermoelectric materials. Phys. Rev. B. 59(24): 15660–15668.

Larson, P., S.D. Mahanti and M.G. Kanatzidis. 2000. Structural stability of Ni-containing half-Heusler compounds. Phys. Rev. B. 62(19): 12754–12762.

Melnyk, G., E. Bauer, P. Rogl, R. Skolozdra and E. Seidl. 2000. Thermoelectric properties of ternary transition metal antimonides. J. Alloy Compd. 296: 235–242.

Mena, J.M., H.G. Schoberth, T. Gruhn and H. Emmerich. 2015. *Ab initio* study of domain structures in half-metallic $CoTi_{1-x}Mn_xSb$ and thermoelectric $CoTi_{1-x}Sc_xSb$ half-Heusler alloys. J. Alloy Compd. 650: 728–740.

Morozkin, A.V. and V.N. Nikiforov. 2005. Thermoelectric properties of ScCoSb, $ScNi_{0.86}Sb$ and MgNiSb compounds. J. Alloy Compd. 400: 62–66.

Nanda, B.R.K. and I. Dasgupta. 2003. Electronic structure and magnetism in half-Heusler compounds. J. Phys.: Condens. Matter 15: 7307–7323.

Nourbakhsh, Z. 2013. Three dimensional topological insulators of $LuPdBi_xSb_{1-x}$. J. Alloy Compd. 549: 51–56.

Nowotny, H.N. 1960. Untersuchungen an metallischen Systemen mit Flusspatphasen. Monatshefte für Chemie 91: 877–887.

Oestreich, J., U. Probst, F. Richardt and E. Bucher. 2003. Thermoelectrical properties of the compounds $ScM^{VIII}Sb$ and $YM^{VIII}Sb$ (M^{VIII} = Ni, Pd, Pt). J. Phys.: Cond. Matt. 15: 635–640.

Offernes, L., P. Ravindran and A. Kjekshus. 2007. Electronic structure and chemical bonding in half-Heusler phases. J. Alloy Compd. 439: 37–54.

Offernes, L., P. Ravindran, C.W. Seim and A. Kjekshus. 2008. Prediction of composition for stable half-Heusler phases from electronic-band-structure analyses. J. Alloy Compd. 458: 47–60.

Ögut, S. and K.M. Rabe. 1995. Band gap and stability in the ternary intermetallic compounds NiSnM (M = Ti, Zr, Hf): A first-principles study. Phys. Rev. B. 51(16): 10443–10452.

Romaka, L., Yu. Stadnyk, A. Horyn, M.G. Shelyapina, V.S. Kasperovich, D. Fruchart et al. 2005. Electronic structure of the $Ti_{1-x}Sc_xNiSn$ and $Zr_{1-x}Sc_xNiSn$ solid solutions. J. Alloy Compd. 396: 64–68.

Romaka, L.P., M.G. Shelyapina, Yu.V. Stadnyk, D. Fruchart, E.K. Hlil and V.A. Romaka. 2006. Peculiarity of metal–insulator transition due to composition change in semiconducting $TiCo_{1-x}Ni_xSb$ solid solution. I. Electronic structure calculations. J. Alloy Compd. 416: 46–50.

Romaka, V.A., Yu.V. Stadnyk, D. Fruchart, V.V. Romaka, Yu.K. Gorelenko, V.N. Davydov et al. 2007a. Conduction peculiarities in $ZrNiSn_{1-x}In_x$ semiconductor solid solution. Ukr. J. Phys. 52(10): 951–957.

Romaka, V.A., Yu.V. Stadnyk, V.V. Romaka, D. Fruchart, Yu.K. Gorelenko, V.F. Chekurin et al. 2007b. Features of electrical conductivity in the *n*-ZrNiSn intermetallic semiconductor heavily doped with the In acceptor impurity. Semiconductors 41: 1041–1047.

Romaka, V.A., Yu.V. Stadnyk, L.G. Akselrud, V.V. Romaka, D. Fruchart, P. Rogl et al. 2008a. Mechanism of local amorphization of a heavily doped $Ti_{1-x}V_xCoSb$ intermetallic semiconductor. Semiconductors 42: 753–760.

Romaka, V.V., E.K. Hlil, L. Romaka, D. Fruchart and A. Horyn. 2008b. Electrical transport properties and electronic structure of RNiSn compounds (R = Y, Gd, Tb, Dy, and Lu). Chem. Met. Alloys 1: 298–302.

Romaka, V.V., E.K. Hlil, O.V. Bovgyra, L.P. Romaka, V.M. Davydov and R.V. Krayovskyy. 2009a. Mechanism of defect formation in heavily Y-doped n-ZrNiSn. I. Crystal and Electronic structures. Ukr. J. Phys. 54: 1119–1124.

Romaka, V.A., D. Fruchart, V.V. Romaka, E.K. Hlil, Yu.V. Stadnyk, Yu.K. Gorelenko et al. 2009b. Features of the structural, electrokinetic, and magnetic properties of the heavily doped ZrNiSn semiconductor: Dy acceptor impurity. Semiconductors 43: 7–13.

Romaka, V.A., Yu.V. Stadnyk, D. Fruchart, L.P. Romaka, A.M. Goryn, Yu.K. Gorelenko et al. 2009c. Features of structural, electron-transport, and magnetic properties of heavily doped *n*-ZrNiSn semiconductor: Fe acceptor impurity. Semiconductors 43: 278–284.

Romaka, V.A., Yu.V. Stadnyk, D. Fruchart, T.I. Dominuk, L.P. Romaka, P. Rogl et al. 2009d. The Mechanism of generation of the donor- and acceptor-type defects in the n-TiNiSn semiconductor heavily doped with Co impurity. Semiconductors 43: 1124–1130.

Romaka, V.A., V.V. Romaka and Yu.V. Stadnyk. 2010a. Intermetallic Semiconductors: Properties and Applications. Lviv Polytechnic Publishing House, Lviv.

Romaka, V.A., D. Fruchart, E.K. Hlil, R.E. Gladyshevskii, D. Gignoux, V.V. Romaka et al. 2010b. Features of an intermetallic *n*-ZrNiSn semiconductor heavily doped with atoms of rare-earth metals. Semiconductors 44: 293–302.

Romaka, V.A., P. Rogl, V.V. Romaka, E.K. Hlil, Yu.V. Stadnyk and S.M. Budgerak. 2011. Features of apriori heavy doping of the *n*-TiNiSn intermetallic semiconductor. Semiconductors 45: 850–856.

Romaka, V.A., P. Rogl, Yu.V. Stadnyk, E.K. Hlil, V.V. Romaka and A.M. Horyn. 2012a. Features of conductivity of the intermetallic semiconductor *n*-ZrNiSn heavily doped with a Bi donor impurity. Semiconductors 46: 887–893.

Romaka, V.A., P. Rogl, Yu.V. Stadnyk, V.V. Romaka, E.K. Hlil, V.Ya. Krayovskii et al. 2012b. Features of conduction mechanisms of the *n*-HfNiSn semiconductor heavily doped with the Co acceptor impurity. Semiconductors 46: 1106–1113.

Romaka, L., Yu. Dovgalyuk, V.V. Romaka, I Lototska and Yu. Stadnyk. 2012c. Interaction of the components in Y-Ni-Sn ternary system at 770 K and 670 K. Intermetallics 29: 116–122.

Romaka, V.V., P. Rogl, L. Romaka, Yu. Stadnyk, A. Grytsiv, O. Lakh et al. 2013a. Peculiarities of structural disorder in Zr- and Hf-containing Heusler and half-Heusler stannides. Intermetallics 35: 45–52.

Romaka, V.A., P. Rogl, Yu.V. Stadnyk, V.V. Romaka, E.K. Hlil, V.Ya. Krajovskii et al. 2013b. Features of conduction mechanisms in n-HfNiSn semiconductor heavily doped with a Rh acceptor impurity. Semiconductors 47: 1145–1152.

Romaka, V.A., P. Rogl, V.V. Romaka, Yu.V. Stadnyk, E.K. Hlil, V.Ya. Krajovskii et al. 2013c. Effect of accumulation of excess Ni atoms in the crystal structure of the intermetallic semiconductor n-ZrNiSn. Semiconductors 47: 892–898.

Romaka, V.V., P. Rogl, L. Romaka, Yu. Stadnyk, N. Melnychenko, A. Grytsiv et al. 2013d. Phase equilibria, formation, crystal and electronic structure of ternary compounds in Ti-Ni-Sn and Ti-Ni-Sb ternary systems. J. Solid State Chem. 197: 103–112.

Romaka, V.A., P. Rogl, V.V. Romaka, Yu.V. Stadnyk, R.O. Korzh, A.M. Horyn et al. 2014. Optimization of parameters of the new thermoelectric material $HfNiSn_{1-x}Sb_x$. Journal of Thermoelectricity 1: 43–51.

Romaka, V.V., L. Romaka, P. Rogl, Yu. Stadnyk, N. Melnychenko, R. Korzh et al. 2014. Peculiarities of thermoelectric Half-Heusler Phase formation in Zr-Co-Sb ternary system. J. Alloy Compd. 585: 448–454.

Romaka, V.V., P. Rogl, L.P. Romaka, R.O. Korzh, Yu.V. Stadnyk, V.Ya. Krayovskyy et al. 2014c. Structure, energy state and electrokinetic characteristics of $Zr_{1+x}Co_{1-x}Sb$ solid solution. Physics and Chemistry of Solid State 15: 563–568.

Romaka, V.A., P. Rogl, V.V. Romaka, Yu.V. Stadnyk, R.O. Korzh, V.Ya. Krayovskyy et al. 2014d. Features of the band structure and conduction mechanisms in the n-HfNiSn semiconductor heavily doped with Ru. Semiconductors 48: 1545–1551.

Rowe, D.M. 1995. CRC Handbook of Thermoelectrics. CRC, Boca Raton.

Rowe, D.M. 1999. Thermoelectrics, an environmentally-friendly source of electrical power. Renewable Energy. 16: 1251–1265.

Rowe, D.M. 2006. Thermoelectrics Handbook: Macro to Nano. CRC Taylor & Francis, Boca Raton.

Savrasov, S.Yu. and D.Yu. Savrasov. 1992. Full-potential linear-muffin-tin-orbital method for calculating total energies and forces. Phys. Rev. B. 46: 12181–12195.

Sekimoto, T., K. Kurosaki, H. Muta and S. Yamanaka. 2005. Annealing effect on thermoelectric properties of TiCoSb half-Heusler compound. J. Alloy Compd. 394: 122–125.

Shklovskii, B.I. and A.L. Efros. 1984. Electronic Properties of Doped Semiconductors. Springer-Verlag, Berlin.

Skolozdra, R.V. 1993. Stannides of rare earth and transition metals. Svit, Lviv.

Sootsman, J.R., D.Y. Chung and M.G. Kanatzidis. 2009. New and old concepts in thermoelectric materials. Angew. Chem. Int. Ed. 48: 8616–8639.

Stadnyk, Yu., Yu. Gorelenko, A. Tkachuk, A. Goryn, V. Davydov and O. Bodak. 2001. Electric transport and magnetic properties of $TiCo_{1-x}Ni_xSb$ solid solution. J. Alloy Compd. 329: 37–41.

Stadnyk, Yu., A. Horyn, V. Sechovsky, L. Romaka, Ya. Mudryk, J. Tobola et al. 2005. Crystal structure, electrical transport properties and electronic structure of the $VFe_{1-x}Cu_xSb$ solid solution. J. Alloy Compd. 402: 30–35.

Stadnyk, Yu., V.A. Romaka, M. Shelyapina, Yu. Gorelenko, L. Romaka, D. Fruchart et al. 2006. Impurity band effect on $TiCo_{1-x}Ni_xSb$ conduction: Donor impurities. J. Alloys Compd. 421: 19–23.

Stadnyk, Yu., A. Horyn', V.V. Romaka, Yu. Gorelenko, L.P. Romaka, E.K. Hlil et al. 2010. Crystal, electronic structure and electronic transport properties of the $Ti_{1-x}V_xNiSn$ (x = 0–0.10) solid solutions. J. Solid State Chem. 183: 3023–3028.

Stadnyk, Yu.V., A.M. Goryn', V.V. Romaka, Yu.K. Gorelenko, L.P. Romaka and N.A. Mel'nichenko. 2011. Structural and thermoelectric properties of $Zr_{1-x}Er_xNiSn$ solid solutions. Inorganic Materials 47: 637–644.

Suzuki, H., T. Yamaguchi, K. Katoh and M. Kasaya. 1993. Physical properties of ternary rare earth compounds RPt(Au)Sb. Physica B. 186-188: 390–392.

Terada, M., K. Endo, Y. Fujita, T. Ohoyama and R. Kimura. 1970. Magnetic susceptibilities of Cl_b compounds CoVSb and CoTiSb. J. Phys. Soc. Jpn. 29: 1091.

Terada, M., K. Endo, Y. Fujita and R. Kimura. 1972. Magnetic properties of Cl_b compounds CoVSb, CoTiSb and NiTiSb. J. Phys. Soc. Jpn. 32: 91–94.

Tobola, J. and J. Pierre. 2000. Electronic phase diagram of the XTZ (X = Fe, Co, Ni; T = Ti, Zr, V, Nb, Mn; Z = Sn, Sb) semi-Heusler compounds. J. Alloy Compd. 296: 243–252.

Tritt, T.M. 2006. Harvesting energy through thermoelectrics: Power generation and cooling. Mater Res. Soc. Bull. 31: 188–229.

Uher, C., S. Hu, J. Yang, G.P. Meisner and D.T. Morelli. 1997. Transport properties of ZrNiSn-based intermetallics. Proc. 16th Int. Conf. Thermoelectrics 485–488.

Uher, C., J. Yang, S. Hu, D.T. Morelli and G.P. Meisner. 1999. Transport properties of pure and doped MNiSn (M = Zr, Hf). Phys. Rev. B. 59: 8615–8621.

Wang, S.H., H.M. Cheng, R.J. Wu and W.H. Chao. 2010. Structural and thermoelectric properties of HfNiSn half-Heusler thin films. Thin Solid Films 518: 5901–5904.

Wu, B., H. Yuan, A. Kuang, H. Chen and Y. Feng. 2013. Spin alignment and magnetic phase transition in Cu-doped half-Heusler compound PdMnSb. Computational Materials Science 78: 123–128.

Wunderlich, W., Y. Motoyama, Y. Sugisawa and Y. Matsumura. 2011. Large closed-circuit Seebeck current in quaternary (Ti,Zr)NiSn Heusler alloys. J. Electronic Materials 40: 583–588.

Xia, Y., S. Bhattacharya, V. Ponnambalam, A.L. Pope, S.J. Poon and T.M. Tritt. 2000. Thermoelectric properties of semimetallic (Zr,Hf)CoSb half-Heusler phases. J. Appl. Phys. 88: 1952–1955.

Young, D.P., P. Khalifah, R.J. Cava and A.P. Ramirez. 2000. Thermoelectric properties of pure and doped FeMSb (M = V, Nb). J. Appl. Phys. 87: 317–321.

Skutterudites for Thermoelectric Applications

Properties, Synthesis and Modeling

R. Carlini,[1,*] *C. Fanciulli,*[2] *P. Boulet,*[3] *M.C. Record,*[4]
V.V. Romaka[5] *and P.F. Rogl*[6]

INTRODUCTION

Skutterudites form a class of promising compounds in the field of thermoelectrics. They are among the most studied thermoelectric materials of the last decade because related to the concept of electron-phonon-glass crystal. In literature there are many studies on Co-based skutterudites filled with Rare Earths (RE) (Ballikaya et al. 2012; Daniel et al. 2015a; Zhang et al. 2015; Lili et al. 2015; Bhaskar et al. 2015) and Fe, Co-based (REFe$_{4-x}$Co$_x$Sb$_{12}$) (Jacobsen et al. 2014; Zhang et al. 2015; Rogl et al. 2010a), while there

[1] Department of Chemistry and Industrial Chemistry, University of Genova, Via Dodecaneso 31, 16146 Genova, Italy.
[2] CNR – ICMATE, Corso Promessi Sposi 29, 23900 Lecco, Italy.
[3] Aix-Marseille University, CNRS, Madirel, 13397 Marseille cedex 20, France.
[4] Aix-Marseille University, CNRS, IM2NP, 13397 Marseille cedex 20, France.
[5] Department of Materials Science and Engineering, Lviv Polytechnic National University, 79013 Lviv, Ustiyanovycha Str. 5, Ukraine.
[6] Institute of Material Chemistry and Research, University of Vienna, Währingerstrasse 42, A-1090 Wien, Austria.
* Corresponding author: Riccardo.Carlini@unige.it

are very few the studies on Fe, Ni–based skutterudites $(REFe_{4-x}Ni_xSb_{12})$ (Tan et al. 2013a; Kaltzoglou et al. 2012; Morimura et al. 2003).

The name "Skutterudite" comes from a small mining town, Skutterud, in Norway, where it was extracted and classified for the first time in 1845. The natural Skutterudite is a cobalt arsenide mineral that contains variable amounts of Ni and Fe, substituting for cobalt, having the general formula: $(Co,Ni,Fe)As_3$. Its orogeny is hydrothermal, in moderate-high temperature; it is often found associated with other minerals such as Arsenopyrite, native Silver, Erythrite, Annabergite, Nickeline, Cobaltite, Silver sulfosalts, native Bismuth, Calcite, Siderite, Barite and Quartz. Its density is about 6.5 g/cm^3 and the Mohs hardness is 6 but these values strictly depend on the composition.

Synthetic skutterudites are now produced in the laboratory. They include compounds having the composition MX_3, where M is a transition metal of the VIII, IX, X group and X represents a pnitogen atom belonging to the XV group. Oftedal in 1928 identified $CoAs_3$ as the first synthetic skutterudites and determined its structure as a cubic body-centred cell and space group $Im\bar{3}$. $CoAs_3$ is now reported as the prototype compound for skutterudites.

Prototype and Binary Skutterudites

The unit cell of $CoAs_3$ contains square radicals of the pnicogen atoms, $[As_4]^-$, situated at the centre of smaller cubes and oriented along the <100> crystallographic directions. Every radical is surrounded by eight trivalent transition metal Co^{3+} cations. The unit cell can be described as constituted of eight small cubes, two of them without the $[As_4]^-$ ring at the centre, keeping the compositional ratio $Co^{3+}:[As_4]^{4-} = 4:3$. A characteristic coordination structure is then obtained having $Co_8[As_4]_6 = 2Co_4[As_4]_3$ composition and 32 atoms per cell.

In electronic terms, each transition metal contributes with nine electrons and each pnicogen contributes with 3 electrons to the covalent bond, giving a total Valence Electron Count (VEC) of 72 for each smaller cube. Considering now one-half of the unit cell and its empty cubes, a general skutterudite formula can be displayed as $\square M_4Pn_{12}$ where $\square$ is the empty cube, M is the transition metal and Pn is the pnicogen atom. Metal atoms are located on 8c sites (1/4, 1/4, 1/4) while the non-metal ones occupy 24 g sites (0, y, z).

Originally Oftedal observed a correlation between y and z, which allowed him to formulate the equation $y + z = 0.5$, the so-called Oftedal's relation. This rule allows to hypothesize that the pseudo planar pnicogen "complex" X_4 is a perfect square (Oftedal 1926), where the edges (d_1, d_2)

have the same length. Nevertheless, Kaiser and Jeitschko (Kaiser and Jeitschko 1999) confirmed that the atomic position of pnicogen atom in the rings fall below this relation but only the unfilled skutterudites include a perfect square X_4; this assumption leads to admit the presence of a small rectangular distortion ($d_1 \neq d_2$).

The experimental values of these parameters, which include the average radius of the voids, are shown in Table 1.

As reported in Table 1, the near-neighbour distances in the rectangular rings are comparable to the nearest neighbour distances in elemental P ($\approx$ 0.22 nm) (Osters et al. 2012), As ($\approx$ 0.25, 0.26 nm) (Ugai et al. 1985) and Sb ($\approx$ 0.28, 0.34 nm) (Lomnytska and Pavliv 2007).

The M-M distance is too large to form a bond, so the only relevant interactions are those among the ions inside the pnicogen ring X_4 (X-X bonds) and those among the M atoms and the pnicogen ions (M-X bonds). Figure 1 illustrates particularly well the binding site of the octahedral metal atom and the formation of inclined MX_6 octahedra sharing vertices with adjacent octahedra. The inclination of these octahedra originates the pseudo planar ring structure X_4. Furthermore, it can easily demonstrate how each pnicogen X is surrounded by two other pnicogens and two M atoms at similar distances. A logical consequence is that the ring is

Table 1: Structural parameters of binary skutterudites.

Compound	a (nm)	y	z	d_1(X-X) (nm)	d_2(X-X) (nm)	R(void) (nm)	Ref.
$IrSb_3$	0.92512	0.3376	0.15365	0.2843	0.3005	-	(Snyder et al. 2000)
$RhSb_3$	0.92322	0.3420	0.1517	0.2891	0.3000	0.2024	(Arne and Trond 1974)
$FeSb_3$	0.92116	0.3402	0.1578	0.2907	0.2944	-	(Möchel et al. 2011)
$CoSb_3$	0.90385	0.3351	0.1602	0.2891	0.2982	0.1892	(Arne and Trond 1974)
$RhAs_3$	0.84507	0.3482	0.1459	0.2468	0.2569	0.1934	(Arne and Trond 1974)
$CoAs_3$	0.82055	0.3442	0.1514	0.2478	0.2560	0.1825	(Arne and Trond 1974)
IrP_3	0.80151	0.3540	0.1393	0.2233	0.2340	-	(Rundqvist and Ersson 1969)
RhP_3	0.79951	0.3547	0.1393	0.2227	0.2323	0.1909	(Arne and Trond 1974)
NiP_3	0.78157	0.3553	0.1423	0.2281	0.2262	-	(Jeitschko et al. 2000)
PdP_3	0.7705	0.3442	0.1514	0.2333	0.2401	-	(Rundqvist 1960)
CoP_3	0.7705	0.3489	0.1451	0.2224	0.2328	0.1763	(Jeitschko et al. 2000)

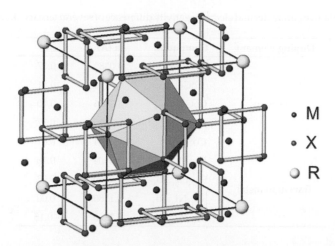

Fig. 1: Crystal structure of a filled skutterudite. The dodecahedral coordination around the X atom and the Sb_4 rings are highlighted.

linked by σ bond where each pentavalent X (ns^2np^3) binds the two nearest X sharing two of its own electrons. The other three valence electrons form, instead, bonds with the two closest M atoms. Due to the six bonds formed between the six X atoms and M atom, the pnicogen contribute to the octahedral MX_6 complex which then has $3*(1/2)*6 = 9$ electrons. This electronic configuration involves the Co-type metal one (d^7s^2) to reach the electronic configuration of the noble gas (18 electrons): this promotes diamagnetism and semiconductive behaviour, the main characteristics of binary skutterudites.

This pattern of bonds and the corresponding electron count poses a constraint on the actual existence conditions of binary compounds within the skutterudite family. For example, totally replacing the cobalt with iron or nickel does not form the skutterudite. Indeed the hypothetical compound FeX_3 having an electron less in the inner d orbitals should be considered a paramagnetic semiconductor. Similarly, for the compound NiX_3, the promotion of an extra non-bonding electron into the conduction band should make it a paramagnetic compound. A substitution of the metals of group IX is possible and, furthermore, the limits of solubility are remarkably large. Particularly, cobalt can be substituted with iron and nickel in accordance with the formula:

$$2Co^{3+}(d^6) \leftrightarrow Fe^{2+}(d^6) + Ni^{4+}(d^6)$$

with the consequent preservation of the total number of electrons.

In the light of these considerations, the existence of ternary skutterudites is expected, as long as it maintains the value of the Valence Electron Count VEC = 72.

Table 2: The lattice parameter and electronegativity difference of several ternary skutterudites.

Doping element	Compound	a (Å)	Δχ
Pnicogen	$CoSb_3$	9.0385	0.17
	$CoGe_{1.5}S_{1.5}$	8.0170	0.41
	$CoGe_{1.5}Se_{1.5}$	8.3076	0.40
	$CoSe_{1.5}Sn_{1.5}$	8.7259	0.37
	$CoSn_{1.5}Te_{1.5}$	9.1284	0.01
Transition metal	$Fe_{0.5}Ni_{0.5}Sb_3$	9.0904	0.18
	$Fe_{0.5}Pd_{0.5}Sb_3$	9.2060	0.03
	$Ru_{0.5}Ni_{0.5}Sb_3$	9.1780	0.00
	$Ru_{0.5}Pd_{0.5}Sb_3$	9.2960	−0.15

Ternary Skutterudites

Starting from the prototype $CoAs_3$, the substitution on both the metal and the pnicogen sites can be assumed as feasible and, indeed, several examples of doped ternary skutterudites are reported in the literature. A small list is shown in Table 2 together with the average lattice parameters (Fleurial et al. 1997) and the calculated Pauling electronegativity difference.

As expected the lattice parameter varies according to the type of substitution, that is, depending on the dimension of the substituted transition metal, pnicogen atom or both.

The investigation on these ternary skutterudites allowed researchers to define, in addition, the existence of more or less extended solid solutions between the parent binary compounds, which generally obey the Vegard' law.

From the perspective of physical properties, the thermal conductivity of ternary skutterudites is sensibly lower if compared to that of binary skutterudites. Unluckily, the decrease of thermal conductivity is consequence of a low carrier mobility: this implies high values of resistivity and small Seebeck coefficient in the ternary skutterudites. The prototype $CoSb_3$ skutterudite, having cobalt as M, has usually a high electrical conductivity and a large power factor but also significant thermal conductivity (10 W/m K) (Bhaskar et al. 2015a). This value can be decreased by filling the voids of the structure with foreign elements that cause rattling with a consequent amplified effect on the phonon scattering.

Filled Skutterudites

The study of the skutterudites structure and the voids identification, as mentioned above, offer the opportunity to add a large number of different species within the lattice. The first "filled" skutterudite $LaFe_4P_{12}$, was synthesized by Jeitschko and Brown in 1977 (Jeitschko and Brown 1977). Shortly after, filled skutterudites were prepared also for the families of arsenides and antimonides. The general formula for the filled skutterudite is $R^{4+}[M_4X_{12}]^{4-}$, where R is an electropositive ion, the exponent is the balance of charge and T the transition metal of filled skutterudite. It's important to notice the difference between the two complexes $[M_4X_{12}]^{4-}$ and $[M_4X_{12}]$: the former has got a tetravalent negative charge, while the latter appears as a neutral unit. This feature greatly affects the chemical and physical properties of the filled compounds.

Thanks to their physico-chemical properties, alkaline earths and rare earths can be considered the main skutterudites fillers. Indeed, according to (Jacobsen 2014), there are two main conditions that must be respected for a filler to enter the structure: the filler must have a size consistent with the host void and the difference of electronegativity (χ) has to be $\chi_x - \chi_{filler} >$ 0.8. Generally, the filler forms weak bonds with x and their delocalization is responsible for the decrease of thermal conductivity.

With respect to the filler properties, it is very important to consider that the lattice parameter and the size of the voids consistently increase passing from skutterudites P-based towards those As-based up to those with Sb. In addition, the lattice constants increase with the increase of the mass of metal within a given family of skutterudites. Figure 1 shows the unit cell centred on the $2a$ position located in (0,0,0), hosting the filler atom. For many of the compounds, the R atoms, frequently rare earths, tend to exhibit exceptionally large thermal parameters corresponding to their rattling in an oversized atomic cage, thus causing a strong reduction of their thermal conductivity (Mochel et al. 2011).

A similar circumstance occurs in the case of $RhSb_3$, where the Rh atoms are replaced by Ru and Pd forming the compound $Ru_2Pd_2Sb_{12}$. The two compounds mentioned above are examples of ternary skutterudites, which are isoelectronic phases with respect to binary compounds. Replacing the pnicogen (as pentavalent Sb) with a combination of a tetravalent (e.g., Ge, Sn, Pb) and a hexavalent (e.g., Se, Te) stable compounds such as $CoGe_{1.5}Se_{1.5}$ could be obtained. On the other hand, if a replacement is carried out on both the metal sites and the pnicogen site, a ternary skutterudites such as $Fe_4Sb_8Te_4$ could be generated. The metals of the group 9, Co-type, are often replaced with metals of group 8 such as Fe, Ru and Os; to date no skutterudite with metals of group 8 were obtained most probably due to the obvious imbalance of charge.

Furthermore, some ternary skutterudites can host varying filler amounts due to the presence of extended solid solutions given by the partial or total substitution of the transition metals. An example is reported by Carlini (Carlini et al. 2016): it is possible to obtain quaternary filled skutterudites having a large range of different compositions. In particular, stable $Sm_yFe_{4-x}Ni_xSb_{12}$ skutterudites were observed for $0 \lesssim x \lesssim 2.4$ and $0.1 \lesssim y \lesssim 0.8$.

The introduction of a suitable filler R could provide the missing electrons, thus saturating bonds and making an electrically semiconductive and diamagnetic neutral structure. The particular "filled" skutterudite having the formula $R^{4+}[T_4X_{12}]^{4-}$ has VEC = 72 with a consequent behaviour, just like the $CoSb_3$ prototype. A difficulty that often occurs in the synthesis of new filled skutterudites is the choice of a suitable filler. In fact, because of the electronic constraints, generally, only trivalent elements provide stable skutterudites. Most of the rare earths are trivalent, some of them (Yb, Sm) show an intermediate valence, while Eu usually has a divalent state; other species of fillers such as alkaline earth metals are only divalent. Accordingly, the total valence number for skutterudites completely "filled" with ions having a valence lower than 4+, will be less than 72 units: consequently, these "full" skutterudites will be considered paramagnetic metals. From the point of view of thermoelectricity, this is not an optimal condition since metals have a very small Seebeck coefficient. To take back this type of compounds in their semiconductive domain (and so, the total valence to 72 units), a charge compensation is required. The adjustment can be obtained mainly in two ways: by modifying the nature of the pnicogen rings (e.g., by partial replacement of x atoms with elements of groups III or IV) or by insertion of electronic-richer metals, such as Ni, Cu, etc., in place of metals belonging to the iron group (VIII) at the 8c site. Both approaches considerably expand the compounds range of the skutterudites family (Rowe 2006).

Nowadays, the most studied compounds are today $Ce_yFe_{4-x}Co_xSb_{12}$ and $La_yFe_{4-x}Co_xSb_{12}$. While the total occupancy of the voids is possible when $x = 0$, the increasing amount of Co implies a decrease of voids occupancy. For $x = 4$ ($CoSb_3$), the ions of Tl lead to an average voids occupation of about 22%, while it increases to 100% using Sn. Some values, for $CoSb_3$ rare earths filled, are reported in Table 3.

The filler amount and the corresponding charge compensation vary depending on the type and the composition of skutterudite, that is, $Sm_yFe_{4-x}Ni_xSb_{12}$ shows a maximum at 49% of void occupancy (Artini et al. 2016).

Table 3: Maximum void occupancy in pure $CoSb_3$.

Filler Ion	Maximum Void Occupancy (%)	Reference
Yb	44	(Rogl et al. 2010a)
Eu	25	(Berger et al. 2001)
La	23	(Nolas et al. 1998)
Ce	10	(Moerelli et al. 1997)

In the case of antimony-based skutterudites, the large size of the "cages" poses a constraint on which type of rare earth (or other species) may be trapped in the structure. Using equilibrium conditions for the synthesis, rare earths up to Europium can enter the network. For the lanthanide contraction, the ions of the heavy rare earths are too small and cannot bind with the atoms that constitute the cage of antimony. Therefore, to get Sb-based skutterudities filled with heavy rare earths, high pressures or non-equilibrium synthesis are generally required.

As reported by (Okamoto 1991) Co_4Sb_{12} is obtained by peritectic reaction at 874°C from $CoSb_2$ and melted Sb; a similar behaviour is found in the other skutterudites. Furthermore, due to the insertion of fillers, often, these materials are characterized by very long syntheses, which frequently involve the use of rather expensive technologies such as the Spark Plasma Sintering (SPS) or the High Temperature High Pressure (HTHP). The Spark Plasma Sintering is a sintering technique which provides for the simultaneous application of pressure and pulsed continued current. Due to the pressure, the particles of the powdered sample are in intimate contact; the current pulses and creates a local rising of the temperature that leads to the melting of the particles' surface; a network of microwelding is produced, making the sample compact.

The HTHP synthesis provides, instead, the heating of the samples at high temperature (usually up to 1000°C) and at high pressure (usually up to 50 GPa). The high pressure on the sample, thanks to the great mobility due to high temperature, favours the filling process by large elements such as rare earth metals or alkaline earth metals. Here are just some examples of the main syntheses reported in the literature.

Ballikaya (Ballikaya et al. 2012) reports a rather complex preparation of skutterudites: the elements are weighed in the respective stoichiometric amounts, a slow heating (1°/min) up to 1373 K is performed, annealing for 10 hr at that temperature to allow the accurate mixing of the constituents, afterwards a cooling with a rate of 4°/min to 1000 K and a hold for 10 d for the reaction to favour the peritectic reaction. The ingot obtained is pulverized and pelletized to be annealed again at 1000 K for another two weeks and finally, re-pulverized to undergo a sintering with SPS. With this synthesis being rather long and cumbersome, however, ZT value result 1.4 at 1073 K, even if the samples are not 100% monophasic.

As reported by Rogl (Rogl et al. 2010b) and Tan (Tan et al. 2013b), the preparation of iron-based skutterudites requires rather long synthetic pathways.

The samples synthesized by Rogl (Rogl et al. 2010b) include the mixing of Fe, Ni and Sb, in the appropriate stoichiometric amounts, at 1230 K and quenching in air. After this, the right amount of didymium is added; the sample is then heated at 873 K for three days and at 995 K for two days. Subsequently, the sample is melted again at 1230 K, tempered again in water and annealed at 873 K for five days. The material obtained finally is pulverized and sintered with the High Temperature High Pressure (HTHP) technique.

Tan (Tan et al. 2013b), instead, uses a simpler synthesis which is still quite long: It includes starting from the pure elements, slow heating and melting at 1273 K, staying at that temperature for 24 h, water salt quenching and annealing at 820–870 K for 7 days. The obtained samples are pulverized and sintered using the Spark Plasma Syntering (SPS) technique.

Other syntheses often reported (Short et al. 2015) involve the "Ball Milling" methods. This method provides synthesis in several steps. The ball milling, through a proper balance between rotation speed and duration, favours the formation of skutterudites with particular features that make them excellent precursors for nanomaterials. However, to obtain dense samples, it is necessary to use the techniques described above.

Some recent research suggest alternative syntheses as the solvothermal route using sodium borohydride ($NaBH_4$) as a reducing agent to obtain $CoSb_3$ nanoparticles (Kadel et al. 2014; Li et al. 2013; Bhaskar et al. 2015b). The Self-propagating-High-temperature-Synthesis (SHS) technique is also applied to synthesize $CoSb_3$ where mixtures of Co and Sb powders are pelletized and ignited from one end (Liang et al. 2014).

Physical Properties

The large number of possible compositions reported for skutterudites corresponds to a wide range of results for the adimensional figure of merit, ZT, as reported in the previous chapter. Each strategy applied to the starting $CoSb_3$ compound acts in term of improvement typically on one of the characteristics of the material, trying to preserve the behaviour of the others. For skutterudites containing elements with low electronegativity differences such as $CoSb_3$, a large number of covalent bonding are formed, enabling high carrier mobility resulting in good electronic properties. However, this strong bonding and simple order leads also to high lattice thermal conductivities, due to the reduced scattering mechanisms acting on phonons. As a consequence of the electronic-crystal behaviour, the main target of materials development is related to the reduction of the lattice thermal conductivity. The introduction of doping in $CoSb_3$ to obtain the carrier concentrations optimizing ZT value, adds enough carriers to significantly increase electron–phonon interactions reducing the thermal conductivity (Fleurial et al. 1997). Another improvement in reducing lattice component of thermal conductivity comes from the replacement of the elements alternatively on the transition metal or on the antimony site: this strategy contributes in creating lattice disorder, an effective way to increase the phonon scattering.

Actually the most used, and probably the most effective, solution to improve phonon scattering in skutterudites is related to the presence of the large voids in the structure. Filling these void spaces with rare-earth or other heavy atoms reduces the lattice thermal conductivity: this approach is widely explored in literature and a large variety of fillers have been used to achieve the larger effects on all classes of skutterudites (Sales et al. 1996). The large database of results obtained allowed researchers to observe a correlation between the size of the filler, and their vibrational motion, and the thermal conductivity of the material: the improvements achieved lead to ZT values up to 1 (Nolas et al. 2006; Sales 1998). A partial occupation of available voids establishes a random network made of filling atoms and vacancies introducing effective point-defect scattering. The large space for the filling atom in skutterudites induce local or 'rattling' modes that lower lattice thermal conductivity. This reduction is so drastic that it overcomes the modest degradation of the electronic properties of the compounds. The mechanism at the basis of this effect is the behaviour of the filler ion which, loosely bonded in a large space, acts as an independent and incoherent oscillator. Its isolated vibration mode is not responsible for heat transfer, but strongly interacts with the phonons of the main framework carrying the heat. This interaction is responsible for the significant reduction of lattice thermal conductivity observed in the filled skutterudites. Spectroscopic and diffractometric experiments, as well as *ab initio* calculations analyses, have been performed to verify and describe in detail the behaviour and the effects related to the filled skutterudites systems (Dordevic et al. 1999; Sales et al. 1999; Singh et al. 1999; Mandrus et al. 1997; Feldman et al. 2000). The results of these developments show that the motion of the filler is harmonic with large displacements. This induces a filler-antimony interaction which modifies the characteristic inelastic vibrational spectrum of the system: the shifts appearing in spectroscopic analyses can only partially be related to the bare filler vibrations, the others, typically at high frequencies, are due to antimony modes movement as a result of the coupled motion of Sb and filler ion. So, the scenario can be summarized as a consequence of a harmonic scattering of phonons by harmonic motions of the cage fillers (Rowe 2003). The employment of different ions to fill the voids, introducing different vibrational modes into the lattice, is at the origin of the multiple filling approach to skutterudites lattice conductivity reduction, resulting in an improvement of the 'rattling' effects. In Fig. 2 (Fleurial et al. 1997; Snyder and Toberer 2008) the effects of the different approaches described above are summarized: it can be seen how the increase of phonon scattering is able to reduce thermal conductivity within the whole temperature range.

Looking at the effects of fillers on the electronic properties of the materials, these filling ions add additional electrons that require compensating cations elsewhere in the structure for charge balance, originating as an additional source of lattice disorder. For the case of $CoSb_3$, Fe^{2+} is frequently used in place of Co^{3+}. A benefit of this partial filling is that the free-carrier concentration may be tuned by moving the composition slightly off the charge balanced composition. This unbalanced charge corresponds to a slight valence unbalance providing metallic carriers to the compound and improving electrical transport. The main characteristic of pure binary skutterudites from the electrical transport point of view, is the high mobility of the carriers, more similar to the one of metals than to the one typical of semiconductors. This condition

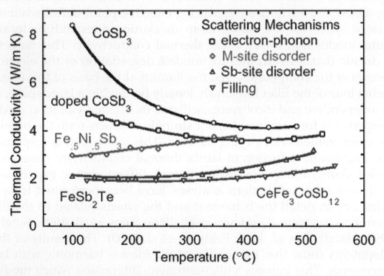

Fig. 2: The effects of the different scattering mechanisms on the thermal conductivity of skutterudites: starting from pure $CoSb_3$, the plots display the decrease of heat transfer obtained introducing scattering sources into the system (redrawn from Fleurial et al. 1997; Snyder et al. 2008).

is the same for both the cases of *p*- and *n*-type compounds and, for given carrier density, the hole mobility is generally higher than electron one (Uher 2002). The high mobility results are limited by different scattering mechanisms deriving by material purity, grain size, phonon or impurity interactions; however mobility exceeds 1000 cm^2/Vs in polycrystalline samples. As already reported, the presence of fillers introduces charge doping into the system: the partial filling easily produces an increase of few orders of magnitude to the carrier density moving the behaviour of the compound from semiconducting to metallic. This effect usually leads to a drastic reduction of Seebeck value, as described in the previous chapter, where the dependence of the thermoelectric parameters on the carrier density is treated. However, the case of skutterudites is peculiar, due to the large effective mass of carriers. In pure binary compounds, Seebeck coefficients are remarkably large with magnitudes of 200–500 μV/K at 300 K typically reported. Even for samples with electron density in the order of 10^{21} cm^{-3}, the Seebeck coefficient usually provides a result of more than 100 μV/K (Rowe 2003; Caillat et al. 1996). Increasing temperature, the conduction regime turns from extrinsic to intrinsic. Consequently, a rapid decrease of the Seebeck coefficient is observed, due to the increase of the available free carriers. The presence of the filler induces a significant change in the electronic bands structure, due to interaction between the fillers' electronic configuration with the one of the rigid cage made of covalent bondings. This change results in an increment of the free carrier density which should affect the Seebeck coefficient of the material inducing its suppression. However, the accessible states at the bottom of the conductive band introduced by the filler, are typically 'flat' in terms of energy dependence: this corresponds to an increase of the effective mass of both the type of carriers (electrons and holes), leading to a reduction of electrical conductivity, and, at the same time, allowing for the avoidance of the drastic suppression expected for the Seebeck coefficient at the new 'metallic-like' carrier concentration. Thus, in spite of the degraded mobility, the enhancement of carrier density, coupled with the large Seebeck coefficient, preserves high power factors.

In order to visualize and better explain the effects described, schematic examples of electrical resistivity and Seebeck coefficient behaviour as a function of temperature are reported in Fig. 3. These plots describe the differences in the parameters varying the level of doping of the material. Start with the undoped material, it displays the lower electrical conductivity (high resistivity) and the highest Seebeck coefficient. Here the passage from extrinsic to intrinsic regimes of conduction due to the thermal activation of carriers is evident in the resistivity behaviour. Gradually increasing the doping level, both absolute values and temperature behaviours change moving towards more metallic-like characteristics: electrical resistivities

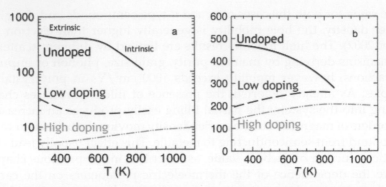

Fig. 3: Electrical resistivities (a) and Seebeck coefficients (b) as a function of temperature used as references to display the effects of doping in a skutterudite (qualitative analyses).

and Seebeck coefficients become more constant with the temperature and the shift between extrinsic and intrinsic regimes practically disappears.

What was reported up to now reflects the high complexity of the scenario related to the class of thermoelectric skutterudites: each attempted strategy for material optimization involves an improvement of some parameters and the degradation of the others. The fillers, acting on charge doping, affect the power factor in a detrimental direction, but, on the other hand, reduce significantly the lattice contribution to thermal conductivity. Thus, the target of the development of this class of materials can be summarized by finding the best balance in the final composition of the material in order to obtain the best ZT value: the high complexity introduced by substitutions and voids filling generally produces greater effects on thermal conductivity than on the final power factor, so recent developments are strongly oriented in the direction of filled and multiple-filled skutterudites, often starting from substituted compositions.

Figure 4 reports a collection of ZT values measured on different compounds (Sales et al. 1997; Sales 2003; Sales et al. 1996; Nolas et al. 2000; Tang et al. 2001) showing the beneficent effect of doping with respect to the pure $CoSb_3$, where the value is strongly limited by the high thermal conductivity. The temperature behaviour follows the dependence on carrier concentration on filled and substituted compositions. ZT is highly increased exceeding in many cases the value of 1 and the temperature dependence is totally changed making these materials feasible for technological employments.

Table 4 reports some of the most recent results obtained on different compounds in terms of ZT; data are collected in order to easily display how the different strategies for materials improvements have been able to lead this class of thermoelectrics to the important target of being competitive with the best materials currently developed.

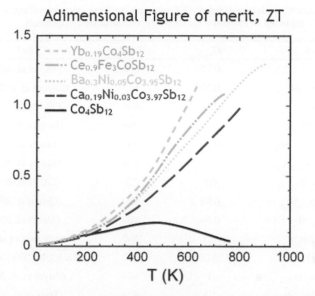

Fig. 4: Adimensional figure of merit for different filled skutterudites starting from undoped and substituted $CoSb_3$. The curves reported, redrawn from Sootsman (2009), show the effects described in the chapter.

In this chapter only the characteristics of the materials related to their thermoelectric performances, such as electrical and thermal transports, have been discussed. However, looking for a technological application of these materials, they only represent a starting point. In fact, in order to produce a device, other characteristics have to be considered. The most relevant example are the mechanical and thermo-mechanical properties, affecting materials' integrity under operating conditions; thermoelectric materials have to work under a simultaneous thermal gradient and mechanical pressure, so they have to sustain a high level of mechanical stress. Despite the fact that more attention has been recently paid to the mechanical characterization of the materials, compiling a state of the art of these characteristics for skutterudites is still difficult due to the lack of data for many of the materials reported and described. Another subject not discussed here is related to the reliability required for thermoelectric technology: this involves a lot of material characteristics like the resistance to the working environment agents or the long term stability, in terms of structure and composition, at operating temperatures. These aspects are sometimes neglected in the study of thermoelectric materials, but they are usually critical for the introduction of novel materials into the technological development.

Table 4: Collection of data related to ZT maxima and the corresponding temperatures reported in literature for different skutterudite compounds.

Compound	zT_{max}	T (K)	Ref.
$CeFe_4Sb_{12}$	0.75	800	(Yan et al. 2014)
$Sm_{0.7}Fe_{2.8}Ni_{1.2}Sb_{12.8}$	0.55	580	(Carlini et al. 2016)
$Mg_{0.4}Co_4Sb_{12}$	0.35	600	(Yan et al. 2013)
$DD_{0.63}Fe_{3.5}Ni_{1.5}Sb_{12}$	1.1	780	(Rogl et al. 2012)
$Nd_{0.61}Fe_{2.98}Ni_{1.02}Sb_{12}$	0.55	550	(Liu et al. 2014)
$Yb_{0.77}Fe_{2.99}Ni_{1.01}Sb_{12}$	0.7	600	(Liu et al. 2014)
$Yb_{0.7}Fe_2Co_2Sb_{12}$	0.8	720	(Liu et al. 2014)
$CeFe_{3.85}Mn_{0.15}Sb_{12}$	0.98	800	(Qiu et al. 2013)
$CeFe_{3.9}Ga_{0.1}Sb_{12}$	0.88	800	(Tan et al. 2012)
$Co_{0.94}Ni_{0.04}Sb_3$	0.5	750	(Katsuyama et al. 2003)
$IrSb_3$	0.15	850	(Slack and Tsoukala 1994)
$Ca_yCo_{4-x}Ni_xSb_{12}$	1	800	(Puyet et al. 2004)
$Ce_{0.28}Co_{2.5}Fe_{1.5}Sb_{12}$	1.1	800	(Tang et al. 2001)
$Ce_xFe_{3.5}Co_{0.5}Sb_{12}$	1.2	870	(Fleurial et al. 1996)
$In_{0.25}Co_4Sb_{12}$	1.2	570	(Chen et al. 2006)
$Yb_{0.8}Fe_{3.4}Ni_{0.6}Sb_{12}$	1	800	(Anno et al. 2002)
$(Ce,Yb)_{0.4}Fe_3Sb_{12}$	1	800	(Berardan et al. 2005)

Property Modelling

Acronyms

DFT	:	density-functional theory
EHT	:	extended Hückel theory
EHTB	:	extended Hückel tight binding
ELF	:	electron localization function
FP-APW+lo	:	full-potential augmented-plane-wave with local orbitals
FP-LMTO	:	full-potential linear muffin-tin orbital
FLAPW	:	full-potential linear augmented-plane-wave
GGA	:	generalized gradient approximation
KKR-CPA	:	Korringa-Kohn-Rostoker with the coherent potential approximation
LDA	:	local density approximation
LDA+U	:	local density approximation with Hubbard corrections
LMTO-ASA	:	linear muffin-tin orbital-atomic sphere approximation

LSDA   :   local spin density approximation
QTAIM   :   quantum theory of atoms in molecules

The modelling of skutterudite properties has been the subject of a large number of publications over the past decades. In this review we will only consider a limited set of results pertaining to Fe- or Ni-based skutterudites. This choice aims at preserving coherence with the experimental section of this chapter that mainly deals with these types of compounds.

Structural and Mechanical Properties

The structural and elastic properties of the ternary skutterudite based on lanthanum (LaFe$_4$A$_{12}$, with A = P, As, Sb) have been investigated by ab *initio* density-functional theory (Hachemaoui et al. 2010) as a function of an applied external pressure. The FP-APW+lo technique as implemented in the WIEN2k code (Blaha et al. 2001) has been used. At a given pressure, the elastic constants C$_{11}$, C$_{12}$ and C$_{44}$ and bulk modulus B$_0$ are found to decrease along the sequence P-> As-> Sb which is opposite to the cell parameters evolution (772.4, 817.9 and 896.3 pm, respectively). A linear increase of these properties is observed when external pressure increases. The compounds are all stable up to at least 40 GPa. In addition, the bulk, shear and Young moduli, the Poisson ratio, Lamé coefficient, sound velocity and Debye temperature have been calculated for an ideal polycrystalline aggregate as estimated from the Voigt-Reuss-Hill approximation (Hill 1952). It is found that these materials are brittle according to the Pugh empirical rule (Pugh 1954). The Debye temperature is decreasing along the sequence P-> As-> Sb (849, 619 and 489 K, respectively) which suggests that the LaFe$_4$Sb$_{12}$ should conduct heat more than the two other compounds. The same properties have been calculated for the UFe$_4$P$_{12}$ filled skutterudite by Ameri and coll. using DFT (Hohenberg and Kohn 1964; Kohn and Sham 1965) and the FP-LMTO approach (Ameri et al. 2014a). The effect of temperature and pressure on these properties has been investigated through the quasi-harmonic approximation. The compound is found to be stable from at least −10 GPa up to 50 GPa. The Debye temperature θ$_D$ increases systematically with the pressure while remaining almost constant within the range of temperatures 0–500 K. At zero temperature and pressure θ$_D$ amounts to about 607 K. DFT combined with the FP-LMTO methodology has been employed by Benalia and coll. to investigate the mechanical properties of the CeFe$_4$P$_{12}$ filled skutterudite at absolute zero temperature (Benalia et al. 2008). The Debye temperature is estimated to be about 366 K for this compound. Recently, an investigation of the elastic properties has been undertaken for the EuFe$_4$P$_{12}$ filled skutterudite by Shankar and coll. (Shankar et al. 2015a). As with the other compounds mentioned above, EuFe$_4$P$_{12}$ is brittle. EuFe$_4$P$_{12}$ possesses bulk and shear moduli of about 116 GPa and 107 GPa, respectively. The Debye temperature amounts to 782.5 K. Finally the $ThFe_4P_{12}$ compound has been investigated using the FP-LAPW+lo technique (Khenata et al. 2007) though only the elastic constants and shear and Young moduli have been reported.

Electronic Structure

Unfilled Skutterudites

The $Co_{4-x}Ni_xSb_{12}$ phases have been investigated by DFT with x an integer parameter varying between 0 and 4 (Christensen et al. 2004). In parallel to this theoretical investigation, an experimental one has been conducted using synchrotron and neutron powder diffraction that showed that the compound $Co_{2.9}Ni_{1.2}Sb_{12}$ is in fact a biphasic system with phases $Co_{3.8}Ni_{0.2}Sb_{12}$ and $Co_{0.4}Ni_{3.6}Sb_{12}$. Nevertheless, the density of states of the $Co_{4-x}Ni_xSb_{12}$ compounds has been calculated using DFT. The results show that Co_4Sb_{12} is a narrow band gap semiconductor. By contrast, all the Ni for Co substituted compounds are found to be metallic with a high electron doping character which, according to the authors, confirm that the Ni for Co substitutions are beneficial for the enhancement of the electrical conductivity of the compounds. The chemical bond nature of Co_3NiSb_{12} has been thoroughly investigated using Bader's QTAIM theory (Bertini et al. 2003; Bader et al. 1994). It has been found that the bonding nature of Co_4Sb_{12} is not fundamentally different from that when Ni is substituted for Co. The antimony atoms are bonded to the two nearest-neighbour metal atoms and to two antimony atoms involved in the pnicogen ring. Metal atoms are not bonded to each other. The metal-antimony bond is ionic in nature: antimony atoms are cations whereas cobalt and nickel atoms are significantly negatively charged which obviously contradicts Dyck and coll. who have stated that either Ni donate an electron to the conduction band at high temperature or retain it at low temperature (Dyck 2002). The ionicity seems to be reinforced when going from cobalt to nickel (substitution). The electronic structure of the NiP_3 skutterudite structure has been investigated using the LMTO-ASA method (Llunell et al. 1996). The band structure and density of states show a strong n-doping character with a small pseudo-gap 0.57 eV wide below the Fermi level. An EHTB analysis of the chemical bonds suggests that this compound is mainly metallic in nature, or eventually covalent. $FeSb_3$ skutterudite has been found to be a stable structure as a ferromagnetic compound with magnetization of about 1 μ_B/Fe (Råsander et al. 2015). The ferromagnetic structure is 0.15 eV more stable than the anti-ferromagnetic one which is itself 0.13 eV more stable than the unpolarized structure. The ferromagnetic $FeSb_3$ compound is described as a near half-metal. The spin-up valence bands are almost completely filled and a direct band gap of 0.26 eV shows up at the Γ point. By contrast, the Fermi level is 0.5 eV below the top of the spin-down valence bands with band gap of 0.7 eV separating the valence bands from the conduction ones. Therefore, the density of states of the spin-down bands near the Fermi level is much higher than that of the spin-up bands. Both spin-up and spin-down conduction bands are mainly contributed by Fe-d orbitals (Råsander et al. 2015; Daniel et al. 2015b) while for the valence bands, spin-down bands are of Fe-d orbital character and spin-up bands are of Sb-p orbital character.

Filled Skutterudites

To date, a large number of investigations have been reported on iron and nickel-based filled skutterudites; as a consequence of which the classification given below follows the type of filled skutterudites, namely, $FeAs_3$, FeP_3 and $FeSb_3$.

$FeAs_3$ Skutterudite. The Ce-filled $FeAs_3$ skutterudite has been found to be a semiconductor with a small, indirect band gap of about 0.238 eV along the Γ-N line (Hachemaoui et al. 2009; Wawryk et al. 2009; Wawryk et al. 2011). Experimental reports are contradictory on this point: Grandjean and coll. (Grandjean et al. 1984) found this structure to be a semiconductor whereas Maple and coll. found it to be a metal (Maple et al. 2008). According to the theoretical results, the valence bands at the vicinity of the Fermi level are mainly contributed by Fe-d orbitals with little mixing with As-p and Ce-d states whereas the conduction bands are contributed by the Ce-f orbitals. Two very flat and well separated (by about 0.2 eV) conduction bands are visible in the band structure (Wawryk et al. 2011). According to the authors, the low-lying bands result from the splitting of the Ce-f orbitals due to spin-orbit coupling.

The band structure as obtained by Harima and Takegahara for the La-filled $FeAs_3$ is very different from that of the previous compound (Harima and Takegahara 2003). $LaFe_4As_{12}$ is found to be a half-metal: some band contributed by the Fe-d orbitals cross the Fermi level. The top of these bands is located at the Γ point whereas the bottom of the conduction bands is between the Γ point and the H point. The Fermi surface around the Γ point is connected through the P point whereas it is closed around the N point.

The magnetism in the hypothetical Na-filled $FeAs_3$ skutterudite has been investigated in detail by Xing and coll. using DFT formalism (Xing et al. 2015). The band structure of $NaFe_4As_{12}$ shows a light band crossing the Fermi level. Hence this compound is a half metal. The maximum of the light band shows up at the G point, and a small gap (about 0.2 eV) separates the top of the valence bands from the bottom of the conduction band. The authors report that an instability in the ferromagnetic state should occur in $NaFe_4As_{12}$ which is evidenced by the high density of states of the Fe-d states near the Fermi level (Stoner theory); The magnetization amounts to about 2.8 μ_B per formula unit.

FeP_3 Skutterudite. To our knowledge most of the filled FeP_3 skutterudites have been investigated using lanthanides and actinides as fillers. Only $NaFe_4P_{12}$ has been reported by Xing and coll. using DFT approach (Xing et al. 2015). It has been shown that $NaFe_4P_{12}$ is a half metal with ferromagnetic ground state and magnetization of 1.6 μ_B per formula unit.

For the lanthanide series the La (Jung et al. 1990; Fornari and Singh 1999; Harima and Takegahara 2003; Grosvenor et al. 2006; Takegahara and Harima 2007; Xu and Li 2007; Saeterli et al. 2009; Flage-Larsen et al. 2010), Ce (Nordström and Singh 1996; Grovenor et al. 2006; Khenata et al. 2007; Benalia et al. 2008), Pr (Galvan 2006; Ameri et al. 2014b) and Eu (Shankar and Thapa 2013; Shankar et al. 2014) atoms have been used as fillers.

In $LaFe_4P_{12}$ the lanthanum gives three electrons to the structure leading to La^{3+}. This is evidenced by the projection of the La-4f orbitals from the total density of states that shows that these orbitals are only located well above the Fermi level in the conduction band (around 3 eV above the Fermi level). The structure may be then described as $La^{3+}(Fe_4P_{12})^{3-}$. Considering the large electronegativity of phosphorus compared to that of iron, phosphorus may be considered as the anion P^- hence giving a Fe^{3+} cationic state for iron. However, the band structure obtained from EHT calculations suggests that the valence state of iron is close to +2 (Jung et al. 1990). Hence, the charge born by phosphorus would be $11/12 \approx 0.92$. The remaining hole could be associated to one out of the twelve phosphorus atoms. Alternatively the hole could be shared among the iron atoms leading to an average valence state $Fe^{2.25+}$. Reported studies tend to show that the hole is at least partially shared among the phosphorus atoms (Jung et al. 1990; Grosvenor et al. 2006; Shenoy et al. 1982). $LaFe_4P_{12}$ is a half metal compound with a light band crossing the Fermi level. Above this band, which has the maximum at the Γ point, a small direct gap is formed (0.098 eV).

By contrast with $LaFe_4P_{12}$, Ce-4f orbitals participate in hybridization with Fe-d and P-p orbitals of the valence band near the Fermi level in $CeFe_4P_{12}$. More specifically, two bands close in energy contribute to the topmost part of the valence band. The highest band is predominantly contributed by the Ce-4f orbitals whereas the second lowest band does not contain Ce-4f contribution. Therefore, the valence state of cerium is not f^0 but rather d^xf^1 with $0.5 < x < 1$ (Nordström and Singh 1996). The consequence of the strong hybridization of the Ce-4f orbitals in the topmost valence band and lowest conduction one yields an overestimation of the Γ-N indirect energy band gap (0.34 eV versus 0.15–0.25 eV from experiments) as calculated with LDA. Probably that LDA+U technique would correct this behaviour but as far as we know, no such calculations have been undertaken to date.

Very few theoretical studies have been reported on $PrFe_4P_{12}$ (Galvan 2006; Ameri et al. 2014b). Using the EHT method, Galvan has found this compound to be a half metal with very small direct band gap of 0.02 eV. However, this is in contradiction with Ameri and coll. findings of 0.442 eV who used FP-LMTO method. The band crossing the Fermi level is essentially contributed by the Pr-f orbitals with little hybridization with Fe-d and P-p orbitals. Hence, as in $CeFe_4P_{12}$ Pr valence state is of the type f^x with $x > 0$.

The Eu-filled Fe_4P_{12} skutterudite has been found to be a ferromagnetic compound with total magnetization of 8.18 μ_B per formula unit (Shankar and Thapa 2013) as calculated with LSDA+U method. The compound is a half metal. The spin-up band structure shows that the Fermi level is crossed by a single, parabolic band at Γ with predominant Eu-f character. This band meets tangentially the conduction band at the Γ point. By contrast, two bands cross the Fermi level and show a maximum at the Γ point though without meeting the conduction band. Hence, a small, indirect gap shows up in the spin-up band structure. The character of these bands is Fe-d and P-p.

Since only thorium and uranium have stable (or very long living) isotopes, only the $ThFe_4P_{12}$ (Galvan et al. 2003; Takegahara and Harima 2003; Khenata 2007; Cheng et al. 2008a) and UFe_4P_{12} skutterudites (Galvan et al. 2003; Ameri et al. 2014a) have been investigated using theoretical approaches (EHT, FP-LAPW). According to the EHT method, both $ThFe_4P_{12}$ and UFe_4P_{12} are metallic compounds which disagree with more elaborated methods such as the FP-LAPW one, where both compounds are described as half-metals with a pseudo band gap located above the Fermi level. For $ThFe_4P_{12}$ it appears that the shape of the band structure and type of band gap (direct or not) is sensitive to the technical details of the method. To highlight this statement we present in Table 5 the results obtained with various methods. Comparing the effect of the exchange-correlation DFT functional (LDA versus GGA) it appears that LDA yields a direct band gap whereas GGA yields an indirect one. Furthermore, the LDA band gap is slightly wider than the GGA one. The inclusion of spin-orbit coupling allows for the splitting of the Th-f orbitals the consequence of which can be observed in the band structure. Irrespective of the technical details, the bands near the Fermi level barely contain Th-f orbitals that expand over the whole conduction band. The valence bands belong to the hybridization of the P-p and Fe-d orbitals. As to UFe_4P_{12} the calculated band structure shows that this compound is a half metal with an indirect band gap at Γ-N of 0.59 eV (as calculated at the FP-LMTO level). The bottom of the conduction band is made of two, flat bands that are separated by a small gap of about 0.1 eV.

Table 5: Band structure features of ThFe$_4$P$_{12}$ obtained with various methods. FP-LAPW: full potential linearized augmented plane waves; lo: local orbitals to treat semi-core electrons if any; SO: spin-orbit coupling.

Method	FP-LAPW+lo+SO with LDA functional[a]	FP-LAPW+lo with GGA functional[b]	FP-LAPW+SO with LDA functional[c]
Type of band gap	direct	indirect	direct
Band gap (eV)	0.52	0.41	0.45
Comment	Splitting of the f orbitals	No-splitting of the f orbitals	Splitting of the f orbitals

[a] From Khenata 2007.
[b] From Cheng et al. 2008.
[c] From Takegahara and Harima 2003.

FeSb$_3$ Skutterudite. The bonding features of NaFe$_4$Sb$_{12}$ and KFe$_4$Sb$_{12}$ compounds have been investigated by Leithe-Jasper and coll. using the ELF function (Leithe-Jaster et al. 2003; Leithe-Jaster et al. 2004; Becke and Edgecombe 1990). The resulting picture is that of a covalent frame of Sb and Fe atoms (with on average 1.76e between two Sb atoms and 2.03e between Sb and Fe) in which sodium cations enter into the cavities. The compound can be described as Na$^{0.98+}$(Fe$_4$Sb$_{12}$)$^{0.98-}$. The same conclusions can be drawn for KFe$_4$Sb$_{12}$. Both compounds have ferromagnetic ground state (Xing et al. 2015). Close to the Fermi level the spin-up density is mainly contributed by Fe-d orbitals whereas the spin-down density is contributed by the Sb-p orbitals. The difference in spin polarization is very close to 1, therefore making XFe$_4$Sb$_{12}$ (X = Na, K) nearly perfect half metals. There are noticeable similarities and differences in the band structure of NaFe$_4$Sb$_{12}$ and KFe$_4$Sb$_{12}$ that have been described in great details by Yang and coll. (Yang et al. 2011) in terms of the band shifting with respect to the Sb-1s core state. The descriptions would be too lengthy to be reproduced in the present review and the interested reader is urged to read the article.

Alcaline-earth metals Ca, Sr and Ba, have been investigated as fillers in FeSb$_3$ skutterudites (Schnelle et al. 2005; Sichelschmidt et al. 2006; Kimura et al. 2007; Yang et al. 2011; Zhou et al. 2011; Yan et al. 2014). These compounds share the same feature of being half metals: the Fermi level is crossed by two, parabolic bands at the Γ point, a light band (antimony) and a heavy one (iron). Below these bands are located several heavy ones around 0.1 eV below the Fermi level. The band gap above the light valence band evolves as 0.47, 0.49 and 0.55 eV for Ca, Sr and Ba containing FeSb$_3$, respectively. The large differences between the conduction bands of the compounds can be exemplified by the energy gap that turns out to change from an indirect one (Γ-N gap) for CaFe$_4$Sb$_{12}$ and SrFe$_4$Sb$_{12}$ to a direct one at Γ for BaFe$_4$Sb$_{12}$.

We treat $InFe_4Sb_{12}$ and $SnFe_4Sb_{12}$ together as they are consecutive in the periodic table. Indeed, they exhibit very similar band structure (Zhou et al. 2011); they are half metals with two valence bands crossing the Fermi level and they have an indirect band energy gap. The two valence bands have maximum at the Γ point and parabolic shape whereas the minimum of the conduction band is between Γ and N. We can notice that the topmost valence band also shows a marked maximum above the Fermi level at the H point where it has a parabolic shape. This feature is not usually observed for other filled $FeSb_3$ skutterudite. The gap between the valence band maximum and the conduction band minimum is very different though: it is much smaller for $SnFe_4Sb_{12}$ (0.028 eV) than for $InFe_4Sb_{12}$ (0.39 eV). However it could be that the band gap of the former compound is underestimated due to the presence of tin. The authors of this review have indeed noticed this feature in other thermoelectric materials such as Mg_2Sn and $Mg_2Si_{1-x}Sn_x$ that are described by DFT methods as metals instead of semiconductors. Nonetheless, this statement needs validation through comparison with experimental data but as far as we know experimental investigations on $SnFe_4Sb_{12}$ have not been reported to date. The hybridization of the orbitals at the vicinity of the Fermi level is as follows. The valence band is as expected mainly composed of the Sb-p and Fe-d orbitals plus a small contribution from the Sn-p orbitals in the case of $SnFe_4Sb_{12}$. The Sn-s orbital contribution is spread over 3 eV of the whole valence band. In $InFe_4Sb_{12}$ no In-p orbitals participate but a small contribution from the In-s orbital appears around 0.5 eV below the Fermi level. The conduction band is mainly composed of the Sb-p and Fe-d orbitals and, in the case of $SnFe_4Sb_{12}$, of the Sn-p orbitals.

Six $FeSb_3$ skutterudites filled with rare-earth metals have been investigated, namely those containing La (Harima and Takegahara 2003; Nouneh et al. 2007; Yang et al. 2011; Xu et al. 2011; Xu et al. 2012; Xing et al. 2015), Ce (Nordström and Singh 1996; Nouneh et al. 2007; Hachemaoui et al. 2009; Xu et al. 2011; Yang et al. 2011), Pr (Yang et al. 2011), Nd (Zhou et al. 2011), Eu (Shankar et al. 2015b), and Yb (Galvan et al. 2003; Schnelle et al. 2005; Sichelschmidt et al. 2006; Zhou et al. 2011; Yang et al. 2011). The band structure and density of states of the La, Ce and Pr containing compounds show that the charge state of these atoms is +3. Cerium- and praseodynium-$FeSb_3$ skutterudites have very similar band structure, both for the valence and conduction bands. There is a slight discrepancy between several reports regarding the band structure of lanthanum-$FeSb_3$ especially near the Γ point where the maximum of the valence band is located. Following Xu and coll. (Xu et al. 2012) the top of valence band is made of a strongly parabolic (very light effective mass carriers) state. This band is intercepted about 0.15 eV below by a heavier state. By contrast, according to Xing and coll. and Yang and coll. (Yang et al. 2011;

Xing et al. 2015) both the heavy and light bands are tangent to each other at the top of the valence band (above the Fermi level). This degeneracy of the states should sensitively increase the density of states and the Seebeck coefficient. Since the methods and technical details used in these studies are of high quality for these types of calculation, the origin of this discrepancy is difficult to trace back. $YbFe_4Sb_{12}$ is also a subject of divergence between the calculations (Zhou et al. 2011; Yang et al. 2011). The authors agree to say that this compound is a half metal and has an indirect band gap but the predicted lines in the Brillouin zone vary. Surprisingly though, the energy band gaps agree with each other and amounts to about 0.42 eV. As expected, the valence state of Yb is +2 in the skutterudite. $NdFe_4Sb_{12}$ is also a half metal compound with heavy p-doping. The band structure, especially the valence band, and the energy band gap (0.37 eV) are very similar to that of $YbFe_4Sb_{12}$ (Zhou et al. 2011). Both in the valence and the conduction bands the Sb-p and Fe-d orbitals are hybridized with the Nd-f states. Hence, the valence state of Nd should be of the type $s^x d^y$ with x and y some real numbers between 0 and 2 and 0 and 4, respectively. $EuFe_4Sb_{12}$ (Krishnamurthy et al. 2007; Shankar et al. 2014) is described as a half metal compound with a small energy gap of 0.1 eV and is ferromagnetic with a total magnetic moment of 5.02 μ_B, 7 μ_B/Eu and –0.5 μ_B/Fe (Krishnamurthy et al. 2007). Hence iron is antiferromagnetically coupled to europium. As with the other filled skutterudites, $EuFe_4Sb_{12}$ possesses a pseudo gap above the valence band. The valence band results from the hybridization of the Eu-f, Sb-p and Fe-d orbitals whereas the conduction band is composed of Eu-d and Fe-d orbitals.

Substituted Skutterudites

A detailed analysis of the bonding structure of the unfilled $Co_{4-x}Ni_xSb_{12}$ skutterudite has been conducted by Bertini and coll. (Bertini et al. 2003) using Bader's QTAIM theory. Each Sb atom is bonded to two metal atoms and to two Sb atoms of the pictnogen ring. The metal atoms are not bonded to each other nor are the antimony ones. The nature of the metal-antimony bond is ionic whereas the Sb-Sb bond is more covalent. The metal-Sb bond is more ionic with nickel than with cobalt, which is in contradiction with the difference of electronegativity between the atoms (i.e., nickel is more electronegative than cobalt). Substitution of Ni for Co leads to a narrowing of the band gap and to an upshift of the Fermi level, which leads to a n-type doping character of Co_3NiSb_{12}, $Co_2Ni_2Sb_{12}$ and $CoNi_3Sb_{12}$ (Christensen et al. 2004). When antimony atoms are replaced by tellurium one, such as in $Co_3NiSb_{11}Te$, the band gap is also narrowed compared to that of $CoSb_3$ (Stiewe et al. 2005) and the Fermi level increases in energy leading to an n-doped compound. When Co_3NiSb_{12} is partially filled with calcium, Ca acts as an electron donor; the empty s states of Ca pin the Fermi level in the conduction band yielding a n-doped compound (Puyet et al. 2007).

Other skutterudites based on mixed metals have been theoretically investigated. We find alkaline-earth filled $(Fe,Ni)_4Sb_{12}$ with Ca, Sr and Ba as fillers (Singh and Du 2010), $Ca(Fe,Co)_4Sb_{12}$ (Thomson et al. 2015) and $La(Fe,Co)_4Sb_{12}$ (Lu et al. 2010), and $A(Fe,Co)_4P_{12}$ with A = La, Y, Ce (Mangersnes et al. 2008). Using the virtual crystal model, which allows to model site disorder within the density-functional theory approach, Singh and coll. have found that in $(Fe,Ni)_4Sb_{12}$ and $(Fe,Co)_4Sb_{12}$ filling the skutterudite with either Ca, Sr or Ba does not change markedly the band structure and the density of states. These compounds are small energy gap semiconductors with energy gap ranging from 0.36 eV to 0.42 eV. The consequence of these observations is that filling with alkaline-earth metals and mixing Ni and Fe at the metal site could lead to good p-type thermoelectric materials. Indeed, the authors found that $SrFe_3NiSb_{12}$ may be a good candidate. The behaviour of $Ca(Fe,Co)_4Sb_{12}$ band structure with varying content of Fe and Co is interesting, as shown by Thomson and coworkers. Indeed, the Fe states are much lower in energy than the Co states. Furthermore, the higher the content in Fe, the deeper in energy the Fe states are. Since the Fermi level is pinned by the Fe-d states, the density of states at the Fermi level increases with the content in iron. According to Lu and coll., in contrast with the previous skutterudite, the Fe-d states in $LaFe_3CoSb_{12}$ are located higher in energy than the Co-d states. In the skutterudites $A(Fe,Co)_4P_{12}$ where A are the rare-earth elements La, Y or Sc, the Fe/Co ratio affects noticeably the density of states and the electronic behaviour of the compound. For pure iron-based skutterudite the compound is a p-doped semiconductor whereas for an equal mixture of iron and cobalt the compound is a n-doped semiconductor.

Thermoelectric Properties

The calculations of electronic transport properties (Seebeck coefficient, electrical and electronic thermal conductivities) have been made possible only recently thanks to the increase of computer power. In effect, for these properties the Fermi surface has to be calculated very precisely, which necessitate an extremely large number of k-points for the Brillouin zone sampling. Hence, studies presenting thermoelectric properties of skutterudites are more seldom than those dealing with electronic structure or mechanical properties.

Chaput and coll. investigated both $CoSb_3$ by combining the FLAPW approach and the Boltzmann transport theory and the inclusion of impurities as dopants (($Ni_xCo_{1-x})Sb_3$, ($Fe_xCo_{1-x})Sb_3$ and Nd_xCoSb_3) using the KKR-CPA approach that allows to account for disordered structures (Chaput et al. 2005). $CoSb_3$ and ($Ni_xCo_{1-x})Sb_3$ are n-type doped compounds whereas ($Fe_xCo_{1-x})Sb_3$ is p-doped. The Seebeck coefficient is largest for the native $CoSb_3$ structure for temperatures below 400 K. The maximum in magnitude is found at around 350 K and amounts to about –420 mV K^{-1}. The same conclusion can be drawn from the experimental findings. It is noted that the calculated maximum of S is about 100 K below the experimental one, but the calculated and measured magnitudes are about the same. Above 400 K, the Nd-doped $CoSb_3$ shows best S characteristics with a maximal magnitude of –400 μV K^{-1} at 500 K followed by Ni-doped $CoSb_3$ (–300 μV K^{-1} at 700 K). The iron doped compound shows worst Seebeck coefficient with values below 200 mV K^{-1} up to 800 K. When cobalt is fully replaced by iron (Daniel et al. 2015b) the compound is strongly p-doped (metallic character) and a high electrical conductivity and low Seebeck coefficient is calculated at the Fermi level (intrinsic compound). A slight doping with, e.g., an excess antimony concentration of 8 10^{21} cm^{-3} leads to an upshift of the Fermi level (n-doping) which increases the Seebeck coefficient substantially (140 μV K^{-1} at 500 K). The power factor is also enhanced due to a slight increase of the electrical conductivity. It is concluded that only the n-doping of $FeSb_3$ leads to good thermoelectric properties for this compound which is due to the specific shape of the band structure. Filled $FeSb_3$ properties have also been investigated with alkaline (Na, K), alkaline-earth (Ca, Sr, Ba) and lanthanide metals (La, Ce, Pr, Nd, Eu, Yb) as fillers by Yang (Yang et al. 2011), Zhou (Zhou et al. 2011) Xu (Xu et al. 2011) and Shankar (Shankar et al. 2015b). Generally, the Seebeck coefficients of rare-earth-filled $FeSb_3$ are very low at temperatures below 300 K and increase markedly with the temperature (e.g., 150 μV K^{-1} for Nd Fe_4Sb_{12} at 750 K). In spite of this increase, the power factor of these compounds is rather small due to a small electrical conductivity. Therefore, the decent ZT values obtained for these thermoelectric materials is mainly caused by low lattice thermal conductivity which results from the phonon scattering by the fillers. Indeed, a strong reduction of the thermal conductivity has been observed by filling the voids in $FeSb_3$ by lanthanum atoms (Bernstein et al. 2010). The lowest value obtained amounts to between 1 and 1.6 W m^{-1} K^{-1} at 900 K, depending on the type of force field parameters used in the molecular dynamics simulations, which is about 2.5 smaller than that for the unfilled skutterudite.

Only a few studies have been reported that deal with the thermoelectric properties of nickel-containing skutterudites. Nickel is present in the structure as a substitutional atom for cobalt or iron or as a filler. The Seebeck coefficient of the unfilled $Co_{4-x}Ni_xSb_{12}$ structure is characteristic of a n-doped materials irrespective of the value of x ranging from 1 to 4 (Christensen et al. 2004). The authors drew the same conclusion for $NiCo_4Sb_{12}$, where Ni fills all the voids. It is noticeable that, the calculated magnitude of the Seebeck coefficient of the $Co_{4-x}Ni_xSb_{12}$ compounds is by far lower than that of the pristine $CoSb_3$ compound for all temperatures except at around 1000 K where $|S|$ is about the same for Co_3NiSb_{12} and $CoSb_3$ (70 $\mu V\ K^{-1}$). $NiCo_4Sb_{12}$ exhibits the smallest Seebeck coefficient magnitude among all the studied compounds. Chaput and coll. (see above) found markedly different results. In particular, above 450 K the $Ni_{0.01}Co_{0.99}Sb_3$ compound exhibit higher Seebeck coefficient than $CoSb_3$. This difference may be due to a different level of doping of the materials. The trends of S with respect to temperature and magnitude for the doubled doped $Co_{1-x}Ni_xSb_{3-y}Te_y$ compound is quite the same as for $Co_{4-x}Ni_xSb_{12}$; $|S|$ increases about linearly with temperature and reaches about 90 $\mu V\ K^{-1}$ at 1000 K (Stiewe et al. 2005). In addition the magnitude of the thermopower decreases as the dopant concentration increases. The best power factor is obtained at 700 K for a ratio Te/Ni of about 0.1. A high magnitude of the Seebeck coefficient (above 200 $\mu V\ K^{-1}$) has been found for the Ca, Ba and Sr alkaline earth-filled Fe_3NiSb_{12} (Singh and Du 2010). In addition, S is high even for high concentrations of fillers. These high values of S are caused by the presence of flat bands near the Fermi level that are contributed by metal d-orbitals. These observations are valid for both n- and p-doped $(Ca,Ba,Sr)Fe_3NiSb_{12}$ skutterudite. However, the advantage of the p-doped compounds over the n-doped ones is that, a mixing of heavy and light bands comes into play in the valence band, which should be favourable for enhancing the carrier mobility. Indeed, a weaker reduction of the Seebeck coefficient due to bipolar conduction has been observed in p-doped compounds than in n-doped ones. For the same reasons, in $Ca_yCo_{4-x}Ni_xSb_{12}$ (Puyet et al. 2007), the electrical conductivity and the thermopower are also enhanced due to the substitution of Ni for Co.

Keywords: Thermoelectric intermetallics, Unfilled skutterudites, Filled skutterudites, Synthesis and characterization, Thermal conductivity, Electrical conductivity, Figure of Merit, Structural and mechanical properties, Property modeling

References

Ameri, M., S. Amel, B. Abidri, I. Ameri, Y. Al-Douri, B. Bouhafs et al. 2014a. Structural, elastic, electronic and thermodynamic properties of uranium filled skutterudites UFe_4P_{12}: First principle method. Mater. Sci. Semicond. Proc. 27: 368–379.

Ameri, M., B. Abdelmounaim, M. Sebane, R. Khenata, D. Varshney, B. Bouhafs et al. 2014b. First-principles investigation on structural, elastic, electronic and thermodynamic properties of filled skutterudite $PrFe_4P_{12}$ compound for thermoelectric applications. Mol. Phys. 40: 1236–1243.

Anno, H., J. Nagao and K. Matsubara. 2002. Proc. 21st Int. Conf. On Thermoelectrics, Long Beach, CA, USA, IEEE, New York, USA, 56.

Artini, C., G. Zanicchi, G.A. Costa, M.M. Carnasciali, C. Fanciulli and R. Carlini. 2016. Correlations between structural and electronic properties in the filled skutterudite $Sm_y(Fe_xNi_{1-x})_4Sb_{12}$. Inorg. Chem. 55: 2574–2583.

Bader, R. 1994. Atoms in Molecules: A Quantum Theory. Oxford University Press, Oxford.

Ballikaya, S., N. Uzar, S. Yildirim, J.R. Salvador and C. Uher. 2012. High thermoelectric performance of In, Yb, Ce multiple filled $CoSb_3$ based skutterudite compounds. Solid. State Chem. Mater 193: 31–35.

Becke, A.D. and K.E. Edgecombe. 1990. A simple measure of electron localization in atomic and molecular systems. J. Chem. Phys. 92: 5397–5403.

Benalia, S., M. Ameri, D. Rached, R. Khenata, M. Rabah and A. Bouhemadou. 2008. First-principle calculations of elastic and electronic properties of the filled skutterudite $CeFe_4P_{12}$. Comp. Mater. Sci. 43: 1022–1026.

Berardan, D., E. Alleno, C. Godart, M. Puyet, B. Lenoir, R. Lackner et al. 2005. Improved thermoelectric properties in double-filled $Ce_{y/2}Yb_{y/2}Fe_{4-x}(Co/Ni)_xSb_{12}$ skutterudites. J. Appl. Phys. 98: 033710 (1–6).

Bernstein, N., J.L. Feldman and D.J. Singh. 2010. Calculations of dynamical properties of skutterudites: Thermal conductivity, thermal expansivity, and atomic mean-square displacement. Phys. Rev. B. 81: 134301 (1–11).

Berger, S., C. Paul, E. Bauer, A. Grytsiv, P. Rogl et al. 2001. Proceedings of 20th International Conference of Thermoelectric. Piscataway. NJ. IEEE Catalog 01TH8589.

Bertini, L., C. Stiewe, M. Toprak, S. Williams, D. Platzek, A. Mrotzek et al. 2003. Nanostructured $Co_{1-x}Ni_xSb_3$ skutterudites: Synthesis, thermoelectric properties, and theoretical modeling. J. Appl. Phys. 93: 438–447.

Bhaskar, A., Y.W. Yang and C.J. Liu. 2015a. Rapid fabrication and low temperature transport properties of nanostructured p-type $CexCo_4Sb_{12.04}$ (x = 0.15, 0.20 and 0.30) using solvothermal synthesis and evacuated-and-encapsulated. Ceram. Int. 41. 5: 6381–6385.

Bhaskar, A., Y.W. Yang, Z.R. Yang, F.H. Lin and C.J. Liu. 2015b. Fast fabrication and enhancement of thermoelectric power factor of p-type nanostructured $CoSb_3$(1 + delta) (delta = 0.00, 0.01 and 0.02) using solvothermal synthesis and evacuating-and-encapsulating sintering. Ceram. Int. 41. 6: 7989–7995.

Blaha, P., K. Schwarz, G.K.H. Madsen, D. Kvasnicka and J. Luitz. 2001. WIEN 2k, An Augmented Plane Wave Plus Local Orbitals Program for Calculating Crystal Properties, Vienna University of Technology, Austria.

Caillat, T., A. Borshchevsky and J.P. Fleurial. 1996. Properties of single crystalline semiconducting $CoSb_3$. J. Appl. Phys. 80: 4442–4449.

Carlini, R., A.U. Khan, R. Ricciardi, T. Mori and G. Zanicchi. 2016. Synthesis, characterization and thermoelectric properties of Sm filled $Fe_{4-x}Ni_xSb_{12}$ skutterudites. J. Alloys Compd. 655: 321–326.

Chaput, L., P. Pécheur, J. Tobola and H. Scherrer. 2005. Transport in doped skutterudites: *Ab initio* electronic structure calculations. Phys. Rev. B 72: 085126 (1–11).

Cheng, Z.-Z., B. Xu and Z. Cheng. 2008a. Optical properties of filled skutterudite $ThFe_4P_{12}$. Commun. Theor. Phys. 49: 1049–1051.

Cheng, Z.Z., Z. Cheng and M.-H. Wu. 2008b. Magnetic properties, electronic structure, and optical properties of the filled skutterudite $BaFe_4Sb_{12}$. J. Magn. Magn. Mater 320: 2591–2595.

Christensen, M., B.B. Iversen, L. Bertini, C. Gatti, M. Toprak, M. Muhammed et al. 2004. Structural study of Fe doped and Ni substituted thermoelectric skutterudites by combined synchrotron and neutron powder diffraction and *ab initio* theory. J. Appl. Phys. 96: 3148–3157.

Daniel, M.V., L. Hammerschmidt, C. Schmidt, F. Timmermann, J. Franke, N. Jöhrmann et al. 2015a. Structural and thermoelectric properties of $FeSb_3$ skutterudite thin films. Phys. Rev. B. 91: 085410 (1–10).

Daniel, M.V., C. Brombacher, G. Beddies, N. Jöhrmann, M. Hietschold, D.C. Johnson et al. 2015b. Structural properties of thermoelectric $CoSb_3$ skutterudite thin films prepared by molecular beam deposition. J. Alloys Compd. 624: 216–225.

Dordevic, S.V., N.R. Dilley, E.D. Bauer, D.N. Basov, M.B. Maple and L. Degiorgi. 1999. Optical properties of MFe_4P_{12} filled skutterudites. Phys. Rev. B. 60: 11321–11328.

Dyck, J.S., W. Chen, C. Uher, L. Chen, L.X. Tang and T. Hirai. 2002. Thermoelectric properties of the n-type filled skutterudite $Ba_{0.3}Co_4Sb_{12}$ doped with Ni. J. Appl. Phys. 91: 3698–3705.

Feldman, J.L., D.J. Singh, I.I. Mazin, D. Mandrus and B.C. Sales. 2000. Lattice dynamics and reduced thermal conductivity of filled skutterudites. Phys. Rev. B. 61: R9209–R9212.

Flage-Larsen, E., O.M. Løvvik, Ø. Prytz and J. Taftø. 2010. Bond analysis of phosphorus skutterudites: Elongated lanthanum electron buildup in $LaFe_4P_{12}$. Comput. Mater. Sci. 47: 752–757.

Fleurial, J.P., T. Caillat and A. Borshchevsky. 1997. Skutterudites: an update in Proc. ICT'97 16th Int. Conf. Thermoelectrics 1–11(IEEE Piscataway, New Jersey, 1997).

Fleurial, J.-P., T. Caillat, A. Borshchevsky, D.T. Morelli and G.P. Meisner. 1996. High figure of merit in cerium filled skutterudites. Proc. 15th Int. Conf. on Thermoelectrics, Pasadena California (IEEE Publishing, Piscataway, NJ, 1996).

Fornari, M. and D.J. Singh. 1999. Electronic structure and thermoelectric prospects of phosphide skutterudites. Phys. Rev. B. 59: 9722–9724.

Galvan, D.H., N.R. Dilley, M.B. Maple, A. Posada-Amarillas, A. Reyes-Serrato and J.C. Samaniego Reyna. 2003. Extended Huckel tight-binding calculations of the electronic structure of $YbFe_4Sb_{12}$, UFe_4P_{12}, and $ThFe_4P_{12}$. Phys. Rev. B. 68: 115110 (1–9).

Galvan, D.H. 2006. Electronic structure calculations for $PrFe_4P_{12}$ filled skutterudite using Extended Huckel tight-binding method. Intern. J. Model. Phys. B. 17: 4749–4762.

Geng, H., X. Meng, H. Zhang and J. Zhang. 2014. Lattice thermal conductivity of filled skutterudites: An anharmonicity perspective. J. Appl. Phys. 116: 163503 (1–6).

Grandjean, F., A. Gérard A., D.J. Braun and W. Jeitschko. 1984. Some physical properties of $LaFe_4P_{12}$ type compounds. J. Phys. Chem. Solids 45: 877–886.

Grosvenor, A.P., R.G. Cavell and A. Mar. 2006. X-ray Photoelectron spectroscopy study of rare-earth filled skutterudites $LaFe_4P_{12}$ and $CeFe_4P_{12}$. Chem. Mater 18: 1650–1657.

Hachemaoui, M., R. Khenata, A. Bouhemadou, A.H. Reshak, D. Rached and F. Semari. 2009. FP-APW + lo study of the elastic, electronic and optical properties of the filled skutterudites $CeFe_4As_{12}$ and $CeFe_4Sb_{12}$. Curr. Opin. Solid State Mater Sci. 13: 105–111.

Hachemaoui, M., R. Khenata, A. Bouhemadou, S. Bin-Omran, A.H. Reshak, F. Semari et al. 2010. Prediction study of the structural and elastic properties for the cubic skutterudites $LaFe_4A_{12}$ (A = P, As and Sb) under pressure effect. Solid State Commun. 150: 1869–1873.

Harima, H. and K. Takegahara. 2003. X-dependence of electronic band structures for $LaFe_4X_{12}$ (X = P, As, Sb). Physica B. 328: 26–28.

He, T., J. Chen, H.D. Rosenfeld and M.A. Subramanian. 2006. Thermoelectric properties of indium filled skutterudites. Chem. Mater 18: 759–762.

Hill, R. 1952. The elastic behaviour of a crystalline aggregate. Proc. Phys. Soc. A. 65: 349–354.

Hohenberg, P. and W. Kohn. 1964. Inhomogeneous electron gas. Phys. Rev. 136: B864–B871.

Jacobsen, M.K. 2014. Enhancement of thermoelectric performance with pressure in $Ce_{0.8}Fe_3CoSb_{12.1}$. J. Phys. Chem. Solids 75: 1017–1023.

Jung, D., M.H. Whangbo and S. Alvarez. 1990. Importance of the X_4 ring orbitals for the semiconducting, metallic, or superconducting properties of skutterudites MX_3 and RM_4X_{12}. Inorg. Chem. 29: 2252–2255.

Jeitschko, W. and D.J. Brown. 1977. $LaFe_4P_{12}$ with filled $CoAs_3$-type structure and isotypic lanthanoid transition metal poliphosphides. Acta Cryst. B 33: 3401–3406.

Jeitschko, W., A.J. Foecker, D. Pashke, M.V. Dewalsky, B.H. Evers, B. Kunnen et al. 2000. Crystal structure and properties of some filled and unfilled skutterudites: $GdFe_4P_{12}$, $SmFe_4P_{12}$, $NdFe_4As_{12}$, $Eu_{0.54}Co_4Sb_{12}$, $Fe_{0.5}Ni_{0.5}P_3$, CoP_3 and NiP_3. J. Inorg. Chem. 626: 1112–1120.

Kadel, K. and W.Z. Li. 2014. Solvothermal synthesis and structural characterization of unfilled and Yb-filled cobalt antimony skutterudite. Cryst. Res. Technol. 49(2-3): 135–141.

Kaiser, J.W. and W. Jeitschko. 1999. The antimony-rich parts of the ternary systems calcium, strontium, barium and cerium with iron and antimony; structure refinements of the LaFe$_4$Sb$_{12}$-type compounds SrFe$_4$Sb$_{12}$ and CeFe$_4$Sb$_{12}$; the new compounds CaOs$_4$Sb$_{12}$ and YbOs$_4$Sb$_{12}$. J. Alloys Compd. 291: 66–72.

Kaltzoglou, A., P. Vaqueiro, K.S. Knight and A.V. Powell. 2012. Synthesis, characterization and physical properties of the skutterudites Yb$_x$Fe$_2$Ni$_2$Sb$_{12}$ (0 ≤ x ≤ 0.4). J. Solid State Chem. 193: 36–41.

Katsuyama, S., M. Watanabe, M. Kuroki, T. Maehala and M. Ito. 2003. Effect of NiSb on the thermoelectric properties of skutterudite CoSb3. J. Appl. Phys. 93: 2758–2764.

Khenata, R., A. Bouhemadou, A.H. Reshak, R. Ahmed, B. Bouhafs, D. Rached et al. 2007. First-principles calculations of the elastic, electronic, and optical properties of the filled skutterudites CeFe$_4$P$_{12}$ and ThFe$_4$P$_{12}$. Phys. Rev. B. 75: 195131 (1–7).

Kimura, S.I., H.J. Im, T. Mizuno, S. Narazu, E. Matsuoka and T. Takabatake. 2007. Infrared study on the electronic structure of alkaline-earth-filled skutterudites AT$_4$Sb$_{12}$ (A = Sr, Ba; M = Fe, Ru, Os). Phys. Rev. B, Cond. Mat. 75: 245106 (1–7).

Kjekshus, A. and T. Rakke. 1974. Compounds with the skutterudite type crystal structure. III structural data for arsenides and antimonides. Acta Chem. Scand. 28: 99–103.

Kohn, W. and L.J. Sham. 1965. Self-Consistent equations including exchange and correlation effects. Phys. Rev. 140: A1133–A138.

Krishnamurthy, V.V., J.C. Lang, D. Haskel, D.J. Keavney, G. Srajer, J.L. Robertson et al. 2007. Ferrimagnetism in EuFe$_4$Sb$_{12}$ due to the interplay of f-electron moments and a nearly ferromagnetic host. Phys. Rev. Lett. 98: 126403 (1–4).

Leithe-Jasper, A., W. Schnelle, H. Rosner, N. Senthilkumaran, A. Rabis, M. Baenitz et al. 2003. Ferromagnetic ordering in alkali-metal iron antimonides: NaFe$_4$Sb$_{12}$ and KFe$_4$Sb$_{12}$. Phys. Rev. Lett. 91: 037208 (1–4).

Leithe-Jasper, A., W. Schnelle, H. Rosner, M. Baenitz, A. Rabis, A.A. Gippius et al. 2004. Weak itinerant ferromagnetism and electronic and crystal structures of alkali-metal iron antimonides: NaFe$_4$Sb$_{12}$ and KFe$_4$Sb$_{12}$. Phys. Rev. B. 70: 214418.

Li, J.Q., Z.P. Zhang, R. Luo, W.Q. Mao and F.S. Liu. 2013. Solvothermal synthesis of nano-sized skutterudite Co$_{1-x}$Ni$_x$Sb$_3$ powders. Powder Diffr. 28: S17–S21.

Liang, T., X.L. Su, Y.G. Yan, G. Zheng, Q. Zhang, H. Chi et al. 2014. Ultra-fast synthesis and thermoelectric properties of Te doped skutterudites. J. Mater. Chem. A. 2. 42: 17914–17918.

Liu, R., J.Y. Cho, J. Yang, W. Zhang and L. Chen. 2014. Thermoelectric transport properties of R$_y$Fe$_3$NiSb$_3$ (R = Ba, Nd and Yb). J. Mater Sci. Technol. 30: 1134–1140.

Llunell, M., P. Alemany, S. Alvarez and V.P. Zhukov. 1996. Electronic structure and bonding in skutterudite-type phosphides. Phys. Rev. B. 53: 10605–10609.

Lomnytska, Y.F. and O.P. Pavliv. 2007. Phase equilibria in the V-Ni-Sb system. Inorg. Mater 43: 608–613.

Lu, P.X., Z.-G. Shen and X. Hu. 2010. A comparison study on the electronic structure of the thermoelectric materials CoSb$_3$ and LaFe$_3$CoSb$_{12}$. Physica B. 405: 1740–1744.

Mandrus, D., B.C. Sales, V. Keppens, B.C. Chakoumakos, P. Dai, L.A. Boatner et al. 1997. Filled skutterudite antimonides: validation of the electron-crystal phonon-glass approach to new thermoelectric materials. Mater Res. Soc. Symp. Proc. 478: 199–209.

Mangersnes, K., O.M. Løvvik and Ø. Prytz. 2008. New filled P-based skutterudites—promising materials for thermoelectricity? New J. Phys. 10: 053004.

Maple, M.B., Z. Henkie, R.E. Baumbach, T.A. Sayles, N.P. Butch, P.-C. Ho et al. 2008. Correlated electron phenomena in Ce- and Pr-based filled skutterudite arsenides and antimonides. J. Phys. Soc. Jpn. Suppl. A. 77: 7–13.

Möchel, A., I. Sergueev, N. Nguyen, G.J. Long, F. Grandjean, D.C. Johnson et al. 2011. Lattice dynamics in the FeSb$_3$ skutterudite. Phys. Rev. B: Condens. Matter Mater Phys. 84. 064302: 1–9.

Morelli, D.T., G.P. Meisner, B. Chen, S. Hu and C. Uher. 1997. Cerium filling and doping of cobalt triantimonide. Phys. Rev. B. 56: 7376–7383.

Morimura, T. and H. Masayuki. 2003. Partially filled skutterudite structure in $Ce_fFe_{8-x}Ni_xSb_{24}$. Scr. Mater 48: 495–500.

Nolas, G.S., J.L. Cohn and G.A. Slack. 1998. Effect of partial void filling on the lattice thermal conductivity of skutterudites. Phys. Rev. B. 58: 164–170.

Nolas, G.S., M. Kaeser, R.T. Littleton and T.M. Tritt. 2000. High figure of merit in partially filled ytterbium skutterudite materials. Appl. Phys. Lett. 77: 1855–1859.

Nolas, G.S., J. Poon and M. Kanatzidis. 2006. Recent developments in bulk thermoelectric materials. Mater Res. Soc. Bull. 31: 199–205.

Nordstrom, L. and D.J. Singh. 1996. Electronic structure of Ce-filled skutterudites. Phys. Rev. B. 53: 1103–1108.

Nouneh, K., A.H. Reshak, S. Auluck, I.V. Kityk, R. Viennois, S. Benet et al. 2007. Band energy and thermoelectricity of filled skutterudites $LaFe_4Sb_{12}$ and $CeFe_4Sb_{12}$. J. Alloys Compd. 437: 39–46.

Oftedal, I. 1926. The crystal structure of skutterudite and related minerals. Norsk Geologisk Tidsskrift. 8: 250–257.

Okamoto, H. 1991. Co-Sb (Cobalt-Antimony). J. Phase Equilib. 12: 244–245.

Oster, O., T. Nilges, F. Bachhube, F. Pielnhofer, R. Weilrih, M. Shoneich et al. 2012. Synthesis and identification of metastable compounds: Black arsenic—Science or fiction? Angew. Chem. Int. Edit. 51: 2994–2997.

Pugh, S.F. 1954. Relations between the elastic moduli and the plastic properties of polycrystalline pure metals. Philos. Mag. 45: 823–843.

Puyet, M., B. Lenoir, A. Dauscher, M. Dehmas, C. Stiewe and E. Muller. 2004. High temperature transport properties of partially filled CaxCo4Sb12 skutterudites. J. Appl. Phys. 95: 4852–4855.

Puyet, M., A. Dauscher, B. Lenoir, C. Bellouard, C. Stiew, E. Müller et al. 2007. Influence of Ni on the thermoelectric properties of the partially filled calcium skutterudites $Ca_yCo_{4-x}Ni_xSb_{12}$. Phys. Rev. B 75: 245110 (1–10).

Qiu, P., X. Shi, R. Liu, Y. Qiu, S. Wan and L. Chen. 2013. Thermoelectric properties of manganese–doped p-type skutterudites $Ce_yFe_{4-x}Mn_xSb_{12}$. Funct. Mater Lett. 6: 1340003 (1–6).

Rogl, G., A. Grytsiv, P. Rogl, E. Bauer, M.B. Kerber, M. Zehetbauer et al. 2010a. Multifilling nanocrystalline p-type didymium-Skutterudites with zT > 1.2. Intermetallics 18: 2435–2444.

Rogl, G., A. Grytsiv, E. Bauer, P. Rogl and M. Zehetbauer. 2010b. Thermoelectric properties of novel skutterudites with didymium: $DDy(Fe_{1-x}Co_x)_4Sb_{12}$ and $DDy(Fe_{1-x}Ni_x)_4Sb_{12}$. Intermetallics 18: 57–64.

Rogl, G., A. Grytsiv, E. Falmbigl, E. Bauer, P. Rogl et al. 2012. Thermoelectric properties of p-type didymium (DD) based skutterudites: $DD_y(Fe_{1-x}Co_x)_4Sb_{12}$ ($0.13 \le x \le 0.25$, $0.46 \le y \le 0.68$). J. Alloys Compd. 537: 242–249.

Rowe, D.M. 2006. Thermoelectrics Handbook—Macro to Nano. CRC, Boca Raton, Fl, USA.

Rundqvist, S. and Ersson, N.O. 1969. Structure and bonding in skutterudite-type phosphides. Ark. Kemi. 30: 103–114.

Rundqvist, S. 1960. Phosphides of the Platinum Metals. Nature 185: 31–32.

Saeterli, R., E. Flage-Larsen, Ø. Prytz, J. Taftø, K. Marthinsen and R. Holmestad. 2009. Electron energy loss spectroscopy of the $L_{2,3}$ edge of phosphorus skutterudites and electronic structure calculations. Phys. Rev. B 80: 075109.

Sales, B.C., D. Mandrus and R.K. Williams. 1996. Filled skutterudite antimonides: a new class of thermoelectric materials. Science 272: 1325–1328.

Sales, B.C., D. Mandrus, B.C. Chakoumakos, V. Keppens and J.R. Thompson. 1997. Filled skutterudite antimonides: electron crystals and phonon glasses. Phys. Rev. B 56: 15081–15089.

Sales, B.C. 1998. Electron crystals and phonon glasses: a new path to improved thermoelectric materials. Mater Res. Soc. Bull. 23: 15–21.

Sales, B.C., B.C. Chakoumakos, D. Mandrus and J.W. Sharp. 1999. Atomic displacement parameters and the lattice thermal conductivity of clathrate-like thermoelectric compounds. J. Solid State Chem. 146: 528–532.

Sales, B.C. 2003. Filled skutterudites. pp. 1–34. In: K.A. Gschneidner, Jr., J.-C.G. Bünzli and V.K. Pecharsky [eds.]. Handbook on the Physics and Chemistry of the Rare Earths, Vol. 33. Elsevier, New York.

Schnelle, W., A. Leithe-Jasper, M. Schmidt, H. Rosner, H. Borrmann, U. Burkhardt et al. 2005. Itinerant iron magnetism in filled skutterudites $CaFe_4Sb_{12}$ and $YbFe_4Sb_{12}$: Stable divalent state of ytterbium. Phys. Rev. B 72: 020402 (1–4).

Shankar, A. and R.K. Thapa. 2013. Electronic, magnetic and structural properties of the filled skutterudite $EuFe_4P_{12}$: LSDA and LSDA+U calculation. Physica B 427: 31–36.

Shankar, A., D.P. Rai, Sandeep and R.K. Thapa. 2014. A first principles calculation of ferromagnetic $EuFe_4Sb_{12}$. Phys. Proc. 54: 127–131.

Shankar, A., D.P. Rai, Sandeep, J. Maibam and R.K. Thapa. 2015a. Investigation of elastic and optical properties of $EuFe_4P_{12}$ by first principles calculation. Indian J. Phys. 89: 797–801.

Shankar, A., D.P. Rai, R. Khenata, J. Maibam, Sandeep and R.K. Thapa. 2015b. Study of 5f electron based filled skutterudite compound $EuFe_4Sb_{12}$, a thermoelectric (TE) material: FP-LAPW method. J. Alloys Compd. 619: 621–626.

Shenoy, G.K., D.R. Noakes and G.P. Meisner. 1982. Mössbauer study of superconducting $LaFe_4P_{12}$. J. Appl. Phys. 53: 2628–2630.

Short, M., F. Bridges, T. Keiber, G. Rogl and P. Rogl. 2015. A comparison of the local structure in ball-milled and hand ground skutterudite samples using EXAFS. Intermetallics 63: 80–85.

Sichelschmidt, J., V. Voevodin, H.J. Im, S. Kimura, H. Rosner, A. Leithe-Jasper et al. 2006. Phys. Rev. Lett. 96: 037406 (1–4).

Singh, D.J. and M.-H. Du. 2010. Properties of alkaline-earth-filled skutterudite antimonides: $A(Fe,Ni)_4Sb_{12}$ (A = Ca, Sr, and Ba). Phys. Rev. B 82: 075115 (1–7).

Singh, Singh, I.I. Mazin, J.L. Feldman and M. Fornari. 1999. Properties of novel thermoelectrics from first principles calculations. Mater Res. Soc. Symp. Proc. 545: 3–11.

Slack, G.A. and V.G. Tsoukala. 1994. Some properties of semiconducting $IrSb_3$. J. Appl. Phys. 76: 1665–1671.

Snider, T.S., J.V. Badding, S.B. Schujman and G.A. Slack. 2000. High-Pressure Stability, Pressure-Volume Equation of State, and Crystal Structure under Pressure of the Thermoelectric Material $IrSb_3$. Chem. Mater 12: 697–700.

Snyder, G.J. and E.S. Toberer. 2008. Complex thermoelectric materials. Nature Mater 7: 105–114.

Sootsman, J.R., D.Y. Chung and M.G. Kanatzidis. 2009. New and old concepts in thermoelectric materials. Angew. Chem. Int. Ed. 48: 8616–8639.

Stiewe, C., L. Bertini, M. Toprak, M. Christensen, D. Platzek, S. Williams et al. 2005. Nanostructured $Co_{1-x}Ni_x(Sb_{1-y}Te_y)_3$ skutterudites: Theoretical modeling, synthesis and thermoelectric properties. J. Appl. Phys. 97: 044317 (1–7).

Takegahara, K. and H. Harima. 2003. FLAPW electronic band structure of the filled skutterudite $ThFe_4P_{12}$. Physica. B. 329–333: 464–466.

Takegahara, K. and H. Harima. 2007. Electronic band structures in $LaRu_4P_{12}$ and $LaFe_4P_{12}$ with modulated skutterudite structures. J. Magn. Magn. Mater 310: 861–863.

Tan, G., W. Liu, S. Wang, Y. Yan, H. Li, X. Tang and C. Uher. 2013a. Rapid preparation of $CeFe_4Sb_{12}$ by melt spinning: rich nanostructures and high thermoelectric performance. J. Mater Chem. A. 12657–12668.

Tan, G., W. Liu, H. Chi, X. Su, S. Wang, Y. Yan. et al. 2013b. Realization of high thermoelectric performance in p-type unfilled ternary skutterudites $FeSb_{2+x}Te_{1-x}$ via band structure modification and significant point defect scattering. Acta Mater 61: 7693–7704.

Tan, G., S. Wang, X. Tang, H. Li and C. Uher. 2012. Preparation and thermoelectric properties of Ga-substituted p-type filled skutterudites $CeFe_{4-x}Ga_xSb_{12}$. J. Solid State Chem. 196: 203–208.

Tang, X., L. Chen, T. Goto and T. Hirai. 2001. Effects of Ce filling fraction and Fe content on the thermoelectric properties of Co-rich $Ce_yFe_xCo_{4-x}Sb_{12}$. J. Mater Res. 16: 837–843.

Tang, X., L. Chen, T. Goto and T. Hirai. 2001. High-temperature thermoelectric properties of n-type $Ba_xNi_xCo_{4-x}Sb_{12}$. J. Mater Res. 16: 3343–3346.

Thompson, D.R., C. Liu, J. Yang, J.R. Salvador, D.B. Haddad, N.D. Ellison et al. 2015. Rare-earth free p-type filled skutterudites: Mechanisms for low thermal conductivity and effects of Fe/Co ratio on the band structure and charge transport. Acta Mater 92: 152–162.

Ugai, Y.A., S.P. Evseeva, A.E. Popov, E.G. Goncharov and O.V. Grigor'eva. 1985. Equilibrium diagram of the $GeAs_2$-As system. Russ. J. Inorg. Chem. 30: 1681–1682.

Uher, C. 2002. Structure–property relations in skutterudites. pp. 121–146. *In*: M.G. Kanatzidis, S.D. Mahanti and T.P. Hogan [eds.]. Chemistry, Physics, and Materials Science of Thermoelectric Materials. Beyond Bismuth Telluride. Kluwer Academic, New York.

Wawryk, R., Z. Henkie, A. Pietraszko, T. Cichorek, A. Jezierski, R.E. Baumbach et al. 2009. Possible metal-insulator transition in the filled skutterudite $CeFe_4As_{12}$. J. Phys.: Conf. Series 200: 012223 (1–4).

Wawryk, R., Z. Henkie, A. Pietraszko, T. Cichorek, L. Kepinski, A. Jezierski et al. 2011. Filled skutterudite $CeFe_4As_{12}$: Disclosure of a semiconducting state. Phys. Rev. B. 84: 165109 (1–11).

Wu, Z.J., E.J. Zhao, H.P. Xiang, X.F. Hao, X.J. Liu and J. Meng. 2007. Crystal structures and elastic properties of superhard IrN_2 and IrN_3 from first principles. Phys. Rev. B: Condens. Mat. 76. 054115: 1–15.

Xing, G., X. Fan, W. Zheng, Y. Ma, H. Shi and D.J. Singh. 2015. Magnetism in Na-filled Fe-based Skutterudites. Sci. Rep. 5: 10782.

Xu, B. and L. Yi. 2007. Optical properties of the filled skutterudite $LaFe_4P_{12}$. Physica B. 390: 147–150.

Xu, B., J. Liang, X. Li, J.F. Sun and L. Yi. 2011. Thermoelectric performance of the filled-skutterudite $LaFe_4Sb_{12}$ and $CeFe_4Sb_{12}$. Eur. Phys. J. B 79: 275–281.

Xu, B., C. Long, Y. Wang and L. Yi. 2012. First-principles investigation of electronic structure and transport properties of the filled skutterudite $LaFe_4Sb_{12}$ under different pressures. Chem. Phys. Lett. 529: 45–48.

Xi, L., Y. Qiu, S. Zheng, X. Shi, J. Yang, L. Chen et al. 2015. Complex doping of group 13 elements In and Ga in caged skutterudite $CoSb_3$. Acta Mater 85: 112–121.

Yan, Y.G., W. Wong-Ng, L. Li , I. Levin, J.A. Kaduk, M.R. Suchomel et al. 2014. Structures and thermoelectric properties of double-filled $(Ca_xCe_{1-x})Fe_4Sb_{12}$ skutterudites. J. Solid State Chem. 218: 221–229.

Yang, J., P. Qiu, R. Liu, L. Xi, S. Zheng, W. Zhang et al. 2011. Trends in electrical transport of p-type skutterudites RFe_4Sb_{12} (R = Na, K, Ca, Sr, Ba, La, Ce, Pr, Yb) from first-principles calculations and Boltzmann transport theory. Phys Rev. B 84: 235205 (1–10).

Yang, J., L. Zhang, Y. Liu, C. Chen, J. Liu, J. Li et al. 2013 Investigation of skutterudite $Mg_yCo_4Sb_{12}$: high pressure synthesis and thermoelectric properties. J. Appl. Phys. 113: 113703 (1–7).

Zhang, Q., C. Chen, Y. Kang, X. Li, L. Zhang, D. Yu et al. 2015. Structural and thermoelectric characterizations of Samarium filled $CoSb_3$ skutterudites. Mater Lett. 43: 41–43.

Zhang, L., B. Xu, X. Li, F. Duan, X. Yan and Y. Tian. 2015. Iodine-filled $FexCo_{4-x}Sb_{12}$ polycrystals: Synthesis, structure, and thermoelectric properties. Mater Lett. 139: 249–251.

Zhou, An, L.-S. Liu, C.-C. Shu, P.-C. Zhai, W.-Y. Zhao and Q.-J. Zhang. 2011. Electronic structures and transport properties of RFe_4Sb_{12} (R = Na, Ca, Nd, Yb, Sn, In). J. Electron. Mater 40: 974–979.

Index

About the Editor

Cristina Artini was born in Genova (Italy) in 1976 and has been working for more than 15 years in the field of physical chemistry. She received degree in Chemistry cum laude in 2000 and Ph.D. in 2004. She worked as a researcher at the Italian National Research Council (CNR), where she studied the chemical-physical aspects of brazing processes, and she is now Researcher at the Department of Chemistry and Industrial Chemistry of the University of Genova. To date she is author of around 50 papers in international journals and books and has been an invited speaker at several international conferences.

Printed and bound by CPI Group (UK) Ltd, Croydon, CR0 4YY
01/11/2024
01782624-0011